THE GIANTS OF STONEHENGE AND ANCIENT BRITAIN

Another essential addition to the ancient mysteries bookshelf. Compelling read from start to finish bringing together everything that could possibly be needed to show that giants once inhabited the British Isles. The authors should be commended for bringing together this fascinating information.

Andrew Collins, author of
Göbekli Tepe: Genesis of the Gods, From the Ashes of Angels, and *The Cygnus Key,*
co-author of, *Denisovan Origins,* and *Path of Souls: The Native American Death Journey*

This is the best book on the legends and myths associated with giants ever put forward to this time. Every page seems to connect and build upon the grand mystery of the entire insular Celtic language region and beyond. I've read the thing twice now, and it appears that, for the first time, a definite connection has been forged regarding the divining of the landscape by true giants, i.e. ancient seers possessing the rarest gift of all, that of a fully functional single or internal eye. Apparently it was they who carefully advised and executed the antique system of 'spirit paths' crossing vast stretches of land through the sacred organ of spiritual vision, crediting forever the ancient landscape patterns to the productions of actual giants. This isn't all they've proven, however, so picking this book up can only add to the 'embarrassment of riches' these two men have already rendered for us all.

Ross Hamilton, author of
A Tradition of Giants, The Mystery of the Serpent Mound,
and *Star Mounds: Legacy of a Native American Mystery*

This book is from two of the most respected and long-standing researchers of the literature on giants in antiquity. The number of giant accounts the authors found and then fully researched is astounding. The book covers giant reports, excavations, and professional journals citing the discovery of giant remains and legends across the UK. Highly recommended for anyone interested in giants and the truth about the ancient world.

Dr. Greg Little, author of
The Illustrated Encyclopedia of Native American Indian Mounds and Earthworks,
co-author of, *Denisovan Origins,* and *Path of Souls: The Native American Death Journey*

Hugh Newman and Jim Vieira once again score a home run with this fascinating collection of British historic fact and associated myth. Absolutely the giants must be brought to the attention of the mainstream, we need to stop hiding from our past, and Newman and Vieira accomplish this comprehensively with style and authority.

Hugh Evans, author of
The Origin of the Zodiac: Cadair Idris and the Star Maps of Gwynedd

Never before have we had such a comprehensive and authoritative discussion of accounts of finds of ancient giants in the UK. The authors take us across the UK and Ireland, uncovering lost finds of giants from newspaper articles and other accounts. Where possible they try to verify the finds, but far from taking an uncritical view, they debunk several finds as forgeries. They uncover remarkable consistencies between the finds, hinting at an ancient culture of nobility, giant-sized rulers. Where possibly they bring in scientific evidence which indeed highlights the possibility for the expression of gigantism in certain related populations today, showing accounts and photos of modern giants, also. The book is extremely well illustrated and will be used for decades by researchers investigating this topic as a primary reference. Account after account of gigantic bones, belonging not to dinosaurs but to humans of a very large size, from ancient burial sites, are produced. I congratulate the authors for having the courage to tackle a difficult topic in such an enterprising and confident manner to produce this well-done publication.

Charles R. Kos, author of
In Search of the Origin of Pyramids and *Confessions of the Gods?*

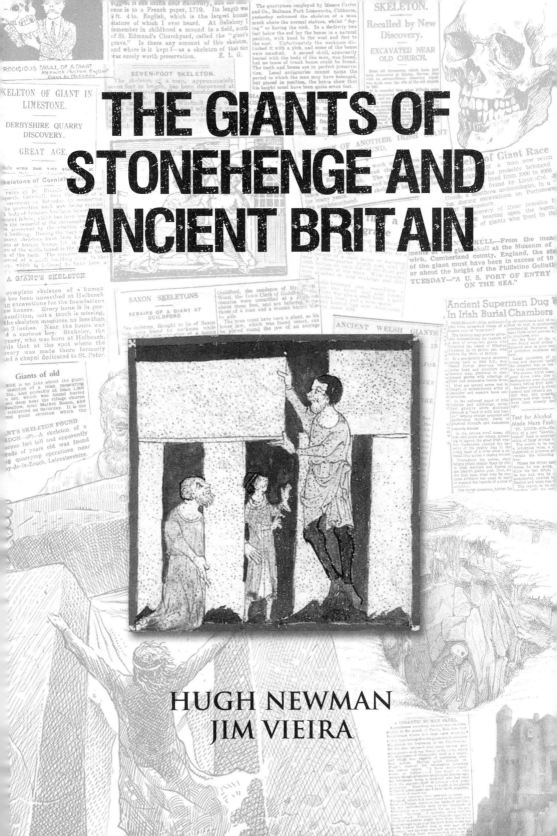

THE GIANTS OF STONEHENGE AND ANCIENT BRITAIN

HUGH NEWMAN
JIM VIEIRA

British Library Cataloguing in Publication Data
Newman, H and Vieira, J
The Giants of Stonehenge and Ancient Britain

Adventures Unlimited Press
One Adventure Place, Kempton, Illinois, 60946, USA

ISBN: 978-1-948803-54-0

Also available in ebook and hardback formats

Front cover artwork and maps by Dan Lish. www.danlish.com
Illustrations by Dan Lish, Emmanuel Martin and Yuri Leitch.
www.yurileitch.co.uk.

Contact the authors directly at: www.facebook.com/giantsonrecord
and more info at www.megalithomania.co.uk/giantsofbritain.html

www.adventuresunlimitedpress.com

About the Authors

Jim Vieira is a stonemason, historical detective and co-author with Hugh Newman of *Giants On Record: America's Hidden History, Secrets in the Mounds and the Smithsonian Files* (2015). He has written for *Ancient American Magazine, The Heretic, World Explorers Magazine* and other publications. In 2012 he created the online blog *The Daily Giant* that showcased a giant report every day for nearly two years. He controversially had his TEDx talk removed from the Internet (much like Graham Hancock and Rupert Sheldrake). He is the star of the History Channel TV shows *Search for the Lost Giants* (2014) and *Roanoke: Search for the Lost Colony* (2015) with his brother Bill, and has featured on *Ancient Aliens* (History Channel), *Legends of the Lost* with Megan Fox (Travel Channel), *Expedition Unknown* (Travel Channel) and many others. He lives in Ashfield, Massachusetts.

Hugh Newman is an explorer, megalithomaniac and co-author with Jim Vieira of *Giants On Record*, as well as the author of *Earth Grids* (2008) and *Stone Circles* (2017), and co-author of *Megalith: Studies In Stone* (2018) *Sensing the Earth* (2021) and *Geomancy* (2021). He is a regular guest on History Channel's *Ancient Aliens, Search for the Lost Giants, UnXplained with William Shatner* and has featured in *The Alaska Triangle* (Travel Channel), *Forbidden History* (Discovery Channel), *Mythic Britain* (Smithsonian Channel), *Ancient Civilizations* (Gaia), and several other TV shows and documentaries. Since 2006 he has been organising the *Megalithomania Conference* and the *Origins Conference* in London since 2013. He runs regular tours and expeditions worldwide and writes for numerous magazines. He has a Bachelor of Arts Honours Degree (BA Hons) in Film and Journalism from London Guildhall University. His worldwide adventures and lectures can be seen at www.youtube.com/megalithomaniaUK. His main website is www.megalithomania.co.uk. He lives next to Stonehenge in Wiltshire, England.

Acknowledgements

Hugh and Jim would like to thank the Newman, Casey, Murray and Gerstner Clans, Bill Vieira, Kyle Vieira, Ross Hamilton ('Godfather of Giantology'), JJ Ainsworth, Linus the cat (RIP), artists Dan Lish, Emmanuel Martin and Yuri Leitch, Chris Kelly, Gary Biltcliffe, Adora Gonzales, Cee Hall, Mark Poyner, Andrew Collins, Greg Little, Robin Heath, Howard Crowhurst, Michelle Bullivant, Nicholas Cope, Peter Knight, Alan Wilson, Giorgio A. Tsoukalos, James Zikic, Martin Morrison, David Hatcher Childress, Jenn Bolm, Hugh Evans, Cliff Dunning, Matt Sibson, Ioannis and Jo at Ancient Origins, John Martineau, Meg Ketch, Stuart Mason, Arianna and Enrico, Yousef Awyan, Keri & Daz, Megalithomania Patrons, Prometheus Entertainment, Like-A-Shot, Wild Dream Films, Gaia TV, and all our fellow explorers, researchers and friends. Special thanks to those who ventured to the other side: John Michell, Anthony Roberts, Chris Grooms, Keith Critchlow, Michael Glickman, Euan MacKie, Edmund Marriage, John Agnew, Sem Meakin and Chuck from cf-apps7865.

Dedicated to
Anthony Roberts
(1940 - 1990)

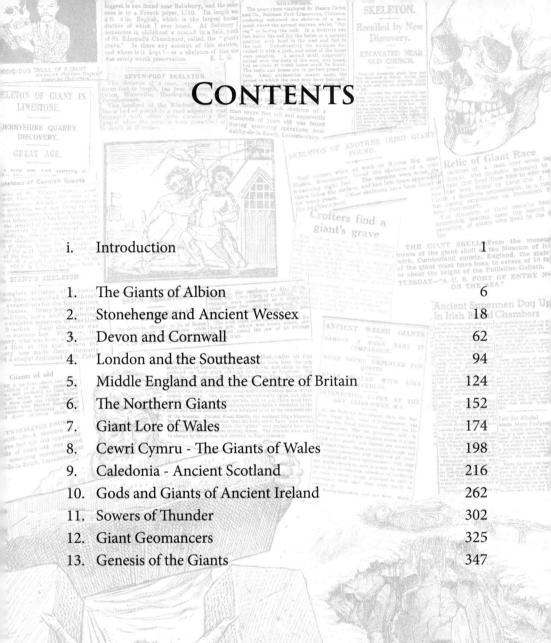

CONTENTS

INTRODUCTION

Stories of giants being involved in the construction of megalithic sites, earthworks and other ancient monuments have been alive in the consciousness of the British for millennia. Legends and creation stories harken back to an earlier age of elemental beings, high magic and giant kings ruling the land. Religious documents, medieval chronicles, oral traditions and origin stories all recount converging tales of giants being an integral part of the founding of the British Isles.

Giant effigies are still paraded around many cities and towns, keeping this ancient memory alive. Mystics, sages and esoteric sources all speak of giants as a literal reality, often originating in a lost sunken realm. Even Stonehenge's creation is attributed to remarkably tall and powerful builders. These were not only giants in stature, but also giants in intelligence, skill and wisdom.

Collating hundreds of historical accounts of massive bones and skeletons being found in the vicinity of sites such as Stonehenge, adds some credence to the idea that age-old myths encoded detailed histories and insights from many thousands of years ago. These were often linked with the secret arts, forgotten sciences and magic from a fabled "Golden Age." The epic annals of Ireland, secret Scottish archives, old manuscripts of Wales and Druidic traditions of England have revealed a lost timeline, a missing chapter in human history that provides evidence of giant human beings inhabiting, ruling and building the megalithic master works of Albion.

The reality of giants existing in prehistoric times is put under the microscope in this book, with the investigation of obscure newspaper accounts, antiquarian diaries, archaeological reports, local history records, newly-translated ancient texts, royal engineering survey data, academic papers, and written evidence from hundreds of sources going back several thousand years. In our previous publication, *Giants On Record: America's Hidden History, Secrets*

1

in the Mounds and the Smithsonian Files (2015), we examined the evidence from across the Atlantic. North America has many legends of the 'Tall Ones' and over one thousand historical, archaeological and news sources support the notion that giants once existed there, so why not in Britain?

It was during our research for *Giants On Record* that we uncovered dozens of accounts from the British Isles. We put them to one side for future reference. After a year of completing other projects, we started to get our teeth into a worldwide book on giants. However, we began by focussing on the chapter covering Britain, and it just kept going on from there, growing and morphing into something bigger. Because of this we turned our attention to Hugh's home country, expanding that already gargantuan chapter, becoming the book you hold in your hands today.

Giants are often said to be the builders of megalithic sites, by Emmanuel Martin.

Our goal in researching this controversial topic is not to be anti-scientific or needlessly provocative, but rather an honest effort to make sense of this profound mystery. We feel that the only difference between our investigative approach and an academic one is that we also believe that there is truth in oral tradition, myth and legend. However, we also understand that some accounts could be hoaxes, mismeasurements of bones, or fabricated news stories to sell newspapers. We have included a comprehensive survey of reports and investigated them to the best of our abilities, but ultimately, we leave it to the reader to decide on their authenticity.

One of our main inspirations for this book was the work of Anthony Roberts, a Glastonbury-based author who wrote the classic *Sowers of Thunder* in 1978. Roberts was a modern-day mystic and seer in his own right, guided by a strong intuition. While researching giant lore, he came to realise that giants were linked with aligning ancient sites, altering the landscape and generally terraforming the land to bring harmony to it and the people; a theory that correlated with the books of John Michell. This all falls under the umbrella of 'Geomancy', or what Roberts' called 'Geomythics'.

Throughout our research we have found many 'leys' (straight alignments of ancient sites) described in giant myths and landscape legends. Throwing large stones into triangular configurations also gets repeated in folklore in different parts of Britain, suggesting giants were measuring and surveying the land. Astronomical principles, geometry and mathematics are all found in the myths of giants. The reader will be introduced to all these themes in more detail during the text but if you want to be fully informed of the geomantic aspect of the giants, jump to the *Geomancer Giants* chapter. As you read through this book, you'll see that Roberts' analysis has now been backed up by 40 years of extra data, much of it difficult to ascertain or simply not in existence when he first published *Sowers of Thunder*.

It seems that the giants, although sometimes misperceived as bumbling oafs, were masters of surveying, astronomy and landscape knowledge to an extremely high degree. Many were remembered as high-kings and rulers of the country, often inhabiting mountain-tops or hillforts. Some are remembered as great shamans and magicians, who could control the weather, and place spells over their burial sites. That being said, there is the dark side to much of this giant lore with dozens of tales of malevolent giants with cannibalistic tendencies, violent attacks and unruly behaviour, just like we find in the biblical stories of the Nephilim and in other ancient traditions worldwide.

The origins of all four countries of the British Isles are intimately connected to ruling giants. Their legacy, when understood properly and investigated with an open mind, can lead to the conclusion that ancient Britain and its giant rulers were held in high esteem by the British populous and quite possibly by the rest of the world.

We hope this book can bring this obscure but fascinating story to life and highlight what a truly immense mystery this is.

Hugh Newman - Stonehenge Cottages, Wiltshire, UK
Jim Vieira - Ashfield, Massachusetts, USA
October 2021

Queen Elizabeth II examining the giant skeleton of Charles Byrne, 'The Irish Giant'. She is apparently an avid giantologist in her spare time.

TOP TEN GIANT DISCOVERIES IN BRITAIN

THE GIANTS OF STONEHENGE AND ANCIENT BRITAIN

FROM THE BEST-SELLING AUTHORS OF GIANTS ON RECORD

HUGH NEWMAN

10. CADAIR IDRIS, WALES - 7FT
9. ST MICHAEL'S MOUNT, CORNWALL - 8FT
8. LUNDY ISLAND - 8FT 7IN
7. GLASTONBURY ABBEY, SOMERSET - 9FT
6. MAESHOWE, ORKNEY - 10FT
5. GLENELG, SCOTLAND - 11FT
4. DUCHUIL, IRELAND - 12FT
3. ST BEE'S GIANT, CUMBRIA - 13FT 6IN
2. STONEHENGE, WILTSHIRE - 14FT 10IN
1. COR GIANT, NORTHUMBERLAND - 21FT

AVERAGE HEIGHT

0 FT 2 FT 4 FT 6 FT 8 FT 10 FT 12 FT 14 FT 16 FT 18 FT 20 FT 22 FT

1

THE GIANTS OF ALBION

Giants are at the heart of national folklore concerning the founding of Britain, and archaic traditions state they have inhabited the country since deep antiquity. Some say they were descendants of Noah, who arrived from Africa about 4000 years ago, dividing the island into three parts.[1] Other traditions state that the giant Albion founded Britain, the son of Poseidon, the Greek god of the sea. Albion represents the archetypal giant figure, being not only of great stature, but also of great wisdom and knowledge, bringing the surviving legacy from a pre-diluvian world. Britain's oldest acknowledged name has uncertain origins that historical author Geoffrey Ashe summarises here:

> Albion is documented quite far back. Greeks writing two or three centuries before Christ apply it to the larger of the two British Isles, Ireland being Ierne. It's meaning is uncertain. A derivation from albus, Latin for 'white', in allusion to the cliffs of Dover... Another etymology relates to 'Alp' and interprets it in Celtic terms as 'high ground', referring not only to cliffs but to hills and mountains on other parts of the coastline. [2]

After defeating the already existing culture with clever strategy and brilliant feats of arms, Raphael Holinshed (c.1525 - 1582), in his *Chronicles of England, Scotland, and Ireland* stated that Albion and the giants were said to have gradually consolidated their position in Britain, ruling the land for hundreds of years. After a long reign Albion went to the south of France (called Gaul at the time) to help his army defeat Hercules, but was defeated and died in battle.

William Blake's art and writing are full of magic and archetypal symbolism that carried the myth of Albion through the centuries and maintained a patriotism and a sacred connection to the British Isles. His poem 'Jerusalem' acted as a love song to the lost golden age:

> *Jerusalem, the Emanation of the Giant Albion! Can it be? Is it a truth that the learned have explored? Was Britain the primitive seat of the Patriarchal Religion? If it is true, my title-page page is also true, that Jerusalem was, and is, the Emanation of the Giant Albion... All things begin and end in Albion's ancient Druid rocky shore.*[3]

William Blake, *Albion Rose*, 1793-96.

The archetypal form of Albion held strong and the inspirational words of William Blake saw to it that he was the key symbol in the spiritual nature of Britain that was succumbing to Christian beliefs, with the Druids and megalithic sites remembered and honoured in his poems:

> *Giant Albion becomes a total archetype both for a primal, magical conception from the collective unconscious, and for a vision of this energizing image running across the surface of the sacredly realized topography.*[4]

The giant race of Britain were said to have continued for six hundred more years, although their numbers decreased and they ended up at the southwestern tip of Cornwall, until the arrival of Brutus after the Trojan wars. However, Britain's original name could also be from a giantess called 'Albina' (also written as Alba, or Albine), who, like the giant Albion, originated in Greece:

> *The Chronicles of Britain, written by John de Wavrin between 1445 and 1455, relate that in the time of Jahir, the third judge of Israel after Joshua, Lady Albine and her sisters came to, and settled in, an island which they named Albion after her, and which afterwards got the name of Britain. While they were living there the devil assumed the shape of a man, and dwelt among the wicked women, and by they had issue great and terrible giants and giantesses, who afterwards much increased and multiplied, and occupied the land for a long time, namely, until the arrival of Brutus, who conquered them.*[5]

The story of Albina has several variations. One version says she was "*a very tall and beautiful girl,*"[6] so potentially had 'giant genes' herself, as her father and mother were also described as taller than average. Albina's father was said to have ruled Greece 3970 years ago, with one account suggesting her father was Diocletian, the King of Syria.[7] Most versions agree that her father had thirty-three wicked daughters, and to kerb their unruly ways he found thirty-three husbands for them. The daughters were displeased and under the leadership of the eldest sister Albina, they plotted to cut the throats of their husbands as they slept. They succeeded in their plan but knew there would be severe consequences:

> *For this crime they were set adrift in a boat with half a year's rations, and after a long and dreadful journey they arrived at a*

great island that came to be named Albion, after the eldest. Here
they stayed, and with the assistance of demons they populated the
wild, windswept islands with a race of giants.[8]

Albina and other daughters of Diodicias (front). Two giants are in the
background, next to a ship carrying Brutus and his men.
French Prose Brut, British Library Royal 19 C IX, 1450-1475.

These dark entities (Incubi) heard the wishes of the thirty-three sisters for males to relieve their sexual needs. They could take on human form and once they had impregnated Albina and her sisters, they disappeared back into the underworld. John de Wavrin, in the 15th century, described the Devil moving in amongst the women, and the results were a particularly powerful and nasty race of giants who were the sole occupiers of the country before the arrival of Brutus. Her first son was thought to have been called 'Albion,'[9] although this could have been added to the story at a later date to explain why Albion was classed as a giant, as the story of Albion arriving from Greece may not have been known at the time.

It should be noted that story of Albina stayed alive in legend until the 19th century, but never reached the popularity of the stories of Brutus arriving from Troy. The tales also parallel the story of the Grigori (or Watchers) of the Bible, who mated with human women and birthed the wicked Nephilim giants, who have similar traits to their British counterparts. Even after they had given birth to a new giant race, there were still no men to partner with the sisters, so they bred with their own sons when they reached an appropriate age. Then the daughters bred with their brothers, and the offspring continued to get larger in stature and more unruly as the inbreeding continued.

However, did they really exist? A rare manuscript from the 15th century hints that the evidence was found beneath our feet:

> *The huge bones which can be found in many places throughout the world are evidence for this; men dig them up in both the country and the city – teeth, legs, ribs, and thighbones four feet long, shoulder-blades as large as a shield. Many people are perplexed by this and wonder if there could ever have been men with such huge skeletons. The giants were hideous to look at because evil spirits had begotten them; their fathers were devils, and their mothers were tall and well-built. They were just the children you would expect from such a union.* [10]

The arrival of Brutus to England and the slaying of giants from Geoffrey's *Historia Regum Britannae*, in Harley MS 1808, fol. 30, British Library. c.1400 and circa 1415.

The giants devolved into what became known as the 'Fairy Race', who multiplied and spread across the land, building great Cyclopean walls and earthen ditches. They preferred the abode of mountains and 'high places'. From there they could dominate their territory and be safe from attack. Many of the huge hillforts of

Britain were thought to be their strongholds. Tradition says that from the time of the arrival of Albina and her sisters, to the time of Brutus was 260 years (or 600 years – depending on which version). During that era the ferocious giants battled between themselves defending their strongholds. Eventually, they were reduced in numbers striving to conquer the whole land, and in the resulting civil wars they slew each other until only twenty-four of them remained in Cornwall, including the infamous Gogmagog.

Geoffrey of Monmouth's influential 12th century *Historia Regum Britanniae* (History of the Kings of Britain)[11] states that hundreds of years after the giants had populated the island, Brutus, the great-grandson of the Trojan hero Aeneas, slowly made his way to Albion. Troy had been defeated by the Greeks and Brutus and his men settled here due to its beauty and fecundity. Geoffrey states that the modern name of Britain came from Brutus. During his trials and tribulations on his epic journey before arriving, the goddess Diana spoke to him in a dream, inspiring his destiny to reach Albion:

> *Brutus,—past the realms of Gaul, beneath the sunset*
> *Lieth an Island, girt about by ocean,*
> *Guarded by ocean—erst the haunt of giants,*
> *Desert of late, and meet for this thy people.*
> *Seek it! For there is thine abode for ever.*
> *There by thy sons again shall Troy be builded*
> *There of thy blood shall Kings be born, hereafter*
> *Sovran in every land the wide world over.* [12]

Geoffrey asserted that he had translated the *Historia* into Latin (in about 1136 AD) from 'a very ancient book in the British tongue,' that was loaned to him by Walter, Archdeacon of Oxford. What this book was, has had scholars debating for centuries, but it could have been the *Historia Brittonum* (History of the Britons)[13] from the 9th century, written by Nennius, a monk from Bangor, Wales. The lost manuscript featured stories of Arthur, giants and knowledge of the Bards and Druids.

An important section of Geoffrey's text has Brutus and his men realising that Albion was already partly populated by unexpectedly tall foes, the descendants of Albina (or Albion): *"It was uninhabited except for a few giants... they drove the giants whom they had discovered into the caves in the mountains."* [14] They arrived on the south coast, sailed up the River Dart and disembarked at Totnes. Brutus stepped ashore on a sacred rock, and pronounced: *"Here I am and here I rest, And this town shall be called Totnes."* [15]

Corineus throwing Gogmagog off the cliffs.

The rock became known as the *Brutus Stone* and what's left of it is on the High Street in Totnes. However, it has been weathered and trimmed to fit the path it was embedded in. Totnes is an ancient sea port and was a settled area stretching back into prehistory, so the Brutus Stone was probably placed hundreds of years previously, possibly by the ancestors of Gogmagog and his followers. [16]

After scaring off the remaining giants and launching regular attacks on the titans, the land was divided up and Corineus was given the southwestern area of Cornwall to rule, named after him. There is no indication that they were willing to learn from the giants, or respect their culture in any way. As Anthony Roberts points out, Gogmagog represented the first patriot in recorded British history, protecting his homeland from invaders:

> *Corineus experienced great pleasure from wrestling with the giants, of whom there were far more there than in any of the districts which had been distributed among his comrades. Among the others there was a particularly repulsive one, called Gogmagog, who was twelve feet tall.* [17]

Other chroniclers state that he was in fact twelve cubits tall, so this would have made him between 18 and 20 feet in height. He was described as being so strong, that he could uproot an oak tree and shake it like a hazel wand. (The hazel wand is a well known dowsing instrument and represents the geomantic techniques mentioned in hundreds of giant myths, that we will explore in more detail in the *Giant Geomancers* chapter).

The original name of Britain's first patriot was "Goemagot", and only later was it refined to "Gogmagog", probably because of the two princes in the Bible called Gog and Magog. The ferocious giant attacked Corineus' camp with twenty of his kin. This turned into an all out battle and Corineus and his men called on their local allies and eventually defeated them in a bloody massacre. Brutus chose to keep one of the giants alive, as he wanted to witness a wrestling match between Gogmagog and Corineus. During the tightly fought match, Gogmagog 'hugged' him and 'broke three of Corineus' ribs' and he was so enraged, he hoisted the giant up on his shoulders with superhuman strength and ran to the cliff where he threw him off to his death. His body smashed into many pieces after hitting sharp rocks and stained the water red, that *"was so discoloured with his blood as to continue tinged with it for a long time."*[18] Curiously, at the Devonshire town of Stoke Fleming, just west of the mouth of the River Dart, *"a very slight churning of the sea produces red waves, as though coloured by a giant's blood."*[19]

> *Across the water, at Torpoint, is a small promontory called Deadman's Point, an interesting coincidence in that a giant of the ogreish, child-eating variety, who formerly enjoyed a bleak reputation in the Gorran Haven District, was bled to death by a wily doctor at Dodman's Point.*[20]

The term *dodman* may be recognisable to 'ley hunters' and followers of Alfred Watkins. It is part of a controversial theory that a dodman was the name given to ancient surveyors, which is represented in a hill figure, the Long Man of Wilmington in Sussex. As we'll discover later in the book, surveying may have been part of the skill-set of the archaic giants, and a further chalk hill figure may also have been present at the location of Gogmagog's final battle. The cliff from which he was thrown became known as *Langnagog* or 'The Giants Leap.' The location was Plymouth Hoe, the legendary place where the wrestling occurred because it was recorded in 1486 that a giant turf-cut figure depicting two foes once existed, one of them being Gogmagog.[21]

South of the chapel in the Royal Citadel was a Giant's Grotto or cave which became the site of a military magazine.[22] It is interesting that in the 1500s, when word came back from Patagonia and North America of 'living giants', it may have inspired revivals in the old legends of the early giants of Britain, and this could have been why the figures were created.

Artist's impression of how the chalk-hill figures on Plymouth Hoe may have looked.

Other traditions state that Gogmagog was thrown *"from one of the rocks not farre from Dover."*[23] However, Plymouth has the strongest claim as a huge jaw and teeth were found in the vicinity, thought to be those of the giant:

> *A 19th-century folklorist was told that, when the foundations of the Citadel at Plymouth were being dug, gigantic jaws and teeth were discovered that were identified as the bones of Gogmagog. Possibly similar finds in earlier times gave rise to the story of a giant's death which was by Geoffrey attached to the legend of Brutus because of his traditional landing place at Totnes.*[24]

Another account relates that a huge femur was discovered in a burial somewhere in Devon, but it is unclear where exactly this was unearthed:

> *A stone coffin in Devonshire contained a thigh-bone belonging to a man eight feet nine inches high.*[25]

The names of Gog and Magog first appear in the Hebrew Bible with reference to Magog, son of Japheth in the *Book of Genesis*, then Gog, the king of Magog, appears in the *Old Testament* in Ezekiel (38:2) as the instigator of a terrible battle. Gog was referred to as being a person and Magog was the land he was from. Similar stories are echoed in the *Book of Revelation* and the *Qur'an*. The tradition is somewhat convoluted as Gog and Magog are presented as men, supernatural beings (giants and demons), national groups, or lands and appear widely in other folklore and mythology. For example, Gogmagog is identified with giants in Edmund Spencer's *Faerie Queen* (1590) and in the medieval legends of Alexander. William Borlase, in *The Dolmens of Ireland*, mentions 'Gig-na-Gog,' the presiding giant of an Irish mound near Beardville.[26] The names even reached Cambridge in eastern England where the low lying hills to the south became known as the 'Gog Magog Hills', where some taller than average skeletons were unearthed in the 1800s and legends of giants prevailed for hundreds of years (see *London and the Southeast* chapter).

After defeating the giants, Brutus travelled all over the country to find a suitable spot to rule from. He decided on the River Thames and founded the city of Troia Nova, or New Troy, which became Trinovantum, which we now know as London, with his captured giant in tow. A later version of the story describes how the giants Gog and Magog were now two giants and were taken prisoner and forced to become porters at the Royal Palace, now the London Guildhall. The effigies of Gog and Magog have been (in various incarnations) the guardians of London since the reign of Henry V. In 1413 Henry V was passing over London Bridge when greeted by the City's first recorded giant. In 1415, after the triumph of Agincourt, he was greeted by a male and female giant at the Southwark gate entrance to London Bridge. The male giant held the city keys, and is the first recorded porter of London. In 1432 Henry VI was greeted by a single giant holding a huge sword. Also, in 1522, Emperor Charles V visited London and was greeted by two giants at London Bridge, notably called Hercules and Samson. In 1554, King Philip of Spain, accompanied by Queen Mary, was confronted by the giants *Corineus Brittanus* and *Gogmagog Albionus*. Although they were burnt down in the Great Fire of 1666, new wooden giants, over 14ft tall and carved by Richard Saunders in 1708, replaced them. In *The Gigantick History of the Two Famous Giants of Guildhall* (1741) it proclaims:

> *Corineus and Gogmagog were two brave giants who richly valued their honour and exerted their whole strength and force in the defence of their liberty and country; so the City of London, by placing these, their representatives in their Guildhall, emblematically declare, that they will, like mighty giants defend the honour of their country and liberties of this their City; which*

excels all others, as much as those huge giants exceed in stature the common bulk of mankind.

Gog and Magog. Wooden effigies in the Guildhall of London after the great fire of 1666, from Volume 1 of *Old and New London*, Illustrated by Walter Thornbury.

Of the two giants, Gogmagog, represented the unruly, savage aspect, equipped with bow and arrow and a spiked globe known as the 'Morning Star'. Corineus, the hero of Troy, was presented as a Roman warrior, with sword and spear. After some 250 years of protecting the city, a German air raid in 1940 destroyed them again, so in 1953 a new set of giants reincarnated as Gog and Magog were set in place. These latest incarnations are nine feet tall and are kilted, bearded and carry lances, prepared to protect the Guildhall at any cost. The Guildhall is London's most venerable reference library, containing books on topography, history, genealogy and heraldry, along with a mighty collection of prints and maps. Furthermore, in the 16th century a 26-inch-long femur bone was unearthed next to the Guildhall that we will investigate in the *London and the Southeast* chapter.

Do these effigies represent the ancient role of giants in Britain not only as protectors of the land, but guardians of ancient knowledge? Wicker versions of Gog and Magog are paraded around London on the Lord Mayor's Show festivities every November, representing giants founding the land of Britain.

To complete this chapter, we visit a fragment of a prehistoric monolith called the London Stone. It is what is left of an ancient menhir thought to have been brought to the capital from Troy by Brutus and incorporated into the Temple of Diana, which was set up shortly before the founding of the city. The original temple may have been at the current site of St Paul's Cathedral as a statue of the goddess was found between Deanery and Blackfriars, and a stone altar dedicated to Diana was found nearby in 1830.

Now embedded in a wall on Cannon Street, the enigmatic stone was once a much more prominent monolith. In 1742 it was six feet high. Before that it was recorded as an established landmark in the twelfth century, where pacts and proclamations were made and was said to be the place where all distances from London were measured. An old saying goes: "*So long as the Stone of Brutus is safe, so long will London flourish.*" Like Gogmagog, the stone is said to protect Britain from invaders. It also has a connection to Albion, the giant founder of Britain, as mentioned by William Blake:

Where Albion slept beneath the Fatal Tree,
And the Druids' golden Knife
Rioted in human gore, In Offerings of Human Life...
They groan'd aloud on London Stone.[27]

Although the founding of Britain is still shrouded in mystery, the stories of the giants seem to go back very far in time. The legends and foundation myths of Britain give some indication that they could be the ancestral memories of real-life giants who ruled here long before Britain emerged as we know it:

The ancient Britons were remarkable for the large stature of their bodies; their eyes were generally blue, which was esteemed a great beauty; and their hair red or yellow, though in many various gradations. They were remarkably swift of foot, and excelled in running, swimming, wrestling, climbing, and all kinds of exercises in which either strength or agility were required. Accustomed to hardships and despising cold and hunger, in retreating they plunged into the morasses up to the neck, where they remained several days. They painted their bodies with a blue dye extracted from woad, and at an early age they were tattooed in a manner the most ingenious and hideous; and in order to exhibit these frightful ornaments in the eyes of their enemies, they threw off their clothes in the day of battle. When advancing to the combat their looks were fierce and appalling, and their shouts loud, horrid, and frightful.[28]

AVEBURY STONE CIRCLE

ALDWORTH CHURCH
GIANT EFFIGIES

Grimsbury Castle,
Newbury, p.40

West Kennet Long Barrow, p.33

Bulford Camp, p.25
Bush Barrow, p.25

STONEHENGE
'THE GIANT'S DANCE'

Salisbury, 9ft 4, p.29

SALISBURY
GIANT

Twyford Down, 7ft,
p.44

Ivychurch Priory,
14ft 10, p.27

Portsdown Hill,
7ft, p.43

Bevis' Grave, Arundel,
p.39

BEVIS STATUE, BAR
GATE, SOUTHAMPTON

MOTE-STONE,
ISLE OF WIGHT

Shanklin Train Station,
p.45

GORM THE GIANT
AT ASHTON COURT

Worlebury Camp, p.55

Loxton Church,
7ft, p.56

Norton St. Philip's,
p.55

Knightstone Island, 9ft, p.59

Wedmore, p.53

Glastonbury Abbey,
9ft, p.46

GLASTONBURY
TOR

Dunster Castle, 7ft, p.57

MAIDEN CASTLE

Sturton Caundell,
p.53

THE CERNE GIANT
HILL FIGURE

Flower's Barrow,
7ft, p.38

Verne Hill Fort, p.37

The Grove, Portland,
10ft, p.38

2
STONEHENGE AND
ANCIENT WESSEX

Wiltshire contains some of the world's most important megalithic sites, and many have giant associations at their core. Numerous Giants' Graves are scattered across this county and various place names have the word 'Og' in them. This title may refer to the biblical giant King Og of Bashan, Gog and Magog from *Revelation*, or the British patriot Gogmagog. Ogbourne St George and Ogbourne St Andrew are two examples, and both have the River Og going through them. In Salisbury, an eighteen-foot giant has been paraded around the town for centuries, and some well-known sites in the Avebury and Stonehenge landscapes were said to be built by these colossi.

Stone circles offer a glimpse into the mysterious world of our ancestors. Druid ceremonies, ancient astronomers and Pagan dances inspire our imagination as to what might have been. Stonehenge is the most famous of all

Imagined Druid festival at Stonehenge, 1815, by C.H. Smith and S.R. Meyrick.

19

stone circles, but it is also unique, in that it has lintels and trilithons in its design. Stonehenge is known for its summer solstice sunrise alignment, but many other circles also have hidden astronomical uses, as well as encoding sophisticated geometry, landscape alignments and ancient measurement systems. Back in the 1600s many local Christians believed they were formed by nature, built by the devil or in most cases, constructed by giants.

Three massive pine poles marked Stonehenge's location on Salisbury Plain 10,000 years ago during the Mesolithic era. Thousands of years later, great earth moving efforts began, shaping the landscape on a grand scale. These included a mighty 1.7 mile-long cursus that was constructed around 3500 BC. In 3100 BC the first structure was constructed with 56 holes around the interior of a circular ditch. These may have contained the bluestones from Wales, that were transported 140 miles. A

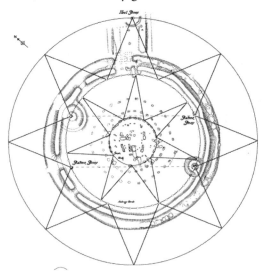

John Martineau's diagram showing sophisticated 8-fold and 7-fold geometry within Stonehenge.

discovery in 2020 revealed 20 large circular pits dating to 2500 BC that formed a massive circle around Durrington Walls measuring 1.2 miles in diameter. The pits (or shafts) were cut out of the chalk and are up to twenty metres wide and five metres deep.[1]

Like Avebury, Stonehenge has a sacred avenue but with no stones along it. At the very end of the sacred pathway, a stone circle once rested on the banks of the River Avon in Amesbury. It was excavated in 2009, with evidence it had bluestones from Wales in its construction, dating from between 3000 and 2400 BC.[2] The great sarsen stones that make up the main circle and the trilithons were erected in 2400 BC, but between the first circular construction and up to 1600 BC, numerous changes were made within the circle with stones being shifted around into various configurations.

The original name for Stonehenge was *Chorea Gigantum* (The Giant's Dance). Stonehenge may be a later Saxon name that roughly translates to 'Hanging Stones' or 'stones supported in the air'. Old English words *stān* meaning 'stone', and *hencg* meaning 'hinge' may also describe the lintels hinging on the uprights.[3] In *Camdens Britannia* from 1772 early names of the monument were discussed with interesting variations:

> *Name of Stonehenge. Leland's opinion that the British one, Choir gaure, should not be translated Chorea Gigantum, a Choir of Giants, but Chorea nobilis, a noble Choir; or else that gaure is put for vaure, which makes it Chorea magna, a great Choir, is probable enough.*[4]

The name *Choir Gaure* (or Gawr) is the Welsh name for the *Giant's Dance* and the relationship to 'choir' is also of interest because recent research in the study of archaeoacoustics has revealed that Stonehenge and the Bluestone site in Wales have pronounced acoustic properties. Another traditional Welsh name for Stonehenge is *Côr y Cewri* or "Council of the Giants."[5]

The earliest illustration of Stonehenge shows two average-sized gentlemen watching a giant lift one of the lintels into place. He is estimated to be about 14 - 16 feet tall based upon the known dimensions of Stonehenge. The twelfth century story that this image comes from is a manuscript called *Le Roman de Brut* by Robert Wace thought originally to be dated to 1155 AD.[6]

Illustration from *Le Roman de Brut* by poet Wace c.1155 AD
showing a giant, Merlin and King Ambrosius.

In *Legendary History and Folklore of Stonehenge*, it states that the illustration is by B. M. Egerton created in the mid 14th century, showing Merlin at work on the trilithons.[7] Whenever it first appeared, it is the earliest illustration of the stones, and appears to show Merlin, King Ambrosius and a giant.

Historia regum Britanniae, (The History of the Kings of Britain) originally called *De gestis Britonum* (On the Deeds of the Britons), is a controversial historical account of British history written around 1136 by Geoffrey of Monmouth. It describes the most famous Arthurian tales and speaks frequently of giants. Geoffrey's *Histories* explains that Merlin relocated and constructed a megalithic monument in honour of 460 slain warriors defeated by Hengist the Saxon, under orders of King Aurelius Ambrosius. The stones, originally transported by giants from Africa, were erected into a stone ring in Ireland. There they stood for thousands of years:

> *If you are desirous, to honour the burying-place of these men with an ever-lasting monument, send for the Giant's Dance, which is in Killaraus, a mountain in Ireland. For there is a structure of stones there, which none of this age could raise, without a profound knowledge of the mechanical arts. They are stones of a vast magnitude and wonderful quality; and if they can be placed here, as they are there, round this spot of ground, they will stand for ever.*

After the 15,000-strong army (led by Uther Pendragon) failed in dismantling and transporting the stones, Merlin stepped in and used what was described as "gears" or "engines". Other accounts talk of the wizard utilising "Magic" and "sleight" to do the job.[8] He effortlessly got them onto boats before reconstructing them on Salisbury Plain in the exact configuration. The stones were said to have been chosen for a very specific purpose, as healing stones of great sanctity. In *Histories*, Merlin states:

> *They are mystical stones, and of a medicinal virtue. The giants of old brought them from the farthest coast of Africa, and placed them in Ireland, while they inhabited that country. Their design in this was to make baths in them, when they should be taken with any illness. For their method was to wash the stones, and put their sick into the water, which infallibly cured them... There is not a stone there which has not some healing virtue.*

Furthermore, in Libya in North Africa, similar megalithic trilithons have been reported and photographed that look strikingly similar to Stonehenge, suggesting there was indeed a prehistoric connection between Africa and Britain, as stated in the text. Numerous legends have North Africa as a haven for the Titans, with stories stretching all the way to Morocco, where a mighty stone circle and burial mound called *Msoura* is thought to be grave of the giant warrior god called *Atlas* (or Hercules).

Old photos of trilithons near Tripoli in Libya.

THE CANGICK GIANTS (PART 1)

The depiction of the giant in Wace's illustration may have been inspired by other traditions within the Wessex landscape. Five hundred years after Geoffrey's *Histories*, Rev. Robert Gay authored *A Fool's Bolt Soon Shot at Stonehenge*,[9] about a ferocious prehistoric culture of giant warriors whose earliest incarnations were as semi-divine beings. The later breed of the *Cangick Giants* became fully human, with their remains found all over the West Country. The *Cangi* were traditionally the builders of Stonehenge. They were based in Somerset, but their territory reached to Wiltshire. The antiquarian views of Gay did not take hold in the British imagination, but how did these particular giants become known as the builders of Stonehenge?

Gay's controversial book was written in 1666 (published in 1725), and provides a unique version of British history. He believed that British people had built Stonehenge and were the "Old Britons" who preceded the "Belgae, Romanes, Saxons, Danes and Normans." Previous to all these old cultures, Gay wrote of a battle that was won by the Cangi tribe against the invading Belgae. However, these giants were not just local folklore, because, Gay wrote: "*some huge bones of men, found amongst others, in the said burrows, as aforesaid, and in other places near Stonehenge.*" We will examine some of the accounts he is referring to in this chapter.

Giants with long hair and clubs. Johan Picard's view of the Iron Age in Britain c1660.

The story gives an interesting history of a powerful group of warrior giants defeating invading armies, who celebrated by building Stonehenge. Moreover, he wrote: "*The Cangick Giants having conquered, triumphed over their enemies… they thought it was expedient to erect this Monument, as their 'Trophie'. That monument was Stonehenge.*" According to the author this is not the only reason why Stonehenge was called The Giants' Dance, as he states "*this triumphant singing and dancing together, as the time and place of Victory, was the common practice of the ancients.*"[10] Descriptions of Stonehenge have said over the centuries that it looks like a circle of giants arm in arm dancing in a victorious circle of celebration.

This could certainly be the memory of an older story that got mixed up when passed down from generation to generation. Further confusion prevailed as seen in this account describing Stonehenge as a sports arena for giants:

> *It commemorates a bloody battle over the Belgae by the Cangick giants; whilst in a manuscript Jones saw with an uncle of his, in the Welsh language, said to be written by Humphrey Llwyd, evidently a piece of ingenious raillery, it is made out to be a play-place of the giant race, where the game was a sort of complicated cricket, and that the holes observable in some of the stones were occasioned by the balls striking against them.*[11]

24

WILTSHIRE

The first two accounts we present here are from the Stonehenge area and are by no means giants. However, the descriptions given suggests the discoverers were shocked by what they found, as these were powerfully built individuals and far taller than the discoverers, and consistently taller than the earlier Neolithic stock who were also being unearthed around this time.

BUSH BARROW: GRAVE OF A TALL CHIEFTAIN

One of the most famous discoveries (in 1808) within the Stonehenge environs was the Bush Barrow Lozenge, a stunning diamond-shaped breast plate made of gold with mysterious inscriptions and geometry carefully carved on it. The owner of this piece was reported to be taller than average, and obviously a Chieftain, quite possibly a ruler of Stonehenge in the Bronze Age.

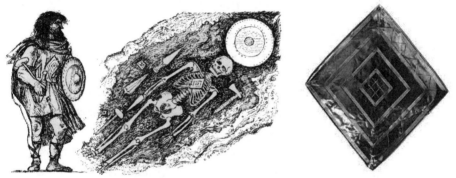

Left: Artist's impression of Chieftain from Bush Barrow near Stonehenge.
Right: The Bush Barrow gold lozenge.

On reaching the floor of the barrow, we discovered the skeleton of a stout and tall man lying from south to north....Near the right arm was a large dagger of brass and a spearhead of the same material, full 13 inches long, and the largest we have ever found.[12]

LARGE BONES FROM BULFORD CAMP MOUND

A similar description refers to a discovery at a mound near Bulford Camp, east of Stonehenge. It is one of three mounds that form a triangle, and they can be seen from Stonehenge behind the New King Mounds in the far distance. In Bulford Village the military camp was built around a mounded landscape, with some on public land in residential areas. The most prominent in the area, that sits higher than all the others, is Bulford Camp Mound, where a powerful individual was interred:

> *Remains of 3 skeletons were found in 1950, 1 in a sitting position and 1 of great size with a low brow. The bones were much mixed.*[13]

Bulford Camp Mounds.

The description is fascinating as it describes a "low brow," a description that we found frequently in our research of the North American giants and also found in the skulls of the *Beaker People*. The next passage outlines the difference between the Beaker People and the earlier Neolithic inhabitants:

> *The invaders differed somewhat from the former inhabitants of the land. The Neolithic folk seem to have been of moderate stature, long headed, oval faced, narrow nosed, with small features. They were not at all a powerfully built race. The new-comers on the other hand - according to Abercromby - were characterised by short square skull showing a great development of the superciliary ridges and eyebrows. The cheek-bones, nose and chin were prominent and the powerful lower jaw was supplied with large teeth. They were a tall, strongly built race and must have presented- at any rate as far as men were concerned - a fierce, brutal appearance. The dead were buried in round barrows, inhumation being practiced. They knew about the use of copper and introduced into England the beaker type pot.*[14]

Ritual burials in a Beaker mound (left) and Neolithic long barrow (right). L. Grinsell.

These human traits have been noted by many commentators, and suggest that powerful interlopers decimated the earlier weaker inhabitants.

Nearby, in the vicinity of Salisbury, further mounds started revealing their secrets in the 16th century.

SKELETON NEAR SALISBURY - 14FT 10IN

Thomas Elyot describes a stunning discovery that made the headlines in the early 1500s and featured in his famous *Dictionary*:

> *About 30 years since I myself, being with my father, Sir Richard Elyot, at a monastery of regular canons (three or four miles from Stonage) [Stonehenge] beheld the bones of a dead man found deep in the ground, which being joined together, was in length 13 foot and ten inches, whereof one of the teeth my father had, which was of the quantity of a great walnut.*

Sir Thomas Elyot (c.1490–1546) was a diplomat, scholar and Member of Parliament for Cambridgeshire. He was best known as one of the first proponents of the use of the English language for literary purposes. Portrait by Hans Holbein the Younger.

In *Journey into South Wales* (1802) George Lipscomb extends the height of the giant skeleton by one foot:

> *...it should be remembered, that Leland, in his "Collectanea", quotes the respectable authority of his friend, Sir Thomas Elyot, as recording, that himself had seen, at some place, near Salisbury, a skeleton which measured fourteen feet ten inches in length.*

A further account from the 1500s (updated in 1883) had more details and agreed with the extra foot in height:

> *An incident which occurred in the course of one such ramble is related by Sir Thomas Elyot in his Dictionary. It appears from this narrative that whilst Richard Elyot and his son were visiting the monastery at Ivy Church, a short distance from Salisbury, some workmen who were engaged in digging stone happened to turn up*

some human bones, which when put together formed a gigantic skeleton measuring no less than 14 feet 10 inches in length.

Thomas Elyot speaks of it as having taken place, about *"xxx years passed."* [15] This puts it at about 30 years before the first edition of Elyot's *Dictionary* which has been dated to around 1538, so this would have been around 1508 AD. John Leland wrote *De Rebus Britannicis Collectanea in Six Volumes* in 1774 and the knighted Sir Thomas Elyot was, as Leland, a respected voice of the time.

Camden evidently alludes to the same incident, though he gives a somewhat different version of it:

...as a tradition runneth, in our grandfathers remembrance was found a grave and therein a corpse of twelve foot and not far of a stocke of wood hollowed and the concave lined with lead with a book therein of very thick parchment all written with capital Roman letters. But it had lien so long that when the leaues were touched they fouldred to dust. Sir Thomas Elyot, who saw it, judged it to be an Historic. [16]

Thomas Elyot observing the giant skeleton, which was found with an ancient book with unusual script on it. A large inscribed metal disc was found nearby. By Dan Lish.

Here we have a 13ft 10in skeleton, also recorded as being 14ft 10in, and finally a 12ft one as well, the final account hinting at strange inscriptions on the mysterious 'book'. Many different heights are mentioned, but one thing is agreed upon, this was a seriously big skeleton. The log coffin in the final account is very similar to one described in Glastonbury Abbey, as discovered in the late 1100s, that we will investigate later in this chapter.

Another fact also mentioned by Camden, in speaking of Stonehenge, points to Elyot's familiarity with this part of Wiltshire and a fascinating artefact that is now lost:

> *I have heard that in the time of King Henry the Eight there was*
> *found near this place a table of metal as it had been tin and lead*
> *commixt, inscribed with many letters but in so strange a character*
> *that neither Sir Thomas Elyot nor Master Lilye, Schoole-maister of*
> *Paules, could read it, and therefore neglected it.*

The description of a "table" of metal and a mixture of "tin and lead" could be connected to the 12ft version of the giant account with the mention of the book with the strange script. Whatever it was, and what was written on it, will never be known.

THE SALISBURY GIANT - 9 FEET 4 INCHES

In *A Theological, Biblical, and Ecclesiastical Dictionary* by John Robinson (1830), it describes a 9ft 4in skeleton unearthed a few miles south of Stonehenge in 1719:

> *Camlet says, that in 1719, near Salisbury in England, a human*
> *skeleton was found, which was nine feet four inches long.*

> Delrio affirms that in 1572, he saw at
> Rohan a native of Piedmont above nine
> feet high. Calmet says, that in 1719,
> near Salisbury in England, a human ske-
> leton was found, which was nine feet four
> inches long. Becanus saw a man near ten
> feet, and a woman full ten feet high.

Another source recounts a further clue as to where exactly this huge skeleton came from. It tells of a local woman remembering some details from her childhood in a local publication. The mound was a *Giant's Grave* next to St Edmund's Church. This particular church was part of the original college founded by the Bishop of Salisbury, Walter de la Wyle in 1269, and is located a few miles from Stonehenge. The earthwork is located just north of St. Edmund's Churchyard, now the *Arts Centre* (founded in 1975), part of the greater Stonehenge landscape. The discovery was published in many sources.[17]

GIANT SKELETON AT SALISBURY.—A French paper on giants gives a list of several, whereof the biggest is one found near Salisbury, and the reference is to a French paper, 1719. Its length was 9 ft. 4 in. English, which is the largest human stature of which I ever heard. At Salisbury I remember in childhood a mound in a field, north of St. Edmund's Churchyard, called the "giant's grave." Is there any account of this skeleton, and where is it kept?—as a skeleton of that size was surely worth preservation.   E. L. G.

THE NEW SALISBURY GIANT

Curiously, in Salisbury, a giant still resides but in a different form. Not skin and bone, but made of textiles and paper mache! Both the Salisbury Giant and Hob-Nob (a hobby horse) were pageant figures created by the Salisbury guild of Merchant Tailors in 1447. The Giant was first recorded in 1496 and even met King Henry VII and his Queen, who were staying at nearby Clarendon Palace. Many similar 'giants' and other Pagan models were destroyed during the reformation, but this one survived, being paraded around Salisbury every year on St John The Baptist's Day, 24th June. Clearly this is connected with the Summer Solstice celebrated at Stonehenge.

The giant and his minions would parade through the city streets accompanied by great crowds with music and singing, with the heavy thud of a bass drum representing the giant's footsteps. Hob-Nob always accompanied him, preceding the giant, clearing the path for him so that he did not topple onto the crowds. The pageant had an unsettling and arcane atmosphere, harking back to Pagan times. From 1746 onwards all the giant's outings were on days of national celebration and he was recorded as being 25 feet tall. In 1784 the giant became known as St Christopher for the first time. In 1869 Christopher the giant and Hob-Nob were given to the Salisbury Museum, purchased in 1873 for £1.10s.0d (£1.50). In recent times the giant has taken part in celebrations for St George's Day (23rd April) that includes

The Salisbury Giant aka St Christopher on display at the Salisbury Museum.

an elaborate re-enactment of St George valiantly slaying a dragon. In 2012 the giant made its last appearance on the streets of Salisbury due to a moth infestation in his garments. He has since been retired and is on display in the museum.

Who really was St Christopher? According to legends of his life he was first named 'Reprobus', and was a Canaanite from the Bible lands. He stood at 5 cubits (7.5 feet /2.3m) tall and was said to have a fearsome look about him. He served under the King of Canaan, but felt this wasn't his true calling and chose to serve "the greatest king there was". This, of course, was Jesus who, as an infant, was famously carried across a river by the saint.

A Canaanite giant? In Salisbury? When the authors realised that a giant who was a Canaanite from the Bible lands had been paraded around the nearest city to Stonehenge for hundreds of years, the reality of the association of giants to the great circle suddenly made sense. Why would a secret brotherhood want to maintain this tradition for so long? Had the founders hidden some secret knowledge lost to us in the modern era? It seems they were holding on to this pageant because it encoded certain truths as to the origins of their beloved Stonehenge. Furthermore, the biblical connection cannot go unnoticed and this repetitive link with earlier foreign giants will be explored a little later.

As we dug deeper we found more information on the enigmatic St Christopher, this example from 1876 stated, *"This Saint is generally represented as of a gigantic stature."*[18] In *The Golden Legend*, Jacques de Voragine described St Christopher, *"He was of gigantic stature, had a terrifying mien, was twelve coudees tall."* A coudee is an ancient measurement of about the distance from the elbow to the end of the middle finger, often referred to as a cubit. The cubit is usually between 18 and 21 inches long. By these calculations St Christopher must have stood from eighteen to twenty one feet tall. He was also described as having a terrible and fearful voice and countenance.[19] Many writers made him out as a dog-headed saint, but "canine" may have got mixed up with "Canaanite," a simple mistake to make, but the association stuck and many paintings

St Christopher, 1524 by Tiziano Vecellio.

31

One of the earliest models of the Salisbury Giant.

and descriptions have him as a dog-headed character.

> *The legend of Saint Christopher quite clearly states that he was a giant. Not just a tall man, but a genuine giant. As the story goes, it was his size that enabled him to easily transport the Christ child across a river and thereby secured his place as patron saint of travellers - that is, until the Church decided that the evidence for Christopher's very existence was entirely legendary and decanonized him. Nonetheless, Christopher remains a popular though unofficial saint, and his image is fixed in our cultural psyche.*[20]

St Christopher is the patron saint of travelling, and is believed to protect people from epilepsy, lightning, storms, pestilence, floods and sudden death. He is depicted in at least 180 churches in Britain. Many soldiers, sailors, mountaineers, and even modern-day surfers pay tribute to St Christopher by wearing his medal or carrying a prayer card with him on it. No doubt, they don't realise he was an 18-foot-tall Canaanite giant.

ADAM'S GRAVE - WODEN'S BARROW

One of the oldest burial sites in Wiltshire is a long barrow near Alton Barnes on the top of a hill. It is located a few miles south of Avebury stone circle. *Adam's Grave* was once called *Wodnesbearth*, which translates to 'Woden's Barrow', confirming a link to the pre-Christian war-god Odin who is usually described as being a giant. Even the name Adam's Grave may have a leaning towards the tradition of the biblical Adam being of super-human size. The archaeologist Leslie Grinsell was told in 1950 that if anyone circles the long barrow seven times, the giant will emerge from it.[21]

WEST KENNET LONG BARROW

Less than twenty miles north of Stonehenge and a few miles from Adam's Grave, stands a magnificent megalithic complex. This includes the world's biggest stone circle, as well as Silbury Hill, massive stone avenues, prehistoric burials, a dolmen and the most impressive megalithic grave in all of England, *West Kennet Long Barrow*. Avebury World Heritage Site covers several square miles and was the megalithic centre of Britain from 3400 BC until Stonehenge took the crown in 2400 BC. West Kennet Long Barrow was built around 3600 BC, making it one of the oldest megalithic sites in Britain. It was placed high upon the hill facing the equinox sunrise. The long barrow was originally 330ft long and in use for over 1000 years. Several skeletons and skulls were discovered during different excavations, but often the 'long bone' (femur) was missing suggesting it may have been taken and used in rituals.

The facade of West Kennet Long Barrow. Courtesy of Dale Haussner.

In 1859, John Thurman, a medical superintendent of the Devizes asylum, encouraged his patients to provide the hard labour telling them excavating the site would be good "occupational therapy." He and his team of inmates were the first to enter the long barrow. Consisting of massive megalithic sarsen stones, the barrow is one of the finest Neolithic achievements in Britain. The side chambers contained the burials and his detailed reports describe six individuals, some of them of large stature:

> Skeleton 1 "had large teeth, which were slightly eroded."
> Skeleton 2: "a large and powerful individual....with a 20in thigh
> bone, aged about 50."
> Skeleton 3: (The skull) "was of a beautifully regular and somewhat
> lengthened oval form."
> Skeleton 4: "had a skull that was elongated." [22]

Signs of artificial cranial deformation are rare in England, but finding some potentially elongated skulls sheds a whole new light on what was going on here in 3600 BC. Firstly, it is an unusual thing to do to an infant, but it is now generally thought it was a sign of high status, and West Kennet was no doubt part of an archaic royal tradition, dominating the south Wiltshire territory.

Harnham Hill and Tilshead Long Barrow near Stonehenge, plus Ballard Down and the Woodyates Complex in Dorset (see next account below) have all revealed evidence of artificial cranial deformation.

West Kennet skeletons showing signs of cranial deformation.

A later culture from Europe came into Britain in the early Bronze Age who were often much taller than average with powerful physiques and a rounder cranium (compared to the Neolithic Britons). This would indicate that *skeleton 2* in Thurman's list was one of the latest internments. The old sites were reused or closed down and their influence was instrumental in the construction of many of the stone circles from around 2500 BC.

There are also numerous sites with 'giant' in their name in Wiltshire, including a 300ft Long Barrow near Pewsey, an Iron Age promontory fort near Oare, a Barrow Cemetery at Aldbourne near Marlborough, and a 180ft Long Barrow at Downton near Salisbury with a nearby Bell Barrow called the *Giant's Chair*.[23] At Luckington a long barrow called the *Giant's Cave* (or Grave) was excavated in 1646 and in it were found five or six small chambers roofed with great stones.[24]

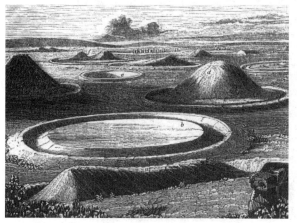

Various types of burial mounds found in Wessex.

DORSET

Dorset is rich in giant legends, and has an 180ft tall chalk hill figure called the *Cerne Giant* in its landscape. At Brockhampton Green, a few miles northeast of the Cerne Giant, a monolith by the side of the road was said to be the one that was hauled up the hill to kill a giant.[25] There is a mound called *Giant's Grave* at Melcombe Bingham, and a *Giant's Grave* and *Giant's Trencher* are two prehistoric earthworks near Swanage. The South Coast and Portland provide an abundance of giant lore that we will also explore, but we will begin our survey with the most iconic, and by far the most shocking image of a giant in all of Britain.

THE CERNE GIANT

Perhaps the most famous giant in Britain is one that can still be seen today, wielding a club, with a huge erect phallus. The Cerne Abbas chalk hill figure in Dorset has been displaying his assets for at least 500 years and fertility rites and rituals are still carried out on his member today. The mighty effigy is rich in mythology. Near the giant, a sarsen stone marks the spot where two local titans once had a rock throwing competition. Several further giant legends surround the club-wielding titan.

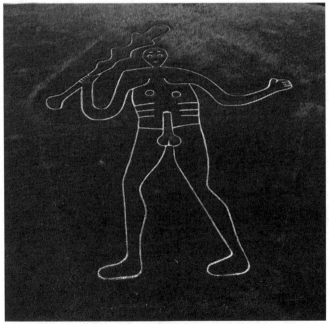

The Cerne Giant.

> *Legend tells of a real giant, a Danish one, who terrorised the district till he ate a number of sheep in Blackmoor and lay down on the hillside to sleep. The peasants beheaded him, and cut the outline around the body as a warning to other giants.*[26]

Other stories say he sometimes awakens to drink from the stream or devour maidens and cooks them in his frying pan (Trendle earthwork looks like a square frying pan above the figure). He also partook in local games with villagers and even acted as referee. He strode between hills and gazed at the setting sun, suggesting his affinity with astronomy and the movement of the stars.

> *Cerne is an ancient British word supposedly taken from the name of the river that runs through the village. However Peter Roberts states that 'Corinaeus' in the Welsh language is 'Ceryn'. Perhaps the giant is a depiction of Corinaeus, his name later corrupted to Cerne, cut as a memorial to his famous battle on Portland.*[27]

THE ISLE OF PORTLAND

Portland is a tiny island off the south coast of Dorset which has numerous giant legends. Author Gary Biltcliffe suggests that the great battle between Corinaeus and Gogmagog took place at the high point where Rufus Castle now stands that was once called *Giant's Castle*. Another tradition speaks of 'mighty men of old', a race of stone-slinging giants who lived in Verne Hillfort:

> *On Portland, a seventeenth century story tells of a large earthquake that caused cracks to open up near the Verne where quarry workers saw giant skeletons within the fissures but failed to recover them. Quarry workers in the nineteenth century also called the mysterious corbeled underground chambers they found in many of the quarries 'giant beehives'. Another Portland tale refers to the discovery of giants' coffins near the Grove, some measuring 10 ft in length. Most were destroyed, apart from a few smaller coffins now in the local museum.*[28]

Biltcliffe also discovered an effigy of a giant in the landscape of Portland. He theorises that this could be the resting place of the Titan Cronus, because in Greek mythology it states that a giant lies sleeping 'beneath' an island off mainland Britain and is bound in chains and guarded by another giant, Briaeus.

This might be referring to a great landscape giant that can only be seen from above, with certain topographical features corresponding to his shape. For example, in his right hand he is holding an ancient hillfort, now Verne Prison. In the area of his head, a rock simulacra exists resembling a huge human face. Local folklore tells of a time when giants arrived on their shore:

> *The old portlanders were renowned for their large stature, and it is possible that these tales allude to ancient memories of a race of tall people who settled here.*[29]

Above: The giant rock face simulacra on the coast of Portland.

Right: The Portland landscape effigy discovered by Gary Biltcliffe.

GIANT SKELETON FOUND IN DORSET HILLFORT

Flower's Barrow hillfort is a multivallate enclosure, encompassing around 5 acres (2 hectares). Though cut in half by coastal erosion, it is thought to have originally been built in the Iron Age. Whoever built it, one of the burials was of a rather tall persuasion:

> *A skeleton 7ft tall, unearthed at Flowers Barrow near Lulworth Cove in Dorset, had its decapitated head facing west, a different orientation to the rest of the skeleton. This story is reminiscent of Bran, whose head was removed to protect Britain. British and Irish myths refer to Bran as one of the beneficent giants and so large that no house could contain him.*[30]

GIANT'S GRAVE AT NORDEN HILL, DORSET

This nugget of folklore was featured in *Giants and Dwarfs*, and once again has geomantic significance:

> *Near Norden-hill, in Dorsetshire, is a lengthy mound which is popularly called the Giant's Grave; and very near to it are two large stones which have probably rolled down from the beds of rock on the side, or from the chalk hill above. A story, says that two giants were once standing on Norden-hill and contending for the mastery as to which of them should hurl the longer distance, the direction being across the valley towards Hanging-hill. He whose stone fell short was so mortified at the failure that he died of vexation, and was buried beneath the mound which has since been known as the Giant's Grave. Pitching rocks across an arm of the sea by way of trying each other's strength was a common amusement of the northern giants.*[31]

Every parish in Dorset seems to have memories of stone throwing giants, although the Devil gets in on the act in numerous stories. This link between two sites may be part of the lost geomantic mystery referencing an alignment between these two hills. Maiden Castle, arguably the largest and most impressive Iron Age hill fort in Europe, as well as Badbury Rings and Spetisbury Rings are all said to have been constructed by giants.[32]

HAMPSHIRE & BERKSHIRE

Hampshire boasts thirty six early Neolithic long barrows which still survive. Many are isolated, but a few occur in groups such as Martin Down and Chilbolton. A few megaliths pop up often related to giants, and the later round barrows of the Bronze Age, of which there are at least one thousand found scattered throughout both Hampshire and Berkshire, some having contained rather large personages. On the slopes of Old Winchester Hill a 180ft Long Barrow named *Giant's Grave* still survives. At nearby Stocks Farm in Meon Stoke, a Long Barrow with the same name and of a similar size exists.[33] Hampshire was part of an ancient kingdom the Celts named Gwent or *Y Went* and during the Roman invasion, they were one of the first to fall before the Druid genocide began. As for giants, we have Bevis and John Ever Afraid. In Silchester a giant called 'Onion' once threw a stone that landed on the common, about a mile away on the Hampshire/Berkshire Border. It is said his giant finger marks are found on what is now called the Impstone.[34] We begin our survey of giants in Hampshire with a giant who was also a giant slayer.

BEVIS THE GIANT SLAYER AND THE GIANT SKELETON

Bevis placed in Southampton's Bar Gate, by Mary Caine.

A legendary giant king called Ascapart is written about in a text from the early 14th century.[35] He was defeated by Sir Bevis of Hampton, also of colossal proportions, as he could carry his horse and his wife under one arm. Their likenesses adorn the famous Bar Gate of Southampton, perhaps representing guardians of the city, like Gog and Magog in London. Bevis defeated Ascapart using a tree as a club, but it got stuck in soft ground. He then saved the giant and made him his squire. The Hampton mentioned in the previous paragraph is thought to be Southampton but Bevis is usually linked with the town of Arundel as the name of his horse was *Hirondelle*.

A remarkable sword, which goes by the name of *Morglay*, is 5ft 9in long and is held in the armoury of Arundel Castle, just over the border in West Sussex. This is supposed to have originated out of the battlements from which he threw the sword to mark his place of burial. This could be the now destroyed *Bevis' Grave* that contained human bones,[36] with one description claiming it contained "*a skeleton of great size*," thought to be either Bevis or Ascapart.[37]

> *Bevis was employed as a warden of the castle and was payed with a whole ox, two hogshead of beer as well as bread and mustard each week. He was said to be able to walk from Southampton to the Isle of Wight without getting his head wet.*[38]

NEWBURY GIANTS AND THE STONE EFFIGIES

Grimsbury Castle is an Iron Age multiple enclosure hillfort located within Grimsbury Wood between Cold Ash and Hermitage in Berkshire where some skeletons of large size were unearthed.[39] Edward William Gray wrote about further skeletons that were uncovered in the area:

...about a foot below the floor ... was found the skeleton of a very large-sized man, the teeth were perfect...and nearly close to the same place, ...they discovered another very large skeleton, about three feet below the surface; the teeth were very fresh and good, and must have belonged to a young person, the jawbone was an inch or two longer than the usual size, the thigh bone was an inch or two longer than that of a person upwards of six feet high, but the most extraordinary circumstance was, that in the back bone was transfixed a flint in the shape of a spear, and which appeared to have been broken off close to the handle. Should this have been the case, the body must have been interred upwards of 1700 years.[40]

When researching this account it said that the mansion *"was the habitation of some of the family of De la Beche."* This name struck a chord with the authors, as a local family with that name own much of the land in the area. In the small church of St Mary's, Aldworth, nine stone statues lie recumbent within its walls. These are said to be life-sized representations of the De la Beche family members, most of them being over seven feet tall. The De la Beche dynasty was well known in the early part of the fourteenth century, as many of them were knights. Some were retainers to the King, Warders of the Tower of London, as well as being Sheriffs of Oxfordshire and Berkshire.[41] Sir Robert, the eldest member of the family, was knighted by King Edward I. The male members of the family rebelled against King Edward II in 1322, but in 1327, King Edward III absolved them, restoring their lands and original positions.

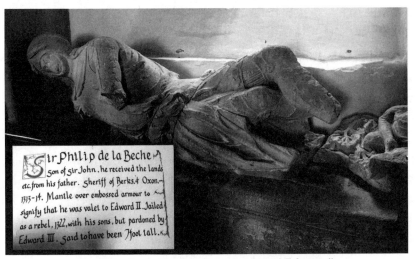

Sir Philip De la Beche who once stood 7 feet tall.

Though the stone monuments can still be seen today, they were vandalized during the Cromwell regime and consequently are badly mutilated...One effigy has disappeared altogether, that of John Ever Afraid.[42]

A tradition existed in this area that maintained its integrity through the centuries relating to the Aldworth giant effigies:

They say that John Ever-Afraid was afraid to be buried either in the church or out, and was consequently buried under the wall, where the arch appears on the outside by the south church door; we understand that the remains of the figure in this arch are bricked up.[43]

The mysterious alcove in the wall where John Ever Afraid was once interred.

The tale claims that he sold his soul to the Devil for earthly riches, and that the Devil could claim John's soul upon his death whether he was buried inside or outside the church. Hence, to stay clear of certain damnation, his body was buried in its walls, his stone effigy being placed in an alcove as described in the previous passage. His skeleton and his effigy are missing and there is a rumour that the statue ended up at Saint Lawrence's Church in Hungerford, where a strangely similar recumbent stone statue of Sir Robert De Hungerford is still on display.[44]

When we looked at local customs, we noticed that a tradition of parading the giant 'John Ever-Afraid' around the village of Aldworth was still happening annually until a few years ago, apparently inspired by the stories of the De la Beche family's ancestors being remarkably tall. Masked puppets would take over the streets and an 18-foot-tall giant would emerge from his slumber, evade the Devil, have minor adventures and return to his burial place after a few days of celebration. It seems giants once existed in this particular area in human, stone, and even puppet form.

To complete this section of a continuing giant legacy in this small Berkshire village, close by are two massive monoliths of note:

I have lately been informed that one of these stones of gigantic size was formerly to be seen in the middle of a field near Kiddington, about a mile west of Streatley. The occupier of the farm removed this immense stone, with a team of eight horses, to a more convenient spot about a quarter of a mile distant, where to this day it still remains. The story that it was thrown hither by one of the Giants is still told, and as implicitly believed by the common people; who say, further, that the print of the giant's hand, made when he grasped the stone, may yet be distinctly seen! [45]

The Berkshire Giant, or
John Ever Afraid on parade.

Poster for the Berkshire Giant
parade from 2012.

RELIC OF GIANT RACE
World's News, 9 April, 1930, p.6

This obscure account of an over seven-foot giant refers to a discovery made in the coastal city of Portsmouth in England.

Relic of Giant Race

THE skeleton of a man over seven feet in height, who probably belonged to a giant race that lived from 3000 to 4000 years ago. has been found by Lieut.-Col. J. H. Cooke. a Southsea archaeologist, in a tumulus during excavations on Portsdown Hill. near Portsmouth.

The discovery of these remains has an interesting bearing upon the legend, still persisting. of giants who lived in the locality

There are other mysteries in this area, including a series of tunnels and reports of strange rumblings coming from beneath the earth.

THE GIANTS' CURSE OF TWYFORD DOWN

In 1992, shortly after a motorway construction team disturbed the graves of eighteen skeletons, some said to be seven feet tall, a number of the road workers were said to have suffered fatal heart attacks and a night watchman dropped dead at his post. A group of protestors called 'Dragon' did their best to protect the area and claim to have witnessed giant skeletons being unearthed from an Iron Age complex close to Saint Catherine's Hill:

> *They dug up graves containing seven-foot skeletons, and there is a definite magical aspect coming into effect…Already four of the building workers and security guards have died of heart attacks. I don't think people should discount the ancient forces they disturbed.*[46]

People protesting against the construction of the motorway pronounced the incident 'the curse of the giants'.

THE GIANT FURZE PLATT HANDAXE

This huge handaxe was produced 400,000 years ago, during the Lower Paleolithic period. It was found at Furze Platt in Berkshire in 1919. It is one of the largest handaxes ever found weighing 2.8kg and is 30.6cm long, with some researchers suggesting it was "*a Neanderthal status symbol.*"[47] However, back then, were people creating status symbols?

The Furze Platt handaxe, photo by James Zikic.

ISLE OF WIGHT

Giants are notorious for choosing small islands as their abode. Just off the coast of southern Hampshire the Isle of Wight not only has a fair number of standing stones, mounds and sacred wells, but was once the residence of a giant written about in folklore. The ferocious cannibalistic giant of Chale lurked in a cave at Blackgang Chine on the south coast of the island. He was eventually defeated by a Holy Man. Another story mentions a megalithic giant connection:

When at the Longstone or Mote-stone which gave its name to Mottistone, in the Isle of Wight, the other day [the writer] was told by an inhabitant of the locality that the two stones were said to have been thrown there from St. Catherine's Down (seven miles away as the crow flies), the larger one by a giant and the smaller by the Devil; and that the giant had to stoop to throw his stone because it was so heavy.[48]

The Mote-Stone,
Illustrated in 1816.

Giant human bones were unearthed during the construction of the train station in Shanklin near the southeastern coast. It appears it could be extremely old, as the account seems to describe a midden (shell) mound that could date back to the Mesolithic era. Similar constructions were found at Culverwell on Portland to the west, and the account seems to be describing archaic skeletal features:

When excavating for building purposes in the Grove Road, a short distance below the railway station, the labourers came upon a 'refuse heap' containing limpet shells and bones, a cooking stone, and various remains of human beings. Among them were jaw and thigh bones which were broken. The lower jaw contained four or five molar teeth, much worn; they were pronounced by an expert to be those of a deformed low-typed savage of gigantic stature. 'The skull was too much decayed for preservation, but the jaw and other bones were placed in the Ventnor Museum.'[49]

Left: Shanklin Train Station where the giant was unearthed.
Right: Antiquarian illustration of the Undercliff.

John. L. Whitehead was a renowned physician with accolades including being a member of the Royal College of Physicians, London, Consulting Physician to the Royal National Hospital for Diseases of the Chest, Consulting Physician to the Isle of Wight County Hospital and Member of the Hampshire Field Club and Archaeological Society. The bones, however, never saw the light of day and appear to have been lost.

SOMERSET & AVON

The delightful counties of Somerset and Avon have several giants hiding within the landscape. Around the town of Glastonbury massive astrological figures are said to be carved in the landscape, and in Glastonbury Abbey King Arthur was said to have been unearthed and found to be significantly taller than expected. According to legend, Gwn Ap Nud, the King of the Faeries, resides in Glastonbury Tor and is often referred to as a giant. The previously mentioned *Cangick Giants* once controlled certain territories of Somerset, who, in some traditions, migrated east and waged a battle at Stonehenge.

In Batcombe at Burn Hill, Somerset, a giant blacksmith is said to appear if you call out his name while standing on this hill.[50] On Exmoor, near the coast of Somerset, Grabbist Hillfort has giants embedded in its mythological landscape, and although no bones were found, various strands of ancient folklore survive:

> *A dip in the hill is said to be the armchair of the Giant of Grabbist. According to legend, he was a very kindly Giant, who amongst other things when waving his hands caused a breeze that dried the villagers' washing.* [51]

THE GLASTONBURY GIANT

The legendary "Isle of Avalon" has a few giants hidden within its records, more often than not linked with the stories of King Arthur. In the year 1191, the monks of Glastonbury Abbey announced a major discovery. The location of King Arthur's Grave was *"revealed by strange and almost miraculous signs,"*(account by Gerard of Wales) and that the burial place of Arthur was between two 'pyramids' in the grounds of the Abbey. They consulted with King Henry II (1133-1189 AD) who was said to have informed them of a long-kept secret of the Royal Family, that Arthur's burial place was at the site of St. Dunstans in Glastonbury, the exact location of the Abbey.

The account by Gerard of Wales said Henry had acquired secret information from 'an ancient Welsh bard' concerning the position of where to dig and at what depth they would find the coffin.[52] These 'pyramids' were first described over sixty years before the discovery by William of Malmesbury, who states that the pyramids were up to 26 feet tall and carved with unusual figures and writings.[53]

Antiquarian illustration of Glastonbury Abbey interior viewed from the east.

During the excavation by the monks, a very controversial object was found at around seven feet below the surface. A lead cross attached to a slab of stone was unearthed that had carved on it in Latin "Hic jacet sepultus inclitus Rex Arthurus in insula Avalonia," meaning "Here lies interred the famous King Arthur on the Isle of Avalon." A hollow oak coffin was discovered at around eighteen feet below the surface between these two pyramids.

Left: A sign marking the original location of the controversial burial in Glastonbury Abbey. Right: An early illustration of the lead cross by Camden.

This skeleton was not alone in its coffin. Alongside it were the remains of a female, and also a plait of blonde hair that crumbled to dust when touched. The discovery became a sensation and it was this series of events which confirmed that the legendary 'Avalon' which featured in the Arthurian epics was indeed Glastonbury and that has stuck to the present day. The male skeleton that was excavated was estimated to be close to 9 feet tall.[54]

The earliest account of the excavation is by Gerald of Wales, who seems to have witnessed the skeletons and the lead cross:

> *...two parts of the tomb, to wit, the head, were allotted to the bones of the man, while the remaining third towards the foot contained the bones of the woman in a place apart; and there was found a yellow tress of woman's hair still retaining its colour and freshness; but when a certain monk snatched it and lifted it with greedy hand, it straightaway all of it fell into dust...the bones of Arthur... were so huge...his shank-bone when placed against that of the tallest man in the place...reached a good three inches above his knee...the eye-socket was a good palm in width...there were ten wounds or more, all of which were scarred over, save one larger than the rest, which had made a great hole.[55]*

A further account was written down by monks at Margam Abbey in Wales in the mid 1200s. It has similarities to the Gerard of Wales version but there are notable differences, that are all featured in the artwork by Yuri Leitch that we feature in this chapter. It states that three coffins stacked on top of each other held the remains and were found by accident when digging the grave for a deceased monk. The first coffin contained female bones and a

The Discovery of Arthur's Tomb, John Hamilton Mortimer, 1767. National Galleries of Scotland.

long lock of hair. The second deepest contained the bones of Mordred, Arthur's nephew, whilst the third enclosed an individual described as *"sturdy enough*

THE GIANT OF GLASTONBURY ABBEY

The burial was located between two 'pyramids' with unusual inscriptions on them

The lead cross

9ft Skeleton buried in an oak coffin (Gerald of Wales)

Three burials with large skleleton at bottom (Margam Chronicle)

Illustration by Yuri Leitch showing the various interpretations of the discoveries at Glastonbury Abbey.

and large." On the lid of this coffin was the lead cross.

Giraldus Cambrensis, a respected historian personally examined the bones and the grave about four years after the discovery and pronounced it a genuine find. In 1278 in the presence of King Edward l and Queen Eleanor, the remains were transferred into a marble sarcophagus with much pomp and ceremony and placed inside the Abbey. Other sources from as far back as 1193 seem to confirm this was a skeleton of gigantic proportions:

> *You must know that Arthur's bones, which were found in that place (Glastonbury), were so big that in them the words of the poet seemed to find fulfillment.*[56]

> *The farmer...marvel at gigantic bones in the upturned graves.*[57]

> *His tibia placed beside that of the tallest man in the place (whom the Abbot pointed out to me), and fixed into the earth by the side of his foot, extended fully three fingers' breadth above the man's knee. His skull bone also was capacious and large enough for a prodigy or a show - so much so that the interval between the eyelids and the space between the eyes might contain the size of a man's palm fully. And in this were seen ten or more wounds, all of which, except one larger than the others and which had made a great gash, and which alone seemed to have caused death, had joined into a firm cicatrix.*[58]

Then, in 1962-63, after carrying out some additional excavations at the grave site, archaeologist Dr. Ralegh Radford "*confirmed that a prominent personage had indeed been buried there at the period in question.*" [59]

What do we make of this? Was it a hoax to attract pilgrims to Britain's first abbey? According to author Yuri Leitch, there was a much wider agenda being played out that involved the Crusades, royal dynasties and the issue of the earlier abbey being burnt down and funding desperately needed to build a new one.[60] The lead cross is likely to be a fraud, possibly left there by the monks to bring fame upon the abbey. The monks also encouraged the nobility to support them in rebuilding the abbey after the fire. This would also help preserve and protect such a sacred spot where England's greatest hero was interred.

Glastonbury is within the territory of the *Cangick Giants*, so perhaps one of their tribal leaders could have been buried here in such a fashion, long before the Abbey was constructed. Whoever was discovered, whether Arthur, a prehistoric giant, or a member of an ancient royal family, it raises questions as to what else may be found under the marshy levels of Somerset.

GIANT 7FT LONGBOW

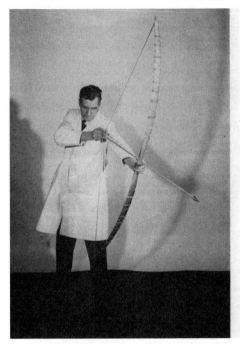

A prehistoric lake village existed very close to Glastonbury at the current location of Godney. It was first constructed during the Iron Age and it was here that hollowed out oak logs were used as boats, just like the coffins described at the abbey. A few miles away more remarkable discoveries were made, dating back to 3500 BC in the Somerset Levels, near Meare.[61]

Preserved for millennia under layers of black peat, two seven-foot longbows made of yew (the sacred tree of the Druids) were unearthed in almost pristine condition. They were beautifully crafted with geometrical patterns. Academics claim they are simply ceremonial, but these appear functional as the game-rich Somerset levels would

Cyril Lilley holding a reconstruction of the 7ft longbow.

have been a prominent hunting ground in early Neolithic times. The *Meare Heath Bow* accords with 20th century principles of scientific design and re-creations of it have proven it was a better weapon than the medieval longbow that appeared several thousand years later.[62] Whomever wielded them must have been significantly taller than Cyril Lilley in the photo.

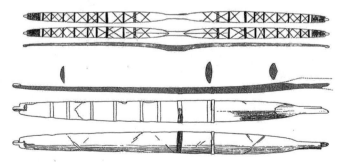

Illustrations of the detail found on the seven-foot longbow.

51

THE CANGICK GIANTS (PART 2)

Long before Glastonbury became a *New Age Mecca*, and before the abbey was constructed, a tribe of powerful warrior giants once ruled the Somerset levels. Known as the *Cangi*, or the Cangick Giants, they lived on in legend and even went down in history as having had a deadly battle in Wiltshire, constructing Stonehenge as their trophy (as outlined earlier in this chapter).

Reverend Robert Gay of Somerset wrote an unusual manuscript on Stonehenge in 1666 called *A Fool's Bolt Soon Shot at Stonehenge*. His research, sarcastic at times, covered much ground and put forward the idea that it was built by ancient Britons, and perhaps influenced by the Greeks before Iron Age Celts from Europe arrived. He clearly influenced John Aubrey (1626 - 1697), the antiquarian who popularised Stonehenge and Avebury in the subsequent decades. Robert Gay lived in Nettlecombe, Somerset and was notorious for a number of reasons, including being involved in an uprising against the courts where he and a local mob set fire to the buildings and he was consequently imprisoned during the Civil War of 1643.[63]

The Cangi he writes about were a real tribe who are remembered to have resided up on the Mendip Hills and surrounding areas where the great Priddy Henges, hundreds of mounds and tin mines are located. He believed their origins were from Flintshire, North Wales.[64] Camden, in his *Britannia*, 1722, also suggested there was evidence of the Cangi in place names such as:

> *...the hundreds of Cannington and Canings, in Wincanton, which is sometimes call'd Cangton; and Kaingsham, as much as to say, the mansion of the Cangi.*

In the text Gay goes into fine details about the origins of place names and suggests the Cangi may have had a far-reaching influence. Some of the place-names he gave were Bishops Canning, All Cannings, Calne, Canford Magna and Little Canford.

The final part of *A Fool's Bolt* focussed on these giants constructing Stonehenge as a monument to their victory. According to Gay: *"By their armour, or pieces of it, (which, when new, was large enough for Giants) found there also"*. In Camden's *Britannia* 1772, which was updated after the publication of Gay's work, it reveals further information:

> *That it was an old Triumphal British Monument, erected to Anaraith the Goddess of Victory, after a bloody battle won by the*

*illustrious Stanings and his Cangick Giants, against Divitiacus
and his Belgæ; and that the Captives and Spoils were sacrificed
to the said Idol in this Temple. An opinion advanced (upon what
grounds I know not) ...written about the year 1666.*

Stanings was the previously mentioned *Stenages* (the tribal leader of the Cangi),
one of ten "principal commanders" of the tribe, all thought to be of giant
stature. Stonehenge was said to have been named after him. Camden stated
that the Canaanites of the Bible also appointed ten commanders in their army.
Were the Cangi also the inspiration for the creation of the Canaanite giant St
Christopher of Salisbury?

The origins of the Cangick Giants were not orthodox, with two types
of colossi described. One group were called 'Giants of Antiquity' who "*had not
been borne into the world by the way of all flesh. For they being heathens, and
not believing any creation, supposed (to be) the first inhabitants of any nation,
were bought forth by the earth...Gigantes.*"[65] This may be a connection to the
Incubi who bred with Albina and her sisters and the monstrous giants were
their offspring. It may also be a reference to the Nephilim of the Bible, as he
made further connections to the Canaanites, not only in name (Canaan =
Cangi), or the fact that both had traditions of sacrificing their victims, but also
descriptions of skeletal remains that were reported upon at the time.

This account from 1810 identifies the most southerly territory of the
Cangick Giants, just over the border in Dorset:

> *At the Lord Sturton's house, in Sturton Caundell, the borders of the
> Cangi; a Giant's thigh bone a full yard, in which instances argue,
> that, as amongst the Canaanites, so amongst the conquering
> Cangi, there were races of Giants.*[66]

In the next report the author reveals a discovery made in Wedmore in Somerset
in 1670:

> *...these Cangick Conquerors were Giants: both of great antiquity
> and ability...it is easy to conjecture at the incredible stature and
> strength of a Cangick Giant, by the top of his skull an inch thick, and
> a tooth of his, which I have, 3 inches long (Ref. H.Hun, lib, Primo
> History) now since the root is broken away, and three inches and
> a quarter round, and three ounces and a half in weight, being full
> four ounces till the roots were broken off; so that, according to this*

instance, the Cangick Giants were very much greater and stronger than Goliath, or any other Giants described in the Scripture.[68]

According to local author John Collinson, the discovery was made while digging a well and found at a depth of about 13 feet.

The Cangi, according to Gay, had metallurgy skills and actually softened, powdered, super-heated and moulded the stones of Stonehenge using "engines." This hints at advanced technology that may have been inspired by Merlin having used 'gears' or 'engines' in moving the sarsens from Killarus in Ireland to Salisbury Plain. The softening and manipulation of stone raises an interesting point because over the last two decades Joseph Davidovits, a French materials scientist has been pioneering the theory that many megalithic sites were not constructed, but rather 'moulded' using powdered stone. He claims to have proven that the blocks of the Great Pyramid in Egypt were created in this way with what he calls geopolymer cement.[69] It may also indicate why the mighty sarsens of Stonehenge have what look like 'scoop' marks on them as though the stone had been softened then carved before they hardened. This 'scooping' technique is also found in Machu Picchu and Ollantaytambo in Peru, and in Aswan Quarry in Egypt.

Striations on two of the stones at Stonehenge (three images top) resemble those from Machu Picchu in Peru (lower left) and Aswan Quarry in Egypt (lower right).

Gay also claimed that the conglomerate monoliths of Stanton Drew Stone Circles in Somerset were moulded from gravel and placed in a circular configuration by these giants. Also, furnaces were known to have been used by the Canaanite giants and the Cyclopes of Greece, who both had a tendency towards metallurgy, warfare, sacrifice and working with stone.

The Cyclopes working with metal and stone in classical illustrations.
Left: Brontes, Steropes and Arges. Right: The Cyclopes depicted in 1670.

THE CANGI OF WORLE HILL

The hut circles of Worlebury Camp Hillfort, Somerset, also shows traces of the Cangi's armour and weapons. This discovery was made by Lord Talbot de Malahide in the mid 1800s:

> ...two skeletons, bearing marks of extreme violence, and apparently of two different races - one a gigantic race, with skull presenting the most uncivilised appearance, and other smaller and more advanced. With them were found iron weapons.[70]

JAW-BONE AND TEETH OF PRODIGIOUS SIZE

This discovery from Norton St Philip was reported once again by John Collinson.[71]

> In digging the stone in the north part of the parish, about the year 1752, some workmen found, at the depth of nine feet beneath a rock, a large quantity of human bones of various sizes, with part of a jaw-bone and several teeth in it of prodigious size.[72]

55

The Three Giants of Brent Knoll

According to tradition, the hillfort of Brent Knoll was created by a shovel full of earth dropped by the Devil when he was digging out Cheddar Gorge. In Arthurian legend it was called the *Mount of Frogs* and three giants resided there.

Brent Knoll, Somerset. Photo by Devon County Council (Bill Horner).

As a new knight, Ider had to pass a test. He was told when at Glastonbury that three giants, 'notorious for their wickedness', lived on Brent Knoll, then known as the Mount of Frogs (Mons Ranarum). King Arthur intended to march against them, and Ider would be required to join him. Young and enthusiastic, Ider galloped ahead and slew all three giants singlehandedly. But unfortunately he was wounded himself, and by the time Arthur arrived Ider lay unconscious and dying. The King returned to Glastonbury blaming himself.[73]

The Loxton Giant Headless Skeleton

A giant skeleton was said to have been unearthed in Loxton Churchyard whose torso alone measured 6ft 3in.[74]

When this discovery was made and how old the skeleton might have been is unclear, but this would have put him at nearly eight feet tall.

Stone Giant Effigies at Chew Magna Church

At Chew Magna Church an effigy of Sir John Loe is represented as being over seven feet tall. The armoured figure is 7 feet 4 inches (2.24 m) long (his true height) and his feet rest on a lion, while those of his lady rest on a dog.[75] This is the same style as the Aldworth Giants who also have lions (and dwarfs) at their feet.

The seven foot tall statue at Chew
Magna Church, on the left.

DUNSTER CASTLE - GIANT SEVEN FEET TALL

An unusually tall skeleton was found manacled to the gatehouse at Dunster Castle. The Castle was built upon a rocky outcrop and is located on the coast between Exmoor and the Quantock Hills dating back to the eleventh century. However, it is steeped in Arthurian myth and is the place where Arthur was supposed to have spent his childhood being trained by a prince called Cadwy.

> *A story, verified by excavations during the late 19th century, confirmed that during the early 1700s an excavation was undertaken of the oubliette which unearthed its macabre secret. A male skeleton was found, some 7 feet tall - a giant of a man, bearing in mind the average height of an adult male at that time would have been around 5 feet 3 inches. The skeleton was discovered manacled to the wall by his wrists and ankles, presumably left to starve to death like so many others. It is said that his remains still lie in-situ at the bottom of the oubliette. The oubliette is now covered and tiled over but just standing there and being aware of what lies below is a little creepy. Haunting cries of men and women have been heard coming from the area of the oubliette.[76]*

The castle is also on the exact same latitude as Stonehenge and the centre of Lundy Island. Geomantic research has revealed its placement in a series of geodetic triangles and sacred measurements across the West Country

Dunster Castle gatehouse.

landscape that we will explore in the *Giant Geomancers* chapter.

One mile west of Dunster lies Grabbist Hill, standing 900 feet high, thickly forested, and with a conical point. It is recorded that a friendly stone throwing giant lived in harmony with the local populous. One day the Devil emerged from the sea and the giant challenged him to a game of quoits. They hurled stones in the direction of Porlock Vale where they still stand today. These two massive boulders were for a long time called the *Hurdlestones* or *Hurlstones*, and are currently known as the Whitstones. Anthony Roberts pointed out that the two granite blocks appear to be shaped and tooled and may have astronomical alignments:

> *The humour and the vigour in these stories is unconsciously concealing and guarding the core factual, geomantic application of a vastly scientific megalithic technology. These are the purest expression of residual geomythic memory.*[77]

THE GIANT OF TARR STEPS, EXMOOR

Tarr Steps is a megalithic 'clapper bridge' across the River Barle in Exmoor National Park near the Somerset border dating back to around 1000 BC.[78] It has been reconstructed many times when the river levels have risen and knocked it over. It is said that a local giant made a wager with the Devil that he couldn't build the bridge overnight. The Devil promptly proved him wrong. The Devil then took ownership of the bridge until a priest drove him away.

Tarr Steps clapper bridge.

Remains of a Giant
Western Flying Post, 15 May 1820

A skeleton which was thought to be nearly 9 feet tall completes this section on Somerset. It was found during the excavation of Knightstone Island just off the coast of Weston-Super-Mare. When excavations for the first baths were made in May 1820, bones and a pottery urn were discovered and assumed to have been of Roman origin. These have now been dated to the Iron Age, and may be related to the previously mentioned discoveries at nearby Worle Hill:

> *Lately, as some workmen were employed in excavating Knightstone Rock, at Weston-super-Mare (an island lately purchased of Mr Howe, of Bristol, for the construction of hot and cold baths), the skeleton of a man of enormous stature was discovered a few feet below the surface, and near it an antique earthen vessel, containing bones of smaller size; the urn and skull of the larger skeleton were, unfortunately, broken to pieces by the carelessness of the workmen, but many of the bones are preserved in the hands of the curious. Conjecture is very busy as to the antiquity of these remains, and the character of the gigantic personage whose frame they once held together. Some suppose they belonged to one of those giants, who, old historians assert peopled this country before the invasion of Caesar; but as there is an old encampment hard by, which, from its construction is probably of British origin, there is little doubt but they are the remains of an aboriginal soldier of distinction. Persons conversant with anatomy infer, from the size of the bones, that he must have been nearly nine feet high.*

Bristol and Beyond

Bristol, in Avon, was the territory of giant cave dwellers who were thought to be responsible for the construction of *Maes Knoll*, a hilltop camp at Norton Malreward. An old snippet of folklore says it was created by a clumsy giant called Gorm (or Goram). He was carrying a huge shovelful of earth and stumbled when on the border with the Cotswolds. The soil fell into the Avon Valley and Maes Knoll was created. After falling over again, he then stumbled along and created the lengthy Wansdyke with the shovel, dragging it across the surface of the earth. This is by far the longest dyke in Britain, originally stretching a staggering 80 miles from the Avon Valley all the way to Andover in Hampshire. Its true age and purpose are unknown, but it is aligned to many prominent prehistoric sites along its path. The disturbance upset the local Lord of Avon, who gave chase, but the giant tripped once more and fell into the Bristol Channel and drowned, his bones forming the two islands of Steep Holme and Flat Holme.[79] Anthony Roberts commented:

This tremendous expression of applied geomancy is certainly 'giant' in conception and the Gorm myth is a perfect encapsulation of its country-spanning appearance.[80]

Sculpture of 'Gorm the Giant' in the grounds of Ashton Court.

Next, we sail to Lundy Island further out in the Bristol Channel, once thought to be the spiritual home of legendary titans.

Left: American researcher JJ Ainsworth with modern giant at Stonehenge on the summer solstice 2018. Right: Aerial View of of a section of Wansdyke.

GAZETTEER OF GIANTS - WESSEX

DATE	LOCATION	SIZE	No.	FEATURES	PAGE
	WILTSHIRE				
1808	Bush Barrow, Stonehenge	6 - 7ft	1	Famous gold lozenge	25
1950	Bulford Camp	6 - 7ft	1	"great size, low brow"	25
1508	Ivychurch Priory, Salisbury	14ft 10in	1	Many famous witnesses	27
1719	St Edmund's Church, Salisbury	9ft 4in	1	'Giant's Grave' Tumulus	29
1859	West Kennet Long Barrow	Large	1	Elongated skull	33
	DORSET				
1600s	Verne Hillfort, Portland	Giant	Many	Also giant coffins	37
1800s	The Grove, Portland	10ft	Many	Site called 'Giant's Castle'	38
1540	Flowers Barrow	7ft	1	Decapitated head	38
1810	Lord Sturton's house, Sturton Caundell	Giant	1	Thigh bone a yard long	53
	HAMPSHIRE				
1770	Bevis' Grave, Arundel	Giant	1	Legendary giant	39
1839	Grimsbury Castle, Newbury	7ft	2	Perfect teeth	40
1300s	Saint Mary's, Aldworth	7ft	Many	Stone effigies, not human	41
1930	Portsdown Hill, Portsmouth	7 - 8ft	1	Tunnels and rumblings	43
1871	Twyford Down, St Catherine's Hill	7ft	18	Cursed	44
1936	Shanklin Train Station, Isle of Wight	Giant	1	Giant skull	45
	SOMERSET				
1191	Glastonbury Abbey	9ft	1	Tomb of King Arthur	46
1670	Wedmore	Giant	1	Massive teeth	53
1861	Worlebury Camp Hut Circles	Gigantic	2	Cangick giant warriors	55
1752	Norton St Philip's	Giant	1	Huge jaw bone	55
1947	Loxton Churchyard	7ft	1	Headless torso	56
1700s	Dunster Castle, Exmoor	7ft	1	Chained skeleton	57
1820	Knightstone Island	9ft	1	Weston-Super-Mare	59

Lundy Island,
8ft - 8ft 7, p.64

MAXIMAJOR
STONE

Huge Femur Bone, p.63

Lellizzick, 7ft, p.74

BOLSTER
THE
GIANT

VIXEN TOR

BOWERMANS
NOSE

St Pirians, 7ft, p.76

Mên Scryfa,
9ft, p.91

Looe Island, p.91

Men-an-Tol
7ft, p.90

Tregony, 10ft, p.75

Trevegean,
p.90

St Michaels Mount
8ft, p.87

3

DEVON AND CORNWALL

It is the quality in the country itself that made it, in the words of Mr. J. O. Halliwell 'anciently the chosen land of giants', the favourite residence, playground and ultimate refuge of the majestic race that inhabited and in some cases constructed the cairns, tors and high places of the west of England before their defeat and extermination by Brutus's Trojan invaders. - John Michell [1]

In the traditions of Brutus and Gogmagog, it is said that many of the giants of Cornwall fled to Dartmoor in Devon and the mountains of Wales to escape the swathe of giant hunters led by Corineus in his final attempt to wipe them out. We have already seen that the areas of Totnes and Plymouth are rich in giant lore contained within the foundation myths of Britain, and we know that *"gigantic jaws and teeth were discovered that were identified as the bones of Gogmagog"* at the Citadel in Plymouth. [2] Further into Devonshire we come across some fascinating folklore and megalithic "giant" sites relating to these surviving titans and their activities on the moor. One report worth repeating is of a huge femur which was discovered in Devon, but it is unclear where exactly this was:

> *A stone coffin in Devonshire contained a thigh-bone belonging to a man eight feet nine inches high.* [3]

Later in the chapter we will also explore the area of Land's End in Cornwall, but before that we investigate the Devonshire Moors, and one particular island that may hold the key to not only the origins of Stonehenge, but also the final resting place of an infamous Greek Titan.

63

LUNDY ISLAND

Lundy is three-miles-long and half-a-mile-wide situated in the approaches to the Bristol Channel, about 11 miles from the North Devon mainland. Its sheer granite cliffs rise from sea level to a fairly level plateau where there is archaeological evidence of habitation from the Mesolithic to the Bronze Age.

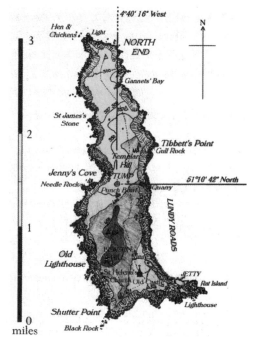

Lundy was once attached to mainland Britain until around 9000 BC when it became an island, reaching its present size by about 7500 BC.[4] By 450 BC, it was being used by the Carthaginians as a trade route for silver.[5] Although the Romans didn't make it as far as Lundy, they recorded an odd reference that *"a specially holy race of men...who refused trade and had visions of the future' resided there."*[6] Were these the Druids? Keepers of the ancient knowledge? Centuries later, Lundy was granted to the Knights Templar by Henry II in 1160. They were a major international maritime force at the time.

The Welsh name for Lundy is *Ynys Elen*, which means 'The Island

Robin Heath's map of Lundy showing the geodetic centre point.

of Elen', or 'The island of the Elbow, bend or right-angle.'[7] Lundy is also the *corner* or *elbow* of a great Pythagorean landscape triangle that links with Stonehenge and Preseli in Wales. We will look more closely at this in the *Giant Geomancers* chapter. The original church on the island was dedicated to St Elen and two recycled megaliths in the graveyard dating to the 5th and 6th century have dedications to Potinus, possibly St Patrick's grandfather, and Vortigern, a renowned British chieftain who features in some of the Arthurian stories.[8] Celtic mythology saw Lundy as an important sacred island, emphasised here by Geoffrey Ashe:

> *In Welsh it has an uncanny character and is supposed to be Annwn or rather a place of access to it. The most important hill is the Tor at Glastonbury, the most important island is Lundy...Annwn's*

inhabitants have human form but are not strictly human. They are immortal - fairyfolk or demons according to one's point of view. Some are Gods thinly disguised. Living humans can enter Annwn, and so are spirits of the dead, but it is neither a heaven or hell in the Christian sense. To a certain extent it resembles Avalon.[9]

The Tigerni memorial stone at Beacon Hill,
showing some inscriptions.

The Celt's saw it as the *Island of the Dead*, where great kings would be ferried to the west to be buried, in the direction of the setting sun.[10] Shrouded in mist, rarely seen from the mainland and hidden from the profane, Lundy was perhaps only known about by an ancient priestly elite.

Left and Middle: Hugh visited Lundy in 2007 and was involved in finding a lost stone circle with Stuart Mason, Robin Heath and the Ley Hunters Group. Right: The layout of the rediscovered 'Lundy Egg' Stone Circle, surveyed by Heath.

Rock-wielding giants who were masters of sorcery are said to be buried at the centre of Lundy Island, who: *"...were driven there from bases in Cornwall and that they angrily hurled rocks back to the mainland."* [11] Another giant lived on Exmoor, Devon (just across the channel on the mainland) who was a great fisherman. He would wade past Lundy followed by fishing boats collecting what he didn't gather. He was a benevolent giant and regularly saved ships in peril, completing his fishing trips by wiping his huge feet on either side of Dunster Castle, so as to avoid splashing the lawn market (see previous chapter).

The medieval chronicler, Richard of Cirencester wrote that Hercules was recognised in Britain long ago, and Lundy Island was where he was buried:

> *Writers are not wanting to assert that Hercules came hither and established a sovereignty. Hercules was one of their gods (the Britons), and Lundy Island was...called Herculea. We make a bee-line for Lundy, and find that it was once rich in the remains of the megalith builders. - tumuli, two tombs of granite blocks locally called the giant's graves.* [12]

Ptolemy also wrote that the passage between Lundy and Hartland Point may have been the original 'Pillars of Hercules', and that it may have even been the final resting place of the Greek Titan Cronus. Plutarch, in his *De Defectu Oralculorum*, recounts that Cronus was said to be asleep under an island off the British coast after he was defeated and banished by Zeus: *"Uranus, as the Greek myths have told us, was the father of the Titans, the giants - one of which was Cronus. Geoffrey of Monmouth links Uranus with England when he states, in the 19th chapter of his third book, that 'Urianus' was a king who reigned in Britain centuries before Arthur."* [13]

On page 156 of *The Wiltshire Archaeological and Natural History Magazine* (1876), the following account discusses a stunning discovery from the southern part of Lundy in an area called 'Bulls Paradise':

> *The skeletons were found on the top of the island, about 2 feet underground, in digging foundations for a wall for farm buildings. The number of the more perfect skeletons was seven, lying in a row with the heads to the West. The first in the row, a male, measured 8ft 5in; by the head were placed two upright stones with the head lying in a little hollow, and protected by a third stone. None of the others had any appearance of coffins by them but great numbers of limpet shells. The one measured was measured by my father, by*

whose orders the remains were buried again, but I am afraid much injured by the workmen in doing so. Some pottery and some beads were found with them, Mr. Etheridge, then curator of the British Institution, showed some of the pottery to a friend of his, an antiquary, I believe, who said it was undoubtedly Ancient British.

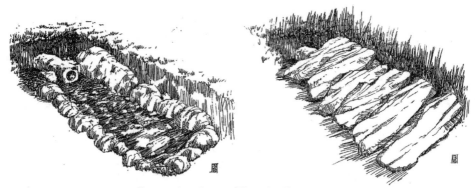

Reconstructions of the giant's grave.

This find was made in May of 1856 and was widely accepted and mentioned in literature spanning over 100 years[14] and although many researchers claim the burials may be Viking, or medieval, it seems that the megalithic grave was indeed prehistoric. One of the capstones from this burial site now rests outside the local tavern and one of the pillow stones is in the porch of the church.[15] Iron Age inscribed monoliths were also discovered dating to 400 - 650 AD, that were found near the giant's graves. In *The Times* it gave more accurate details: *"The measurement of the grave from the top of the skull to the feet gave eight feet seven inches."* Another account, indicated a second giant skeleton was unearthed:

LUNDY ISLAND.—A private letter from Lundy Island, received at Bristol, says, " A curious circumstance has happened here since our last post-day. In digging for the foundation of a wall the workmen have discovered several skeletons about two feet under the surface, one of which is that of a man of gigantic stature, and probably a man of note besides, for his grave had been enclosed, whereas not one of the others had anything to mark where they were deposited. The bones had evidently been under ground for many centuries, for, on being touched and exposed, nearly all of them fell to pieces; besides, as there has been ground consecrated on the island prior to the time of Edward III., one may presume the burials would have taken place in the churchyard if they had been of anything like recent date. Nearly all the bodies were covered with limpet-shells, perhaps originally buried with the fish in them. All the burials must have taken place nearly at the same time, for the skulls were lying in nearly a straight line. A stone about three feet long, roughly squared, about 18 inches through, and hollowed at one end to the depth of about two inches, like a shallow basin, contained part of the side of the skull of the large skeleton. The heads of all were laid to the west. The spot around shows traces of former habitation, such as pieces of pottery, peat-ashes, &c.

The first report in the *Taunton Courier, and Western Advertiser*, 21st May 1856.

The smaller cist, which also contained a skeleton, was but 8 ft. long and differed from the other in having no head or foot rest.

Mr. Heaven was sent for and the skeletons carefully measured. The larger had a stature of 8ft. 2in. Mr. Heaven was present the whole time and not only saw the measurement taken, but, as he himself told me, saw one of the men place the shin bone of the skeleton against his own, when it reached from his foot half-way up his thigh, while the giant's jaw-bone covered not only his chin but his beard as well. Close by seven other skeletons were discovered but these were of ordinary stature, and buried without stone coverings.[16]

We did locate the original announcement of the discovery within the same month (May 1856) of "*a man of gigantic stature.*"[16] Yet another account described the giant, then an unfortunate bovine accident on the same page!

> **Digging foundations for cattle-house at Lundy, found ancient grave covered with tile and layer of mussel shells. Remains of 8 ft. giant. Skull and bones in good preservation. Bullock grazing cliff near Hartland Quay fell over 50 ft. not much hurt. Same place two killed previously.**
>
> Hartland and West Country Chronicle, 14 September 1935

Finding two skeletons between 8ft and 9ft tall buried in stone-lined graves is a sensational discovery, suggesting these were very important people. Why they were on Lundy in the first place raises more questions as to who these giants really were.

DEVONSHIRE

Dartmoor has hundreds of rocky outcrops called *tors*, and natural features of balanced rocks called *Logan Stones*, that to ancient folk were alive with stories and legend. These folk tales were passed on from generation to generation. Certain landscape configurations would be noted, such as a sleeping giant, and rock faces often had giant human features called simulacra. Perhaps, like the indigenous people of Australia, the ancient Britons believed these natural features were the resting places of giants, spirits, and their ancestors. Walking across the country would become a three-dimensional story or quest, merging names of places and legends into a mythologised landscape. Wales is particularly rich in elemental and ancestral folklore where nearly every field has a name and every lane, mound or boulder has a story.

The tors were deemed the domain of the giants, mythical anchor points of knowledge and ancestral history. During Neolithic and Bronze Age times, the construction of stone circles and other megalithic sites wrote a new story, that of man's presence in and domination of the landscape. The natural outcrops became a smaller part of a re-mythologised topography with the stone circles as the focal point. Over the centuries bones of enormous magnitude have been unearthed throughout Dartmoor and Devon, often coinciding with legends of these ancient titans.

Joseph Bligh's illustration of giants moving massive rocks around in Devon.

THE MAXIMAJOR STONE AND THE GIANT'S GRAVE

The giant Maximajor lived in an elevated stronghold and could see across into Wales and down to the South Devon coast. The giant had a strange relationship with bovines and bullocks. He was something of a *cow whisperer* and could understand them in a way no other man could.[18] However, one day, two local children were on the moor and found Maximajor stone cold dead. The villagers were alerted, and a grave covered with stones was constructed. The grave regularly became surrounded by cows, as though they missed their giant human friend. Some time later, when checking to make sure his grave had not been disturbed, the locals noticed that it had, but from the inside!

The Maximajor Stone with a bullock.

The giant had risen from the dead and footprints led to a huge standing stone on the moor. The thing is, the menhir had never been seen before, as though the giant had turned into stone by a magical spell. No one knows what really happened, but cattle often congregate around the stone. Later the Christian folk of the moor renamed the stone 'The Headless Cross', to stop Pagan stories of giants rising from the dead doing the rounds, as it was a bit too similar to their saviour's story at Easter.

THE GIANT OF DINGER TOR & THE PHOENICIAN

The giant Blunderbus (also known as Blunderbore) escaped to Dartmoor from Cornwall and made a home on *Dinger Tor*.[19] The tor is a small rocky outcrop on Dartmoor's North Moor, not far from Okehampton, but in the vicinity of various megalithic ruins. An alternative version of the story calls him *Moran The Mighty*, a giant king who ruled over several tors in Cornwall and Devon. The two names of the giant tell the same story with slight variations. In both versions it involves Phoenician pirates, many giant wives, reconciliation and the origin of clotted cream! It is worth noting the Phoenician connection at this juncture, because the earlier incarnation of their people were the biblical Canaanites, who may be connected to the Cangick Giants and St Christopher.

VIXANA, THE GIANTESS

Vixana was an evil giantess who lived in a cave at the foot of *Vixen Tor*, which had been constructed at her command by the earth gnomes over whom she had power. She would call up a mist to confuse travellers who would get stuck in the boggy ground. After clearing the mist she would watch them suffer and slowly sink to their demise in the bog, laughing at their misfortune.

Old illustration of Vixen Tor.

BOWERMAN'S NOSE, DARTMOOR

On the very edge of the vast moorland, startling every eye,
A shape enormous rises! High it towers
Above the hill's bold brow, and seen from far,
Assumes the human form; a granite god, –
To whom in days long flown, the suppliant knee
In trembling homage bowed. The hamlets near
Have legends rude connected with the spot,
(Wild swept by every wind) on which he stands
The giant of the Moor! by N. T. Carrington

Bowerman's Nose, by Baring-Gould, 1907.

71

Bowerman was a local giant on the east side of Dartmoor who enjoyed hunting with his hounds. His derogatory comments about some local witches, eventually got on their nerves so badly that they tricked him, and turned him and his loyal canines to stone, becoming Bowerman's Nose granite outcrop.

CORNWALL

I find, over a tract of country extending, from the eastern edge of Dartmoor to the Land's End - and even beyond it, to the Scilly Islands - curious relics of the giants. This district is in many respects a peculiar one. - Robert Hunt, 1903.[20]

Classic image of how antiquarians saw ancient Cornwall, by W. Borlase, 1754.

A strange story about the origins of the Cornish giants discusses a race of "*inferior*" people who "*went about on all fours*" and "*spoke a strange tongue.*" These were called the *Lizard People*.[21] A large ship was one day wrecked off the coast and some giant survivors settled amongst the Lizard People, gradually interbreeding with them over the generations. Their offspring became renowned in the area, being very tall, intelligent and superior in many ways. It is said that in this area of Cornwall, local residents still often reach over six feet tall.

Rock formations, mounds, stone circles, tors, cairns and dolmens are often attributed to these mighty beings with a persistent presence in the minds of the people in this part of England. These stories concentrate in the small triangle of land encompassing Land's End, Penzance and St Ives. The stories appear to represent a movement through the land in a precise yet poetic way, more often that not following straight lines.

Around us, on the hill-sides and up the bottoms - huge boulders of granite are most fantastically scattered. All these rocks sprang from the ground at the call of the Giants.[22]

The Giants' Quoits (Dolmens)

Dolmens are specifically associated with the Cornish titans such as *Trethevy Quoit* at St Cleer, which is also called 'The Giant's House'. The famous *Zennor Quoit* is also known as the 'Giant's Quoit', and the surrounding countryside is littered with granite boulders of all shapes and sizes.

Trethevy Quoit in 1872.   Towednack Chun Quoit in 1871.

Zennor Quoit, by W. Borlase.

Cornwall's most famous dolmen is the impressive *Lanyon Quoit* at Madron. Legend says that the bones of a giant were found in the tomb, hence its alternative name of 'The Giant's Table'.[23] Half a mile away, *West Lanyon Quoit* is located in a sloping meadow. During its discovery in the 19th century bones were found and *"the excavator suffered the fate commonly met by those who destroy megalithic sites. His cattle died and his crops failed, 'which left a warning impression on the minds of his neighbours.'"* [24]

Sperris Quoit is a ruined Neolithic portal dolmen in open moorland near Zennor Quoit. The capstone is missing and all that remains are one upright stone and three fallen stones. It is also known as 'The Giant's Rock of Tregerthen'. *Carwynnen Quoit* near Camborne, is a rare example of a portal dolmen outside the Penwith peninsula. It is also known as 'The Giant's Quoit' and 'The Giant's Frying Pan', due to its flat shape. Near the stunning *Boscawen-Un* stone circle, an extremely large footprint supposedly from a giant who was walking up country from the Scilly Isles was recorded in the mid 19th century:

> *On the way back to the carriages [from Boscawen-Un] the party visited Careg-Tol, a fine pile of granite rocks not far from the Circle, commanding an extensive view. Thereon are some shallow rock basins, the outline resembling a human foot, and which, being of superhuman size, are locally called giant's or devil's footprints.*[25]

The footprint in Celtic cultures often symbolised the beginning point of a journey or survey, and were often inauguration stones of local chieftains.[26]

An Iron Age hillfort not far from Tintagel in North Cornwall is said to be the resting place of a giant who was defeated by another giant. The fort is one of the largest earthworks in Cornwall. It is an oval enclosure with an area of about 19 acres. It is known as 'The Giant's Grave' or 'King Arthur's Grave.'

Tolmen Rocking Stone, by W. Borlase. Cornwall is known for its balanced rocks or 'Logan Stones' that were said to have been constructed by giants. The Cornish colossi tended to reside in these kinds of 'high places' away from the humans.

THE LELLIZZICK GIANT

Chapel Field is located near a small church in Lellizzick in North Cornwall. Here, an unusually tall skeleton was reported in the 1780s, in an area where many burials were found in stone coffins. It only got a brief mention, as though it was not such a strange thing to find:

> *In the adjoining ground many human bones were digged up, and*

on one occasion a whole human skeleton was discovered, that measured nearly seven feet in length.[27]

THE GORAN HAVEN GIANT AND DODMAN POINT

In the parish of Goran is an elongated earthwork running from cliff to cliff. It is 20 feet wide, and 24 feet high in most places. Folklore assures us it was constructed through magical means by a giant in a single night. This fortification has long been known as *Thica Vosa*, and the *Hack and Cast*. The giant ruler, who lived on the promontory, was feared by all those residing in the area, but when the giant fell ill through eating some food (possibly children, as he was a cannibal) his roars and groans were heard for miles. A reluctant doctor arrived and, like the stories of Jack-The-Giant-Killer, he told him that he must be bled. The giant submitted, and the doctor told him to ensure relief, a large hole in the cliff must be filled with the blood. The giant lay on the ground, relieved by the loss of blood but unaware the hole continued through the other side of the rock, he permitted the stream to flow on, until he, at last became so weak that the doctor kicked him over the cliff and killed him. The promontory of *The Dead Man*, or *Dodman*, is so called after the deceased giant. The spot on which he fell is called the 'Giant's House.'[28]

Anthony Roberts noticed that this particular legend encoded many esoteric clues. The giant used 'magic' to create the earthwork. He was a local ruler, suggesting great antiquity, and the sacrificial death of a so-called king to allow the rightful adept to come to power is also hinted at. Of course, to all the ley hunters, whenever a dodman is mentioned, it reminds them of the name for ancient surveyors of the megalithic world as discovered by Alfred Watkins.

GIANT OF TREGONEY - TEN FEET TALL

This account comes from the *Annual Register of Cornwall for 1761* about a discovery in one of the many tin mines in the area. Tregony is located east of Truro a few miles from the Roseland Heritage Coast. It was said to contain "*a skeleton of a man estimated at 10 feet.*"[29]

The Annual Register for 1761 tells us that in March of that year, as a miner was working at Tregoney, in Cornwall, in a new mine, he accidentally discovered a stone coffin, on which were some inscribed characters. Within it was the skeleton of a man of gigantic size, which, on the admission of the air, mouldered into dust. One tooth, two inches and a half long, and thick in

75

proportion, remained whole. The length of the coffin was eleven feet three inches, and its depth was three feet nine inches.[30]

The inscribed characters were very badly worn, but may hint that this was an important person. If it was one of the ancient legendary giants featured in this chapter, it would give some indication of burial practices and the fact they had some form of writing.

SKELETONS OF CORNISH GIANTS
Western Times, 30 August 1910, p.6

This fascinating report we include here from a seemingly scrunched up 110 year old newspaper!

Skeletons of Cornish Giants

The ruins of St. Piran's Oratory, near Perranporth, Cornwall, were the scene of a public ceremony on Saturday in connection with the protective work now being carried out. A body of trustees has been formed, and the ancient fabric, which has been brought to view by the removal of the surrounding sand, will be preserved by the erection of an enclosing building. During the work of excavation many skeletons have been found, some of them of human beings 7ft. in height. A feature of the skulls found is the perfect condition of the teeth. The remains have been discovered of a small building close to the oratory, which is believed to have been a priest's residence.

*The ruins of St Piran's Oratory, near Perranporth, Cornwall, ...
During the work of excavation many skeletons have been found, some of them of human beings 7ft in height. A feature of the skulls found is the perfect condition of the teeth.*

St Piran was an old Irish saint who arrived in England on the coast and built an oratory in the Irish style. He is *the* patron saint of Cornwall to many and founded the earliest Christian building in this part of the country. Two Celtic stone heads of a man and a woman were also discovered that may have been part of a prehistoric head-cult found all over the British Isles.[31] Who these giants were, or what period they came from is unclear.

The stone head.

John of Gaunt at Carn Brea

Carn Brea Neolithic hilltop settlement towers over Redruth, and dates from as early as 3900 BC. Neolithic remains, Bronze Age axes, gold, Celtic and Roman coins, and later artefacts were unearthed here. Inca-style terracing and Cyclopean walls were also part of the geomantic design of this 738-foot-high granite outcrop. According to legend, the giant John would often throw boulders at neighbouring giants, who would toss them back at him, hence the hillside is covered in large rocks. The giant's remains are said to be buried somewhere on the site.[32]

CARN BREA, DRUIDICAL ALTAR STONE.

Old postcard of Carn Brae's 'Altar Stone'.

Another version mentions one of the natural springs in the area. A well can be found on the northern slope of the hill, near the castle called the 'Giant's Well.' A local folk tale says that a giant lived on Carn Brea, and engaged in a fight with a giant from St Agnes Beacon (see next section). The Carn Brea Giant lost the fight, and the rock formations that litter the hilltop are his petrified bones.[33] Other stories tell that it was an important meeting point of the local titans:

> Some tales say the giants regularly gathered within its ramparts
> to hold great contests among their strongest members, watched
> over quite naturally by the equally ubiquitous fairies or spriggans.
> These tutelary guardians were said to have had a vast hoard of
> treasure buried in a hidden cave under Carn Brea.[34]

BOLSTER AND ST AGNES

This Cornish myth combines Christian and Pagan motifs by having a giant fall in love with a saint. Bolster was a tyrant, who battled with other giants, ate children and treated his wife badly. He forced her to bring rocks up from the base of the St Agnes Beacon to use as ammunition against his foes. When he witnessed St Agnes evangelising her new Christian beliefs to the locals, he started following in her footsteps attempting to woo her. However, the terrified saint reminded him of his wife and his immoral actions, so gave him a range of difficult tasks to complete. She asked him to fill a hole in the ground up with his blood to prove his love to her, but (like the Goran tradition) the hole never filled as it led down to the sea, and he eventually died. All along the coast, the sea ran red for days and even now the cliffs around Chapel Porth retain the crimson stain from the giant's lost blood. She was hailed as a heroine who later become canonised. However, the loving gaze of the giant Bolster remained with her until her dying day.

Left: Old illustration of Bolster striding over six miles between Carn Brea and St Agnes. Right: Photo of the pageant that takes place every May.

This is one of thousands of stories of Christianity defeating Paganism as the giants were regarded as symbols of the archaic religions. The early names of this area also echo themes that we find with biblical giants:

> *Bolster had his stronghold on the hill known as Carn Bury-anacht or Bury-anack, said to mean 'the sparstone grave'. The words anacht and anack have affinity with the biblical terms Anak or Anakim, the race of martial giants encountered by the early*

*Hebrews. The hill later became the celebrated St Agnes Beacon,
where the fire of annual rituals regularly blazed forth to mark the
passing of the seasons.*[35]

In legend, Bolster must have been a truly enormous figure as the curious account
of the *Giant's Stride* has him stepping over six miles between St Agnes Beacon
and Carn Brea. These two Beacon Hills are what Alfred Watkins called 'initial
points' marking the genesis of a survey. Another tradition has his wife carrying
large stones in her apron to the top of the hill. This is yet another snippet of
folklore that is found all over Britain that has a female giantess striding across
the landscape dropping piles of stones, or even large monoliths, and where
they fall marks ancient megalithic sites and cairns. The name Bolster is also still
attached to a linear earthwork that runs from the foot of St Agnes Beacon to a
nearby tin-mining area.

Bolster has been revived in a modern pageant that has a 28ft tall effigy
of the giant parading through the area of St Agnes. It has been staged each
May Bank Holiday weekend for the last 20 years on the cliffs at Chapel Porth
Beach.[36]

The Giant of Carn Galva

Carn Galva is the omphalotic *world centre* for the Penwith area, and is one of
many rocky outcrops said to be an early residence of giants. A peaceful giant
called Holiburn protected the people of Zennor and Morvah from the giant
bandits of Trencrom Hill, who pillaged, robbed and terrorised the locals.
After a few games of quoits (stone throwing), with a human friend from
nearby Choon, the giant patted his friend's head in a playful way, unfortunately
forgetting his own strength, his fingers piercing the human's head killing the
man instantly. The giant died of a broken heart seven years later.[37]

The Giant of Trecrobben (Trencrom) Hill

Trencrom Hill is a large prominent hill inland from St Michael's Mount, known
in the early Cornish legends as *Trecrobben Hill*. Giants were said to reside in a
stronghold made of huge, carved megalithic blocks:

*On the summit of this hill, which is only surpassed in savage
grandeur by Carn Brea, the giants built a castle - the four
entrances to which still remain in Cyclopean massiveness to attest
the Herculean powers by which such mighty blocks were piled*

upon each other. There the giant chieftains dwelt in awful state. Along the serpentine road, passing up the hill to the principal gateway, they dragged their captives, and on the great flat rocks within the castle they sacrificed them. Almost every rock still bears some name connected with the giants - 'a race may perish, but the name endures.' The treasures of the giants who dwelt here are said to have been buried in the days of their troubles, when they were perishing before the conquerors of their land. Their gold and jewels were hidden deep in the granite caves of this hill, and secured by spells as potent as those which Merlin placed upon his 'hoarded treasures.' They are securely preserved, even to the present day, and carefully guarded from man by the Spriggans, or Trolls.[38]

One of the giants was revered as a great builder, and Cormoran, the lord of the sea mount, was a colleague of his. They assisted each other in their endless reshaping of the landscape. The two are said to have thrown boulders to one another as recreation. In one version, they had only one large hammer to do their megalithic work. They would throw the hammer to one another over great distances between their respective strongholds.[39] Unfortunately, Cormelian, the wife of Cormoran, leaned out of her window one day and was struck dead by the flying hammer. The subsequent cries of anguish from her husband were said to have caused a violent storm that cut off the mount from the mainland for days. She was later buried beneath Chapel Rock, on the shore facing the island.

Again, the geomantic significance is interesting here. Throwing hammers in straight lines over distances is an analogue of 'leys' and when something gets in the way, this line of force can have a catastrophic effect by interrupting the geomancy of the landscape. The prehistoric earthworks that encircle the hill were once thought to be the location where numerous giants played elaborate games of bowls, with huge granite boulders, to mark out their territories. One now stands in a garden at the base of the hill that is marked by a National Trust Plaque as 'The Bowl Rock'.

Even this simple tale has undertones of the old, markings of the countryside with the patterns of geomantic measurement, forming a religious geometry that was an imposed infrastructure on a natural terrain.[40]

An old postcard of Bowl Rock.

Another huge boulder sits on top of the stronghold on a flattened area that marks a classic ley that goes through St Michael's Mount to the south and the spire of the hilltop church of St Ives to the north. Nearby is an earthwork called the *Giant's Grave* near Varfell, and if you continue south to Ludgvan towards St Michael's Mount, a church sits within an earlier prehistoric earthwork. A medieval painting inside the church is a curiosity to giantologists. Saint Christopher, the patron saint of pilgrims (and a giant in his own right) is represented in a wall painting above the north door. He is holding a flowering staff with a serpent coiled around it. It was at this church that famed antiquarian William Borlase was a rector, who preserved and recorded much of the megalithic landscape around Cornwall.[41]

BLUNDERBORE THE GIANT

A Cornish giant who was said to have lived in the area of Ludgvan was called *Blunderbore* (also recorded as B*lunderboar, Thunderbore, Blunderbus,* or *Blunderbuss*). He used to terrorise locals who travelled north to St. Ives and appeared in the Jack-the-Giant-Killer stories.[42] Jack gets trapped by the giant after three lords and ladies were captured by Blunderbore. The giant invited his friend Rebecks to help eat Jack but Jack somehow makes some nooses with rope he finds and manages to put them round the giants' necks and slit their throats. The giant also found his way into the stories of Tom the Tinkeard (or *Tinkard*, a tin miner), a local version of Tom Hickathrift, who is an East Anglian version of Jack who often battles giants.

Old illustrations that accompanied the tales of Blunderbore the Giant,
also featuring Jack-The-Giant-Killer.

JACK-THE-GIANT-KILLER

The violent chronicles of Britain's most famous giant hunter stretch far back into history. According to researchers at Durham University and the Universidade Nova de Lisboa, stories such as "Jack and the Giants" are thought to have originated more than 5000 years ago.[43]

A Chapbook of *Jack the Giant Killer* from the early 19th century.

Mainly based in Cornwall, Jack's exploits lingered around campfires across the whole of Britain. He was presented as a clever young man trained in sports and warfare, who often outwitted his gargantuan foes. He appeared to have replaced King Arthur as the fresh, new, giant hunter and he even found his way into a modern Hollywood movie in 2016. Other Jack-the-Giant-Killer stories continue in this vein, and it was only when the printing press was developed in the Victorian age that the story was toned down, transforming into the children's classic *Jack and the Beanstalk*.

Jack is described in folklore as an early Cornish (Celtic) warrior carrying out typical heroic feats, especially murdering his foes in original and cruel ways. His first kill was Cormoran on St Michael's Mount. The giant's famous outcry in the Jack stories has alone become the stuff of legend. "*Fee! Fie! Foe! Fum! I smell the blood of an Englishman*" appears as "*Fy, Fa and Fum*" in the pamphlet *Haue with You to Saffron-Walden* (published in 1596) written by

Thomas Nashe who made it clear that the rhyme was already old and its origins lost in time.[44] It also appears in William Shakespeare's early 17th century *King Lear* as "*Fie, foh, and fum, I smell the blood of a British man.*"[45] The earliest known printed version of the Jack-the-Giant-Killer tale appears in *The History of Jack and the Giants* in 1711:

> *Fee, fau, fum, I smell the blood of an English man, Be alive, or be he dead, I'll grind his bones to make my bread.*[46]

In one tale Jack is challenged to an eating contest by a giant rival but the hero concealed a bag under his garment and dropped the food into that. The giant is bemused further when Jack plunges a sword into his upper-body that releases the contents telling the giant that it had no ill effects on him. The giant does the same thing to himself, committing suicide.

When he was an old man, his penultimate battle took place near the Scottish border. The hero was said to have shot his final arrow at a giant whilst hiding behind an ancient monolith. Unfortunately, when Jack went to check to see if his enemy was dead, the giant leaped up and killed him, and the giant died from the arrow wounds. *The Giant's Stone* is said to mark Jack's burial.[47] The monolith forms part of a group of three standing stones situated on both sides of the road leading to Fruid Reservoir from Tweedsmuir village.[48]

THE GIANTS OF ST MICHAEL'S MOUNT

St Michael's Mount is one of the most prominent landmarks in England, located off the coast of Marazion in South Cornwall. It is a conical granite mountain

St Michael's Mount. Aerial photo by H. Newman.

with a monastery on top, surrounded by beautiful gardens and a population of 35 people. When the tide is low, it's possible to walk across the landbridge. In the days of the giants however, it was some six miles from the coast and was known as the *White Rock in the Wood*, or in Cornish, 'Carreg luz en kuz.'[49] This may represent a folk memory of a time before Mount's Bay was flooded and when woodland surrounded it. Carbon dating has established that a forest was submerged between 1700 and 2500 BC and remains of fossilised trees have been seen at low tides.[50] The area has been inhabited from at least 4000 BC and Mesolithic flints were found on the lower eastern slope (circa 8000-3500 BC).[51] The Mount is one of several candidates for the island of *Ictis*, described as a tin trading centre in the *Bibliotheca Historica* of the Sicilian-Greek historian Diodorus Siculus, writing in the first century BC.[52]

In 1275 an earthquake hit this part of Cornwall that disturbed an already existing Christian structure on the Mount. During the 14th century the chapel was rebuilt and it was around this time that a giant skeleton was probably first discovered. Before we examine the archaeological context of this find, let's take a closer look at the intriguing stories of the legendary giants who once inhabited this island.

Cormoran was the famous giant who featured in Jack's stories with some accounts putting him at 18 feet tall.[53] His name translates from Cornish as *'The Giant of the Sea'* - *Kowr-Mor-An*. He and his wife lived on the existing promontory and constructed the Mount, labouring for many months. They brought large chunks of white granite from a nearby shore, but Cormoran would frequently get her to do the bulk of the carrying in her apron. They were said to have walked from shore to shore, when the sea level was lower. One day, when Cormoran was sleeping, his tired wife decided to bring some greenstone rocks that came from a much closer location, but he awoke, and was enraged that his beloved granite was not being transported, so he kicked at her and broke her apron strings causing the boulders to fall to the ground. Today, the partially submerged causeway has a large green stone marking its entrance on the shore. Geomancy is certainly present here and Anthony Roberts noted the significance:

> *The 'broken apron strings' motif appears throughout British giant-lore as well as in Scandinavian, Indian and American legends, usually in conjunction with alignments of megalithic stones or, sometimes, linked to single free-standing boulders.*[54]

White quartz was a prized material for megalithic builders, not only for its glowing aesthetic, but because of its magical qualities, electromagnetism and piezoelectric properties. The piezoelectric effect may also be partly responsible for numerous sightings of glowing apparitions, thought to be St Michael himself above the Mount. In 495 AD, some fishermen swore they saw the archangel blazing alight on the summit.

The mount has been revered as 'sacred' throughout the Christian era and long before. It was once known as 'Dinsul' (Dunn-sol), or 'the mount of the sun', and is aligned to Beltane sunrise (May Day) and Samhain (Hallows Eve) sunset, along the great St Michael line (discovered by John Michell in 1969). This Mount bisects another 2500 mile alignment linking a series of hilltop sanctuaries dedicated to St Michael and Apollo across Europe, so it clearly has been much visited and had a prolonged place in the consciousness of seekers and pilgrims going back millennia.

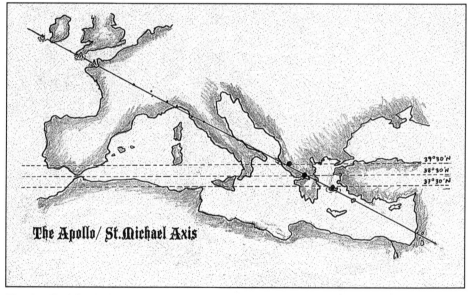

The Apollo / St Michael Line that stretches 2500 miles from Skellig Michael in Ireland, through St Michael's Mount in Cornwall, over to Mont Saint-Michel in France and many other Michael and Apollo sanctuaries in Europe, including 'Oracle' sites such as Delphi, Delos and Dodona in Greece, that are exactly one degree of latitude apart.

Giant legends are also found at Mont Saint-Michel in France, another hilltop island sanctuary very similar to its English counterpart that also sits upon the St Michael/Apollo axis. In fact, St Michael's Mount monastery was an offshoot of the French establishment. These 'high places', were previously the abodes of

85

giants, who had a fondness for elevated strongholds:

> *Mont Saint-Michel has its own giant traditions, in which a terrible titan inhabited the summit and ceaselessly preyed upon local residents. When he captured a lady of noble birth, King Arthur came over from Cornwall and put an abrupt end to his career by killing him upon the hilltop. A neat gathering-together of all the main giant motifs.*[55]

The oldest recorded name of Mont Saint-Michel was 'Mount Gargan' and the original inhabitant was the father of the Giant Gargantua, who is still said to be buried there.[56] Traditions also exist of giants throwing boulders between the coast of Normandy and St Michael's Mount in Cornwall, mythologically identifying the great Apollo/Michael axis.

Arthur finds a Giant, from *Roman de Brut* (mid 14th century).

Returning to Saint Michael's Mount, in the most famous Jack-the-Giant-Killer story, he succeeded in defeating Cormoran in a unique and elaborate manner. One morning he woke the sleeping giant by blowing a horn loudly in his ear. The alarmed titan came rushing out, was blinded by the sun,

Jack in action killing Cormoran the Giant.

and fell into a deep pit that Jack had prepared and covered with twigs and straw. Other versions say it was a thin layer of stone, or even full of water so the giant drowned. He was then hacked to death by Jack in his usual ultra-violent way. Jack was dubbed 'Jack-the-Giant-Killer' for this feat and received not only the giant's wealth, but a sword and belt to commemorate the event from the Marazion town elders. The belt was embroidered with: "*Here's the valiant right Cornishman, Who slew the giant Cormelian.*" [57] Cormoran (sometimes spelt Cormelian) was reported to have had six fingers on each hand and six toes on each foot.[58] He was also a 'Cyclops' with one eye and battled 'white magicians' for power and land, both losing and gaining wealth in the process.[59]

GIANT SKELETON OF ST MICHAEL'S MOUNT

In the early 1800s an 8ft skeleton was unearthed from an old underground chamber on St Michael's Mount. From *The Age*, Jan 24th, 1955, it reads:

The evidence: Some 200 years ago, the skeleton of a man who, in life, stood eight feet tall, was found in a narrow dungeon cut in the solid rock of the island.

The Killer

ST. Michael's Mount, in Cornwall, the island which Jack the Giant Killer wrested from an oversize man named Cormoran, has been given to the British National Trust, to be preserved as an historical site and tourist attraction.

There may be more truth—and history—than fiction in the old nursery tale.

The evidence: Some 200 years ago, the skeleton of a man who, in life, stood eight feet tall, was found in a narrow dungeon cut in the solid rock of the island.

To this day, the visitor can see, beside the steep cobbled path leading up to the castle, the pit where Jack trapped his foe.

The first published account of this giant skeleton was published in 1824:

A low Gothic doorway was discovered, closed up with stone in the southern wall...a passage was found to descend, containing ten steps, which led to a vault of stone under the church, about nine feet long, six or seven broad, and nearly as many high. In this vault was found the skeleton of a very large man, without any remains of a coffin.[60]

The National Trust's *Archaeological Evaluation of St Michael's Mount* (1993), confirmed the account and gave some interesting details referencing various contemporary sources:

> *Giant's skeleton: One of the Mount's several quaint stories relates to the reported discovery in the early years of the 19th century (probably 1804 or 1811 when the church was renovated) of the skeleton of a larger than normal person within a stone vault beneath the church. The door to the steps down to the vault had been blocked with stone and the body was not in a coffin. 'The discovery, of course, gave rise to many conjectures, but it seems most probable, that the man had been there immured for some crime' ('A Physician' 1824, pg.59). 'The bones of the unhappy victim were taken up from the dark abode, in which they had been immured for unknown ages, and interred within the body of the church' (Thomas, 1831, pg.18). According to Canon Taylor, writing about 100 years later, the body whose shin 'was larger by one-half than that of an ordinary man' was re-buried in the north court and in 1864 was moved again, down to the island's cemetery (Taylor, 1932, pg.6). Its location within the cemetery is not known. Was the discovery the cause of the association of the Mount with giants? Or was it used to revive earlier traditions?* [61]

Numerous lines of evidence come together to confirm the authenticity of the discovery of an eight-foot-tall skeleton on St Michael's Mount.

THE GIANT SORCERER OF TRERYN HILL

The fortress on Treryn Hill, close to Land's End, is a prime example of Cyclopean architecture built by the power of enchantment and thought to be a remnant of the lost land of Lyonesse. The chief giant was the strongest of the tribe and also a necromancer who ruled with sorcery during the creation of the fort. The huge walls and the outer earthworks are still partly visible today:

> *He (the chief giant) sat on the promontory of Treryn, and by the power of his will he compelled the castle to rise out of the sea. it is only kept in its present position by virtue of a magic key. This the*

giant placed in a holed rock, known as the Giant's Lock, and whenever this key, a large round stone, can be taken out of the lock, the promontory of Treryn and its castle will disappear beneath the waters. There are not many people who obtain even a sight of this wonderful key. You must pass at low tide along a granite ledge, scarcely wide enough for a goat to stand on. If you happen to make a false step, you must be dashed to pieces on the rocks below. Well, having got over safely, you come to a pointed rock with a hole in it; this is the castle lock. Put your hand deep into the hole, and you will find at the bottom a large egg-shaped stone, which is easily moved in any direction. You will feel certain that you can take it out, but try! Try as you may, you will find it will not pass through the hole; yet no one can doubt but that it once went in…no one has ever yet succeeded in removing the key of the giant's castle from the hole in which the necromancer is said to have placed it when he was dying.[62]

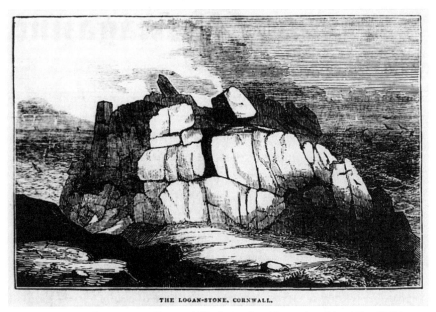

THE LOGAN-STONE, CORNWALL.

Cyclopean Fortress on Treryn Hill, *Saturday Magazine*, Nov 7th, 1835.

Further legends talk about his tragic end, when his wife was coveted by a younger giant who resided below the Logan Rock at Trereen. He crept up on the chief who was sitting in his rock-cut chair or throne and stabbed him in the stomach, while his wife watched on and did nothing to save him. The young titan threw him in the sea, and he and the deceased chief's wife were said to have lived happily ever after. Again, the 'sacrificial' tones of this tale echo others, and may have more meaning that just a simple murder.

THE GIANTS OF MÊN-AN-TOL

This account is from the early 1700s in the proximity of *Mên-an-Tol*, a wonderful megalithic site featuring a holed stone that has many legends attached to it, the main ones being based around fertility and healing. However, it seems that giants also resided in the area too:

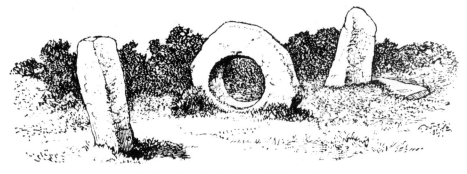

Antiquarian illustration of Mên-an-Tol.

In the village of Mên, near the Lands End, a farmer, in the year 1716, removing a flat stone seven feet long and six wide, discovered underneath it a cavity formed by stone, two feet long at each end, and on each side another stone twice as long. In the middle was an urn, full of black earth, and round it were some very large human bones irregularly dispersed. In some sepulchres have been found bones much larger than those of the human body, which are therefore thought by the vulgar to have belonged to the giants.[63]

TREBIGGAN THE GIANT

Trevegean (also known as Trebegean) is the name of a village near Land's End, also known as 'the town of the giant's grave' in Cornish. A local titan once had the name Trebiggan. He was a vast man, with arms so long that he could take men out of the ships passing by Land's End, and place them on other ships. He was also fond of eating children, who he would fry on a large flat rock a short distance from his cave. A skeleton of colossal proportions was unearthed in the area in the early 1800s that gave the legend some life:

> *Carew mentions a little village called Trebeyan, or, the town of the giant's grave, not far from the Land's End, "near which, and within memory, certain workmen searching for tynne (tin), discovered a long square vault, which contained the bones of an excessive big carkas (carcass), and verified this etymology of the name.* [65]

THE GIANT KING OF MÊN SCRYFA

Mên Scryfa is an inscribed stone, one of several standing stones along the Tinners' Way (a modern long distance footpath) close to Mên-an-Tol. There is an inscription in Latin reading *Rialobrani Cunovali Filii,* meaning 'Royal Raven, son of Cunoval'. It is thought to be the tombstone of a king killed in the 6th century Battle of Gendhal Moor. The stone is nine feet high, and the legend is that the king was as tall as that. He is said to be buried under the stone, complete with his treasures and weapons.[64] The site has never been excavated, so the skeleton may still be entombed below the great menhir.

THE GIANT OF LOOE ISLAND

An archaeological excavation carried out by 'Time Team' (on British TV's Channel 4) at St Michael's Chapel on Looe Island (also once called *Lammana*) discovered a prehistoric standing stone in the grounds of a church once built by the monks of Glastonbury Abbey in the thirteenth century. During an antiquarian excavation in the late 18th century a *"remarkably large human skeleton"* was discovered. A second inhumation and a stone cist burial were also found. In 1823 a certain T.L. Bond wrote that *"on the top of it are the remains of some building, which goes by the name of the Chapel. Some years since a remarkably large human skeleton was found in it."*[66] These remains are probably the same individual that was found in c.1783.[67]

There is also a stone circle at Stonetown Farm, Dunloe, round barrows south of Wilton Mill, as well as as site called Mabel Barrow and a tumulus at Bin Down. A large bronze ingot found by divers south of the Island, has put forward the theory that this may be the legendary Ictis, the tin trading island seen by Pytheas in the 4th century BC and recalled by Diodorus Siculus in the 1st century BC.[68]

GIANTS OF THE ISLES OF SCILLY

Giant's Castle at St Mary's on the Isles of Scilly was occupied by a nameless giant who was said to have constructed this Iron Age cliff-castle or promontory fort, that has stone-built ramparts and ditches cut out of solid rock. Another giant, lived on Buzza's Hill (now Buzza Hill) at Hugh Town. He was called Bosow, but as James Orchard Halliwell-Phillipps observed in 1861: *"It is of the rarest occurrence to hear the name of a giant mentioned in the recital of any oral tradition in this district."*[69] William Borlase opened some 'Giant's Graves' on the island and had a 'sowers of thunder' experience that we will look at more closely in the chapter of the same name.

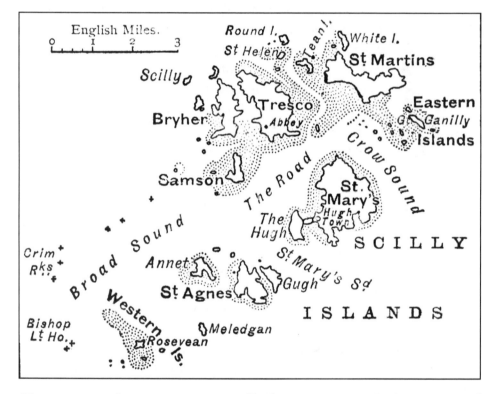

The next step of our giant journey will take us east, towards the counties of Sussex, Kent, Surrey, East Anglia and into the nation's capital, London. It is here that Gogmagog reappears far from his home county of Cornwall, and in a different guise, becoming the principal giant protector of Britain.

More antiquarian illustrations of Cornish giants Blunderbore and Cormoran battling their heroic foe, Jack-the-Giant-Killer.

GAZETTEER OF GIANTS - DEVON & CORNWALL

DATE	LOCATION	SIZE	No.	FEATURES	PAGE
		DEVON			
1808	Unknown location	Huge	1	Giant femur & stone coffin	63
1856	Lundy Island	8ft-8ft 7'	2	Stone-lined graves	64
		CORNWALL			
1780s	Lellizzick	7ft	1	Found in churchyard	74
1761	Tregony	10ft	1	Carved inscription	75
1600s	St Piran's Oratory, Perranporth	7ft	1	Perfect teeth	76
1804	St Michael's Mount	8ft	1	Found in vault	87
1716	Mên-an-Tol Holed Stone	Giant	1	Urn full of 'black earth'	90
1824	Trevegean, Land's End	Giant	1	Legendary giant in area	90
1810	Mên Scryfa Standing Stone	9ft	1	Possibly still buried there!	91
1823	St Michael's Chapel, Looe Island	Giant	2	Possibly two burials	91

Sheringham Hand-Axe, p.122

Holbeach, 7ft 2, p.121

Burgh Castle, 7ft 4, p.120

Peterborough, 7ft, p.118

WANDLEBURY CHALK FIGURES

TOM HICKATHRIFT

War Ditches, Cambs, p.115

Heydon Ditch, p.117

Wrabness, Essex, p.119

Weston Church, p.107

Walton, Essex, p.119

Dunstable, 7ft, p.108

Finchingfield, p.119
Brent Pelham, p.111

St Albans, 8ft, p.108

Nazeing, 7ft, p.118
King John's, 7ft, p.98
Guildhall, p.97

Thundersley, 7ft, p.119

Canonbury, p.100

Brentford Ends, 7ft, p.100
Aldeuman Buri, 11ft, p.97
Woolchurch, p.98

Bethnal Green, 7ft, p.99
St Paul's, p.99

ST PAUL'S CATHEDRAL

Deal, p.103

Guildford Jaw, p.102

Rotherhide, 10ft, p.100

Cuxton Handaxe, p.104

Charterhouse, 8ft, p.101

Mitcham, 7ft, p.101

Wotton, 9ft 3, p.101

Dover, p.103

Hythe Church, 7ft, p.102

Boxgrove, p.105

LONG MAN OF WILMINGTON

GOG AND MAGOG

4

LONDON AND
THE SOUTHEAST

The giant Gogmagog transformed into *Gog and Magog* after he was captured and taken to London. Chained to the city gates, the two titans were placed there to protect the capital from invaders. Gog and Magog are still on display at the Guildhall and they are also a prominent part of the Lord Mayor's Show every November. We have already explored the singular Gogmagog's origins and significance in the opening chapter, but there is a tradition that their captor, Brutus, was in fact buried in a mound at the location of the Tower of London, "*on a spot known as Bryn Gwyn or The White Mount.*" [1] However, a legendary giant king may have also been buried there.

Brân the Blessed (Welsh: Bendigeidfran or Brân Fendigaidd, literally *Blessed Crow*) was a renowned giant and ancient ruler of Britain. He is written about in several of the Welsh Triads, but his most significant role is in the Second Branch of the Mabinogi, *Branwen ferch Llŷr.*[2] He is a son of Llŷr (God of the Sea) and Penarddun, a human female. As we saw in the previous chapter the name Brân in Welsh is usually translated as crow or raven:

> *In Celtic mythology, Bran appears as a semi-humanized giant residing at Castell Dinas Bran, the later home of the later Kings of Powys. Though Bran himself was supposed to have been an early King of the Silures tribe of Gwent.*[3]

In his final battle against the Irish he is mortally wounded, so sacrifices himself by asking that his head be cut off and taken to various places for specific periods of time. After some strange twists, such as the head continuing to talk and getting lost in a time warp, his cranium is eventually taken to Gwynfryn,

The White Tower, 1872. Illustration by Gustave Doré.

Raven at the Tower of London.

the 'White Hill' which is traditionally the location of the Tower of London. It is then buried facing France to ward off any further invasion. King Arthur disinterred Bran's head to assert his supremacy, but it quickly backfired, as soon after this the Saxons invaded Britain. This gesture is referred to as one of the three *unfortunate disclosures of Britain*. The tradition survives with the Ravens currently kept at the Tower of London. It is said that if they were ever to leave, then Britain would fall to invaders. Their wings are wisely kept clipped so they have no choice but to fulfill their duty.

There is a connection to Bran in Cornwall. In West Penwith the name Bran is associated with Men Scryfa, a significant standing stone featured in the previous chapter, which records: *Rialobrani Cunovali Filii,* meaning 'Royal Raven, son of Cunoval,' suggesting a local leader carried the name of the hero, the son of a Cynfawl. Could this standing stone mark the renowned burial place of Bran?

The Celts believed Bran belonged to a race of giants called the Tuatha which preceded mankind. Bran was a semi-humanised giant. [4]

We will explore who the 'Tuatha' were in the chapter on Ireland.

A gigantic suit of armour is on display in the Tower of London standing at nearly 7ft tall. Some researchers speculate that it was the armour of John Gaunt (1340 - 1399) or the Duke of Lancaster, a soldier whose great height was legendary.[5] Perhaps it was created to represent Bran, or one of the other giant protectors of London.

There are several reports of giant bones and skeletons being unearthed in London, from prehistoric times right up until the early 1900s. We start at the home of Gog and Magog at the Guildhall.

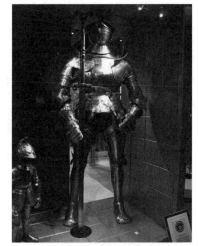

This 7ft suit of armour at the Tower of London.

THE GIANT BONE AT THE GUILDHALL

This report was written in the 16th century translated from old Welsh:

> *And the thigh bone of a man is still to be seen in the church of Saint Laurence close to Guild Hall or the Debating House in London; and this bone in times past was twenty six inches in measure of length and height.*[6]

Described here was a giant human bone tantalisingly close to the Guildhall, the residence of Gog and Magog. The 26 inch femur would be from someone well over seven feet tall.

COLOSSAL 32 INCH THIGH-BONE

The next account from the same manuscript describes a remarkably long thigh bone. At thirty-two inches, it may once have been part of an eleven-foot-long skeleton. The area is now called Aldermanbury in the City of London:

> *And another also is to be seen in Aldermari Buri or Aldeuman Buri, measuring thirty two inches long and the statue that was made of it through the art of perspective, and was put in the eastern end of that church's cloister and not far either from that bone itself, shows the complete image of a man eleven feet in length or height;*

this bone was found within the cloister to the ground as far as its foundations, carrying away also the stones from that cloister to the Strand for building his own home from those stones.[7]

MASSIVE BONE NEAR THE LONDON STONE

Another remarkable discovery was made close to the famous *London Stone* (see *Albion* chapter) just after the Great Fire of London:

After the great fire of 1666, upon pulling down the church of St. Mary Wool-Church, and making the site into a market-place, was found a huge thigh-bone, which was assumed to have belonged to a woman, and which was afterwards to be seen at the King's Head Tavern at Greenwich.[8]

The church is just a few hundred feet from the London Stone and had an early foundation dating from the time of William the Conqueror, during the 11th century. The area around Cannon (formerly Candlewick) Street also hides remains of a huge pre-Roman wall, although John Cotton, of *The Museum of London* believes it is Roman. The

The London Stone as it looks today.

wall extends to at least fifteen feet deep, whilst six feet is the average Roman level. London historian Rob Stevenson told the authors this: *"There's always an element of truth in these things. If the wall is twenty two feet thick, under the station, it must pre-date the London Wall. There must have been a civilisation that existed here in the distant past".*

CARTLOAD OF GIANT BONES AT SOMERSET HOUSE

This next account is in the vicinity of the London Stone near the River Thames. It is at least 2000 years old as Romans were living in this area at that time.

In February, 1830, during the excavation of ground on the eastern side of Somerset House, now the site of King's College, a cartload of human bones was found; and among them were two skulls and several bones of extraordinary dimensions, which must have belonged to persons of great stature.[9]

There is no indication of how deep the bones were buried, but just down the river near Vauxhall, Mesolithic activity has been recorded.

HUGE THIGH-BONE OF A GIANT NEAR ST. PAUL'S

St Paul's Cathedral was Sir Christopher Wren's finest achievement. There is much debate as to whether a Pagan Temple of Diana once stood on the site of Ludgate Hill under the building. Writing on St Paul's around 1600, the respected historian William Camden claimed there were many connections to the name *Diana* in the close surrounding area and he conjectured a temple dedicated to her once stood there. The discovery of hundreds of ox skulls emphasised the notion that it was an ancient ritual site. E.O. Gordon, author of *Pre-Historic London*, even suggested that "*Silhouetted against the sky, (was) the mighty unhewn monolith of the druidic circle,.....no trace of the circle remains.*" Rob Stevenson believes there was some sort of circular foundation that Wren outlined in his more obscure notes, which may indicate a ring of stones.

This account found in the immediate area of St. Pauls gives no indication of size, but "huge" seems like it was something beyond the norm:

The Neios, for June 2nd, 1664, advertised that at the Mitre, near the west end of St. Paul's, was to be seen a rare collection of curiosities, among which was the 'huge thigh-bone of a giant.'[10]

Lud (Welsh: *Lludd map Beli Mawr*) was said to have ruled Britain before the Romans. Founding the City of London, he was buried at Ludgate. During his reign he built cities and refortified Trinovantum (London), suggesting he may have been involved in the mysterious subterranean city walls. Geoffrey of Monmouth suggested the name London came from *Caer Lud* or *Lud's Fortress*. Variations of the legendary king feature in both Irish and Welsh mythology.

SEVEN-FOOT SKELETON NEAR BETHNAL GREEN

In March, 1813, was found at King John's Palace, Old Ford, a stone coffin containing the remains of a skeleton, which, from the length of the thigh-bone, must have belonged to a person nearly seven feet long.[11]

All indications of this discovery near Bethnal Green point to this being authentic as the historical record confirms a stone coffin was unearthed at this time.[12]

THE 10FT TALL ROTHERHIDE GIANT

London-based newspaper *The Weekly Packet* from late December 1717, tells us of a colossal discovery just in time for Christmas:

> *Last week, near the new church at Rotherhithe, a stone coffin of a prodigious size was taken out of the ground, and in it the skeleton of a man ten foot long.*

A GIANT'S SKELETON - BRENTFORD END
Nottingham Evening Post, 5 January 1895

A GIANT'S SKELETON.

On Thursday some men who were making excavations at Brentford End found, at a depth of 10ft., the skeleton of a man "about 7ft. in length," and later they came upon a hip and thigh bone. Mr. Field, of the post office, has taken charge of the remains. The ground where the discovery was made had not been disturbed for half a century. Immediately over the right ear there is a clean-cut hole, such as might have been made by a rifle bullet, and directly over the left temple there is another hole of the same size.

This curious report described what appears to be a very ancient 7ft skeleton because it was buried so deep, but it was found to have had two bullet holes in his skull. This could indicate an example of 'trepanning', a technique of drilling holes in the skull for therapeutic and ritual purposes that although a rarity in Britain, is featured in several accounts in this book.

NOTES FROM THE ISLINGTON BELLS
The Islington Gazette, 9 June 1904

This brief mention of a giant discovery was originally part of a lecture given at Canonbury Tower to the *London and Middlesex Archaeological Society*, and featured in a column of the *Islington Gazette*:

> *The giant skeleton was discovered by the Brothers Nichols in 1770.*

Who these brothers are, or details about what they actually discovered are unclear as we could find no other follow up to this.

GIANT SKELETON
Daily Gazette for Middlesbrough, 31 March 1914

We finish the section on London on the border with Surrey, with a description of a seven foot tall skeleton possessing well-preserved teeth in Mitcham.

> **Giant Skeleton.**
> While making a new road for the Brighton Railway Company near Morden-road, Mitcham, workmen unearthed three human skeletons in perfect condition, the teeth being exceptionally well preserved. One is that of a man who stood more than seven feet high. It is suggested that the skeletons are those of men who fell at the Battle of Merton fifteen hundred years ago

SURREY

Surrey is situated just southwest of London, adjoining the River Thames. The county is bordered to the northwest by Berkshire, to the northeast by the Greater London conurbation, to the east by Kent, to the south by Sussex, and to the west by Hampshire. Paleolithic activity and Neolithic and Bronze Age remains are spread across the county. Before the Roman era the county was occupied by the Atrebates tribe, centred at *Calleva Atrebatum* (Silchester) in Hampshire. Eastern areas were ruled by the *Cantiaci*, based mainly in Kent, who also controlled the southern bank of the Thames River.

THE WOTTON 9FT 3IN SKELETON

John Evelyn, the famous philosopher, was buried in Evelyn Chapel in St John's Church, Wotton. In the same churchyard another remarkable human being was also buried:

> *About one hundred and sixty years ago, as some workmen were digging in the church-yard, they found an entire skeleton, which, on being measured proved to be 9ft. 3in long. It was lying between two boards of the coffin; but on the workmen endeavouring to take it out, it fell to pieces.* [13]

MASSIVE THIGH-BONE AT CHARTERHOUSE

This account from Surrey describes another enormous internment:

In May, 1763, while some workmen were digging a vault under the master's apartment in the Charterhouse, they discovered a perfect human skeleton, of a surprising length; the thigh-bone measured two feet two inches, and the other bones were in proportion.[14]

This would have made the height of the person in question over eight feet tall.

SAXON SKELETONS: REMAINS OF A GIANT AT GUILDFORD
Portsmouth Evening News, 12 December 1932

SAXON SKELETONS

REMAINS OF A GIANT AT GUILDFORD

Two skeletons, thought to be of Saxon
origin, were found by workmen while
engaged in the construction of a tennis
court in the garden of Upton, Guildown,
Guildford, the residence of Mr. C. H.
Wood, the Town Clerk of Guildford. The
remains were unearthed at a depth of
about three feet, and are believed to be
those of a man and a woman buried side
by side.
The man must have been a giant, as his
lower jaw, which was found intact, can
be placed round the jaw of an average
modern man.

Two skeletons were unearthed, thought to be Saxons, but one of the traits of the male skeleton was the massiveness of the jaw that was easily *"placed round the jaw of an average modern man."* This is a repetitive story that appeared in hundreds of reports in America. There is also a larger than expected percentage in the British Isles.

KENT

The county of Kent is famous for a series of megalithic sites in the Medway area, including Kits Coty Dolmen and Coldrum Burial Chamber dated to between 2500-1700 BC. A majority of the featured skeletal discoveries are thought to be Danish or Jutish warriors from the Iron Age, when they are known to have arrived. Possibly the most important discovery relating to giantology was two remarkable and very ancient handaxes that were of massive size and date to a staggering 300,000 years ago. We begin our survey in the Cathedral city of Canterbury.

GIANT WARRIORS OF HYTHE, KENT

In a vault under the church of Hythe, Kent, has long been preserved a huge pile of several thousand skulls, and arm, leg, and thigh bones, some of which are very large, and are said to be the remains of the Danes and Britons who were killed in a battle near the place. One of the thigh-bones must have belonged to a being nearly seven feet high.[15]

GIANT JUTISH WARRIOR'S SKELETON FOUND
The Dundee Courier, 27 December 1933

A huge skeleton was unearthed at Deal in Kent that caused a local sensation.

GIANT JUTISH WARRIOR'S SKELETON FOUND

SPEAR AND KNIFE IN GRAVE.

While digging foundations for the new Roman Catholic Church in the mining colony at Deal, workmen unearthed another Jutish burial ground.

Some four feet below the surface a complete skeleton of a Jutish warrior, with spear by the right side and knife on the left, was found.

Local archæologists state that the skeleton is that of a Jutish giant and that the date of burial would be approximately 500 A.D.

The remains were found on high ground overlooking the sea, and were apparently those of an old man.

GIANT SKELETON AT DOVER
Cheltenham Chronicle, 9 March 1901

Originally in the *Dublin Evening Telegraph* from 4th March 1901, this later account gives more details of the discovery in Dover:

The skeleton of a gigantic man, some glazed tiles, Roman bricks, and part a Roman jug are among the remains.

GIANT SKELETON AT DOVER.

Some interesting finds have just been made in a part of Dover once covered by the sea, now being excavated for the enlargement of the premises of Messrs. Leney and Co., brewers. The skeleton of a gigantic man, some glazed tiles, Roman bricks, and part of a Roman jug are among the remains, while above the sand level was also found a small quantity of coal, lying in thin strata two yards long.

PALEOLITHIC GIANT HANDAXES

One of the most interesting archaeological discoveries in the county was not of any bones or teeth, but two massive handaxes that date to the Lower Paleolithic period, which lasted from 600,000 to 250,000 BC. The first excavations took place in 2003, and produced a total of 206 handaxes.[16] In 2006 the University of Southampton excavated twenty more flint handaxes at a site at Cuxton, including two of exceptional size and sophistication: *"The axes - one of which measured 307mm (1ft) in length - were dug up from old sand deposits in a front garden."*[17] The larger axe was the second biggest ever found in Britain (the largest being the Furze Platt axe found in Berkshire) and has caused debate among the academic community as it is simply much too sophisticated to have been produced at this time, long before modern humans were thought to have existed in England.

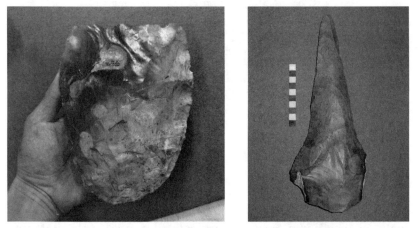

Left: The gigantic 'rounded' axe being held to show the size. Right: The elongated 'giant' ficron handaxe, c.300,000 years old. Photograph by Thomas Wynn.

Preserved in almost mint condition, the handaxe displays exquisite, almost flamboyant, workmanship in addition to its extreme size. Another giant handaxe was also uncovered beside it, this time a cleaver, 179mm long by 134mm wide.[18]

The quality and sophistication is not the only question here, because the sheer massiveness of these axes suggests very tall and powerful humans were creating and wielding them hundreds of thousands of years ago. If so, who were these prehistoric titans?

SUSSEX

The name 'Sussex' derives from the *Kingdom of Sussex*, founded by Ælle of Sussex in 477 AD. There is evidence that there were standing stones in the Hove area which is rich in sarsen, the same type of stone found at Stonehenge and Avebury. In Alfriston and Ditchling there are remnants of stone structures and at Philpots Camp there are some ancient carvings, and Mesolithic rock shelters.[19] In the Neolithic period, Britain's largest and most important flint-mine, near Worthing was in operation. This was at Church Hill in Findon, carbon-dated to between 4500 BC and 3750 BC, making it one of the first in the country. Some of the flint tools from here have been discovered as far away as the Eastern Mediterranean.[20] At a site called The Beedings near Pulborough archaic tools have been discovered that date from around 35,000 years ago, which were manufactured by either the last Neanderthals in Northern Europe or incoming groups of Cro-Magnons.[21]

Sussex was also once the residence of some of Europe's earliest recorded hominid activity, whose remains have been found at Boxgrove dating to half a million years ago. These were the bones of a male *Homo heidelbergensis*:

It is believed that Boxgrove Man was extraordinarily large and robust; from measurements of the unearthed tibia he is believed to have been nearly six feet tall, and the teeth measure nearly 1.5 times the length of those of modern man.[22]

Boxgrove Quarry. A reconstruction by artist Ivan Lapper.

105

Homo heidelbergensis have been recorded at over seven feet tall, with one fossilised bone "twice the size of a normal person" reported in South Africa.[23]

THE LONG MAN OF WILMINGTON

The Long Man of Wilmington hill figure in Sussex is cut into a soft chalk base. He stands around seventy metres tall holding a 'staff' in each hand and is thought to represent any one of several legendary colossi:

> *Flint pits on the hill are said to be craters formed by missiles hurled by the giant on Firle Beacon...at the giant of Windover Hill. This could be a sighting legend. Another version says that the giant fell over the steep north edge of the hill and broke his neck - the Long Man on the north slopes if the outline drawn round the giant's dead body. Beneath the outline a Roman is supposed to be interred in a golden coffin. A third version of the legend hints a memory of the ley being described here: it says that the giant was killed by pilgrims on their way to Wilmington Priory, our fourth ley point.[24]*

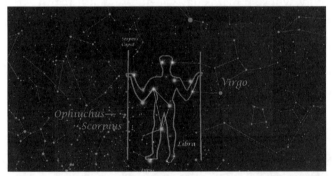

The Long Man of Wilmington and associated constellations by R. Hamilton.

The legend mentioned here describes two great giants who lived on two hills: Windover Hill and Firle Beacon. After a long period of joyfully throwing earth and boulders around for fun, they got agitated with each other and began fighting. After a long battle the Windover Hill Giant was defeated. The Firle Beacon Giant was devastated as he had now lost his old friend. The villagers drew a line around his body, removing the grass to reveal the chalk surface below:

> *It is said in local tradition that if you ever find yourself in Wilmington, on a dark and stormy night, try and listen: between the whistle of the wind and the roar of the rain, you can hear the*

rumble of the fighting giants. Sometimes you can even make out the weeping of the Firle Beacon giant at the death of his friend by his own hand.[25]

The Cerne Giant may represent many characters in ancient folklore including Beowulf, Woden, Thor, Apollo or Mercury, but to an avid 'ley hunter' he is what Alfred Watkins called the 'Dodman', with his sighting staves in hand. Further geomantic connections are evident in the area. A phantom black dog is also reputed to lurk above the figure (who is also said to move in straight lines) and various 'leys' have been noted going through the figure.[26]

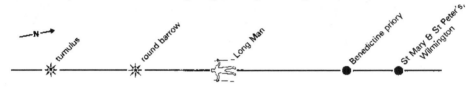

The Long Man of Wilmington Ley

HERTFORDSHIRE

We now jump slightly north to the county of Hertfordshire, just above London. The name Hertford is derived from the Anglo-Saxon *heort ford*, meaning deer crossing (of a watercourse). Mesolithic and Neolithic activity, extending up to the Bronze Age has been found in this area, with a few notable giant discoveries, such as this one called Jack.

THE LOST THIGH-BONE OF JACK O' LEGS

Just within the gate of Weston Churchyard in Hertfordshire are two blocks of stone 14ft apart, said to mark the burial place of *Jack o' Legs*, the Weston Giant. The stones are meant to mark the head and the foot of the legendary titan's grave. In 1728 Nathaniel Salmon wrote about Jack in his *History of Hertfordshire*:

> *The Giant, called Jack of Legs, as fame goes, lived in a wood here, he was a great robber, but a generous one, for he plundered the rich to feed the poor.*

However, he got caught by the Baldock Bakers who gouged out his eyes and voted to hang him, but he was granted one last request to fire an arrow and be buried there. It landed in Weston Churchyard. The legend held strong in the local area and one of his bones was said to have been unearthed in the 1600s:

> *Moreover, the parish clerks of Weston would for a tip exhibit what they claimed was Jack's thigh-bone, a very long bone which used to be kept in the parish chest.*[27]

The Grave of Jack O'Legs with the two stones on the left and right in the churchyard.

EIGHT-FOOT GIANT MEASURED BY ANATOMIST

This next account is from the Roman town of Verulamium in St Albans. A detailed report in the *Philosophical Transactions of the Royal Society* in 1712 by a well known medical anatomist revealed something extraordinary:

> *The circumference of the skull length-wise was twenty-six inches, and its breadth twenty-three inches. The greatest diameter of each os innominatum was twelve inches. The left os femoris was twenty-four inches long, and the right one was twenty-three inches in length. Each tibia was twenty-four inches long. If all the parts bore a due proportion, this man must have been eight feet high. The bones were found near an urn, inscribed "Marcus Antoninus," on the site of a Roman camp.*[28]

At first glance it appears to be an 8ft tall Roman warrior, and if that is the case, it is known that the Romans often hired the tallest recruits from territories they invaded.

BEDFORDSHIRE
UPWARDS OF SEVEN FEET LONG

This account from Dunstable in Bedfordshire, again involves the Romans, as it is known they sometimes buried their dead in lead coffins. However, only 300 of this type of coffin have been found in total in Britain so this was still a rare find, but not as rare as discovering two giant skeletons in one place.[29]

> *Singular Discovery.* — A short time ago, Mr. Fossell, of the Square, Dunstable, being about to erect a wall round his premises, employed some labourers, who commenced digging, when they discovered the remains of what were supposed to be the monasteries, which in "olden times" existed in that neighbourhood. On digging further, they discovered a leaden coffin, containing the skeleton of a human being, upwards of seven feet long, and proportionately large in its dimensions, the hair being perfect, and the nails and teeth undecayed. The labourers continued their exertions, when they succeeded in finding another leaden coffin, containing a skeleton of a similar description. The circumstance excited considerable interest and curiosity in the neighbourhood, and the relics are now in the possession of the above-named gentleman.

EAST ANGLIA

In this final part of the chapter we explore the mystical realms of East Anglia, comprising Cambridgeshire, Suffolk, Norfolk and Essex. Before the English Channel flooded the area between England and Europe, a landmass called Doggerland joined England to Europe before 6000 BC. Water played a role in shaping the land and the mindset of early people, and some of the oldest inhabitants of Europe lived here dating to around 80,000 years ago. There is even a report of human remains in Ipswich that date back 330,000 - 400,000 years ago.[30] Most of East Anglia has been occupied since the Paleolithic, with Mesolithic, Neolithic and Bronze Age activity in all areas.

Folklore was widespread throughout East Anglia, with stories of phantom black dogs called Schucks, Willow the Wisps, ghostly apparitions, and of course, a few giants. Most famously, the legendary warrior-queen Boudica (or Boadicea), who ruled the Iceni tribes of various parts of Norfolk and Suffolk, went down in history for her final battle against the Romans in 60 AD. The 3rd century Roman historian Dio Cassius wrote of her:

> *She was enormous of frame, terrifying of men, and with a rough, shrill voice. A great mass of bright red hair fell down to her knees;*

109

she wore a huge twisted torque of gold, and a tunic of many colours, over which was a thick mantle held by a brooch. When she grasped a spear, it was to strike fear into all who observed here.[31]

An engraving of Bouidica by William Sharp published in 1793.

PIERS SHONKS DRAGON SLAYER

A renowned giant, this time just over the border in East Hertfordshire, was reported to be 23ft tall, who was one of the last 'Dragon Slayers'. His name was Piers Shonks and he lived beneath an oak tree on an island in Peppsall Field. He single-handedly rid Brent Pelham of demons, but one day, whilst out hunting with his three winged hounds, he shot the Devil's favourite dragon with an arrow and it died. Satan was enraged and swore that he would possess his soul whether he was buried inside or outside the holy church. Years later, on his death-bed, he fired one last arrow (like Robin Hood and Jack o'Legs) to divine the site of his grave (and not dissimilar to Tom Hickathrift - see next section). The arrow went through the church's south window and lodged itself in the northern wall, where he was later buried, therefore defying the vow of Satan.

Today you can see the coffin-slab that shows dragons, angels and winged beasts on it with a date of 1086 AD. Similar legends appear at Tolleshunt D'Arcy and Daffyd Ddu in Flintshire and show how widespread the tradition of dragon slaying was in ancient England, highlighted by the stories of St George and the Green Man.

Left: The grave of Piers Shonks inside the wall at Brent Pelham Church.
Right: Celtic Cross with huge megaliths at its base, part of an earlier megalithic site.

All around the base of Brent Pelham Church are numerous stones that appeared to have been part of an earlier megalithic structure. There is also tantalising evidence from an account in 1861 that the giant Piers Shonks once really existed:

> *Support is said to have been given to the tradition that Piers Shonks was a giant in 1861 when the tomb was opened and unusually large bones were found.*[32]

TOM HICKATHRIFT - GIANT OF THE MARSHES

The most famous giant-killer of the Eastern Shires was called Tom Hickathrift. He is a legendary figure similar to his western counterpart, *Jack-the-Giant-Killer*, although he is often represented as a giant. He famously battled a giant in Wisbeach (Cambridgeshire). Various versions of the tale exist, but Tom is said to have lived in the Smeeth area of Norfolk and fought off a fierce ogre one day while travelling through the marshland. An elaborate moulded plasterwork (pargeting), showing his other famous battle against the Wisbech Giant, decorates the Old Sun Inn pub in Saffron Walden, Essex.

Left: Tom Hickathrift battling a giant on the frontage of the Sun Inn in Saffron Walden.
Right: Old illustration of Tom battling the Wisbech Giant.

One of the most interesting adjuncts to the Hickathrift mythos was the proposed burial of one of his slain giants, which stood at the Smeeth: "*It is called the giant's grave, and the inhabitants relate that there lie the remains of the giant slain by Hickathrift.*" [33]

There are also geomantic associations with Tom, as he is said to have thrown a massive hammer great distances, and even kicked a stone ball six miles distant in a straight line. One version of the story said the hammer or ball landed at Tilney All Saints Church, where he was eventually buried. His gravestone marks a burial 7 feet 6 inches long and in legend, it is the exact stone he threw from all those miles away. Since 1631 AD the stone in this churchyard is claimed to mark the burial place of the giant. It is a simple slab of unadorned granite on an east-west axis. His covert Pagan slant is provided in the stories of him throwing missiles of stone at churches, attempting to damage or destroy them.

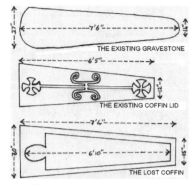

The various grave covers.

He is often referenced as a 'god of the marshland' of the Eastern Wash,[34] often depicted with a wheel for a shield and an axle-tree for a sword. His name may be connected to Boudica's Iceni tribe that ruled parts of East Anglia. *Hickathrift* or *Hicafrith*, according to Tom Lethbridge, translates as "the trust of the Hiccas' or 'Iceni'. [35]

One of the giant hill figures discovered by Tom Lethbridge at Wandlebury in the 1950s (see next section) may represent Hickathrift. The chalk carving of a 100ft tall warrior holds a round shield and weapon above his head. Lethbridge also claims that some bronze offerings found near Belsars Hill in Willingham (where St Christopher the giant is depicted in the local church) are of the giant Hickathrift.[36]

CAMBRIDGESHIRE
GIANTS IN THE GOG MAGOG HILLS

A few miles southeast of Cambridge, the flatlands of the East Anglian countryside are interrupted by the Gog Magog Hills, rising 236ft (72m) above sea level, housing a complex of tumuli, megaliths, earthworks, ring-forts, chalk hill figures and sacred springs. The original name of the hills still remains a mystery, but its relationship to giants may be how they got this name. The first recorded reference to the hills was in Michael Drayton's (1563-1631) *Poly-Olbion* verse called *one and twentieth song*: *"Old Gogmagog, a Hill of long and great renowne, which neer to Cambridge fet, o'ormlookes that learned Towne."* The rest of the song tells a story of how the giant fell in love with a nymph, although however hard he tried to woo her by tempting her with presents and changing his appearance, she still rejected him.

Another similar tradition reveals more (which a certain H. C. Loft, heard from an elderly man living in the neighbourhood):

I have never seen or heard of it being anywhere in print. It asserts that previous to the formation of these hills (Which are three in number), and near to the same spot, was a very large cave, which was inhabited by a giant and his wife (a giantess) of extraordinary stature, whose names were Gog and Magog. They did not live very happily together, for scarcely a day passed by without a quarrel between them. On one occasion the giantess so outraged the giant, that he swore he would destroy her life. She instantly fled from the cave, he quickly pursued her, but she running faster than her husband, he could not overtake her. Gog, in his anger, stooped down, took up a handful of earth and threw at her, it missed her, but where it fell it raised a hill, which is seen to the present day. Again the enraged giant threw earth at his wife, but again it missed her, where it fell it was the cause of the second hill. Magog still kept up her pace, but again the giant, in his rage, threw more earth at his wife; but this time it completely buried her alive, and where she fell is marked by the highest hill of the three. So runs, the local tradition respecting the origin of the Gog-Magog Hills.[37]

In the 1950s children from Cherry Hinton were warned to stay away from the hills because they might wake the giants. There are also the giant chalk hill figures at Wandlebury, rediscovered in the 1950s by Tom Lethbridge who had heard about giants being buried in the hills.

T.C. Lethbridge's complete plan of what he believed to be carved into the chalk at Wandlebury. The large figure on the right may be a representation of 'Hickathrift'. Only a segment was revealed before the excavation was halted, c.1955.

THE WAR DITCHES SKELETONS

The nearby Cherry Hinton 'War Ditches' was a circular earthwork about half the size of Wandlebury that is now completely destroyed by chalk quarrying. On May 27th 1854, a short item appeared in the *Cambridgeshire Chronicle* that described a find by workmen who were in the process of preparing the land for the first reservoir for Cherry Hinton. Nine skeletons were found:

> *Several of them were of large size, and were evidently the remains of men who reached to a greater height than ordinary men in the present day.*[38]

Some of the skeletons discovered at the War Ditches.

According to Archaeologist Michelle Bullivant, "*this sparked local interest as to a possible race of giants having once lived upon the hills.*"[39] Unfortunately, this is the only record of that particular discovery but it gave some clout to the 'giant' legends and place-names that persist around the Gog Magog Hills.

THE GIANT'S GRAVE - CHERRY HINTON

A fresh water spring in Cherry Hinton at the base of the Gog Magog Hills is referred to as *The Giant's Grave* (or *Magog's Grave*). It is a natural spring (also called Springhead, Springfield and Robin Hood Dip), that possibly once had a round barrow next to it. In the centre of the pond is an island twenty-five feet in width that, in legend, is thought to be the burial place of a giant.

In the car park of the *Robin Hood and Little John pub*, lies a mysterious lonely monolith. The sarsen stone is about three feet across and has an unusual man-sized 'footprint' deeply carved into it. It may have been a Celtic 'Kingship Stone' that tribal leaders would place their foot in before embarking on a journey or to be proclaimed leader of the tribe. The pub's name, again references a British giant, *Little John* of Robin Hood fame.

Left: The 'Kingship Stone' in the car park of the Robin Hood and Little John pub in Cherry Hinton. Right: The Giant's Grave or 'Magog's Grave'.

THE DEVIL'S DYKE

In a small village northeast of Cambridge, not far from the fabled Fens of Ely, a great 'dyke' emerges smoothly from the ground and rises to over two stories high. This incredible earthwork stretches for nearly eight miles. This mighty bank and ditch are assumed to be of Anglo-Saxon origin and comprise the most complete of all the dykes in Britain. It is located near the most northern route of the ancient Icknield Way, the oldest road in Europe, having been in use since the Stone Age.

Illustration of the Devil's Dyke in Cambridgeshire, c.1853.

In *Legends of the Fenland People*, Christopher Marlowe (Published by C. Palmer, 1926) describes an old story associated with the dyke saying it was constructed by, "*...a race of giants, renowned alike for cunning, strength and ferocity.*"

116

THE GHOSTLY GIANTS OF HEYDON DITCH

Heydon or Bran's Ditch, also in Cambridgeshire, has eerie ghost stories of giant warriors attached to it, and when it was excavated in the 1950s several taller-than-average skeletons were reported to have been unearthed, some of them decapitated. These were originally thought to be the skeletal remains of the ghostly Saxon warriors who had haunted the area for centuries. The great rampart is about three miles long, and although it has been decimated by agriculture, some of the banks are still over five feet high. It runs against the paths of the prehistoric Icknield Way (much like Fleam Dyke, Devil's Dyke and other local earthworks), suggesting it was for defensive purposes. It is likely to be much older than the official Saxon date of between 600 - 700 AD.[40]

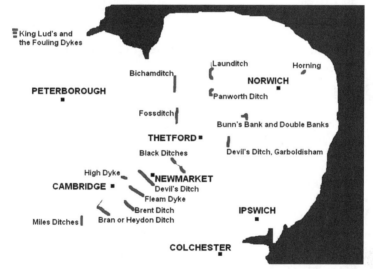

The major dykes of East Anglia including Heydon Ditch and the Devil's Dyke.

THE HISTON GIANT

Hugh grew up in the village of Histon, a few miles north of Cambridge. Little did he know during his first nine years of life that it was once the home of a famous local titan. A sarsen stone sits within the garden of The Boot Inn public house. *The Histon Giant* was Moses Carter, born in 1801 and died in 1860. He was nearly seven feet tall and weighed twenty-three stone. He was renowned for his strength and in 1847, for a bet, he carried a massive stone from a ballast pit in Park Lane to The Boot Inn, and it is still there today.

Left: The village sign of Histon, showing Moses Carter carrying the monolith. Right: The remains of the stone now in the garden of The Boot Inn.

THE PETERBOROUGH GIANT
The Gloucester Citizen, 12 August 1922

This short account was tucked away in a local newspaper without fanfare, but shows that an unusually tall skeleton was unearthed near Peterborough, coming in at seven feet tall. Whether it was prehistoric burial, or a very tall drunk man who had stumbled over a beer keg one fateful night is unclear.

> A seven foot skeleton of a man has been unearthed in his yard by the landlord of the Windmill Inn, Orton Waterville, near Peterborough.

ESSEX

There are very few giant legends in Essex compared to other counties. Author of *Secret Essex,* Glyn Morgan reveals that in fact *"there is a chronic shortage,"*[41] although Tom Hickathrift turns up on the front of the Sun Inn in Saffron Walden. Skeletal remains of giants are also scarce, but what is quietly ample is the number of monoliths, earthworks, and ancient trackways. Even though the Essex giants are a rarity, the ones that did get remembered are impressive examples that Tom Hickathrift would have been proud of.

A FAMILY OF SEVEN FOOTERS
Reading Eagle, 4 February 1953, p.19

This account tells us that five skeletons were unearthed in 1953 in a small town called Nazeing near Epping in Essex. This seems to indicate something remarkable because they were unearthed in modern times and to find this many oversized skeletons in one place points to an anthropological mystery. Was this the warrior class of the local tribe, or perhaps a family of giants?

> Five human skeletons all more
> than seven feet tall have been
> found during excavations on a
> nursery site at Nazeing, England.

LARGE TEETH AND BONES OF EXTRAORDINARY BULK

Two accounts from over 300 years ago tells us of further bones of massive size and teeth that were much larger than the norm discovered at Wrabness in the Essex countryside:

> *In 1701 some large bones were found at Wrabness, near. Harwich, in Essex; and Camden says that in the times of Richard II. and Elizabeth, large teeth and bones of extraordinary bulk, which were accounted to have belonged to giants, were found at the eastern promontory of the same county, meaning probably at Walton.*[42]

GIANT SKELETON UNEARTHED
Eastern Evening News, 24 June 1905

A 7ft skeleton was discovered in June 1905 in Finchingfield, Essex with a trait of perfect teeth we find in much of our research on giants:

> *A skeleton seven feet in length has been unearthed at Finchingfield, Essex. The skull possessed a perfect set of teeth.*

THE THUNDERSLEY GIANT

This strange account discusses the discovery of a small standing stone with what appears to be 'eyes' carved on it at the 13th century church of St Peter in Thundersley, Essex. It was unearthed during the First World War and reported in a booklet called *The Ancient Parish of Thundersley* by Rev. E.A.B. Maly (1937):

> *A week or two after, whilst digging another grave adjoining we came across the skeleton of a man seven feet in length.*

He noted that the skeleton with burn marks on it and that it was lying north-south, rather than the usual east-west, which got him thinking this was a prehistoric burial linked with the stone. The village's name was recorded as *Thunreslea* in the *Domesday Book* of 1086 meaning, "Sacred grove belonging to the god Thor." It appears that the church was built upon an earlier prehistoric site or Druid grove, with ancient connections to the 'sowers of thunder' mythos.

119

NORFOLK

Norfolk has a very ancient heritage where some of Britain's most important archaeological discoveries are located. These include priceless metal artefacts and the now-destroyed circle of upturned tree stumps called 'Seahenge' that dates back to the Neolithic era. Before that, the Icknield Way and Peddars Way ancient tracks were spreading trade and communication in the late Mesolithic. Earlier still, the oldest human footprints ever discovered outside of Africa were found in Happisburgh, dating to an astonishing 800,000 years old. Mighty handaxes seemingly yielded by giants were also unearthed and there are numerous legends of giants in the county.

Legendary history states that Somerleyton was once the hunting area of a giant, who is said to briefly appear on the evenings of 17th July or 19th November. He stumbled across a plot to murder him, so struck first and killed the would-be assassin. The assassin's blood is said to rain down from the sky on 19th November.[43]

THE BURGH CASTLE GIANT

A massive skeleton was found near a Roman fort within a 7th century Saxon Castle in Norfolk, England. The height of the man was approximately 7ft 4in (2m, 23cm) he was around forty-years-old at death. This is *old* for a Saxon of this era and it was revealed that this was not a pathological case of pituitary gigantism. Dr. Rideout, of the Department of Radiology at Princess Margaret Hospital, Toronto, described the skeleton as a once healthy, strong, and long lived individual for his era. His bones showed uniform growth, were very strong, and his skull showed no signs of pathology. The paper even discussed trepanned skulls from the same area suggesting there were advanced surgeons in 7th century England.[44]

The 7ft 4in skeleton from Burgh Castle.

THE RUDHAM DIRK GIANT BRONZE AGE DAGGER

The Rudham Dirk dagger

An unusually large bronze dagger was discovered on a farm in Norfolk, a few miles northwest of Burgh Castle. It is at least three times as large as any other yet discovered in Europe. Was it wielded by a mighty Bronze Age warrior or was it a ceremonial object? Whatever its purpose, it is one of the most significant Middle Bronze Age objects ever found in Europe and displays one of the earliest uses of metal in the British Isles. The 1.9kg (4lb) dirk is made from bronze, which is nine-tenths copper and one-tenth tin. The nearest source for the copper is North Wales, while the tin probably came from Cornwall.

It remained in the ground for over 3500 years, and was eventually ploughed up in a farmer's field and used as a door stop for twelve years before being identified. The dirk is now on display at Norwich Castle Museum.[45] *"It is about three times the size of a normal Bronze Age dirk, making it impractical for use as a weapon,"* unless of course you are a 'three times the size' human giant.

THE HOLBEACH SKELETON
The Singleton Argus, 3 September 1903

This noteworthy account comes from just outside Norfolk, in Lincolnshire, but it seems like it may be connected to the other reports from this area. It is just a few miles west of Rudham. This is from one of many newspapers across Britain that reported on this:

> **A GIANT'S SKELETON.**
>
> THE complete skeleton of a human giant has been unearthed at Holbeach during excavations for the foundations of new houses. Every bone is in per-

A Giant's Skeleton: The complete skeleton of a' human giant has been unearthed at Holbeach during excavations for the foundations of new houses. Every bone is in perfect condition, not a tooth is missing, and the skeleton measures no less than 7 feet 2 inches. Near the bones was found a curious key. Stukeley, the antiquary, who was born at Holbeach, records that at the spot where the discovery was made there formerly existed a chapel dedicated to St Peter.

GIANT HAND-AXE FROM SHERINGHAM

This account outlines the discovery of a huge handaxe, similar to the Furze Platt and Cuxton examples. Again, this part of north Norfolk seemed to be the haunt of giants, but this time from a much earlier epoch. It was first reported in 1935 in the scientific journal *Nature*:

An altogether remarkable and gigantic hand-axe, discovered embedded in the beach below Beeston Hill, Sheringham, by Mr. J. P. T. Burchell, has been figured and described by Mr. J. Reid Moir. The implement measures in its greatest length 15 inches, in greatest width 6 inches, in greatest thickness 5 inches. Its weight is approximately 14 lb.[46]

This article came from the *Syracuse Herald* from July of the same year:

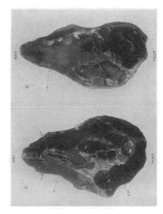

That a race of giants roamed the earth ages ago in the early Pleistocene era seems to be indicated by the discovery of a huge stone ax at Norfolk, England, which apparently was wielded by prehistoric man.

As the great ax was found in strata that underlies glacial deposits, scientists are sure it was produced and used by a race that antedated the Ice Age.

It was confirmed as dating to the early Pleistocene epoch and was championed as an amazingly advanced piece of tooling for its age. We are looking at between 2.5 million years ago and one million years ago, although more recently it has been reclassified as Paleolithic at 500,000 BC to 10,000 BC.[47] Whatever era it came from this final comment hints at the elephant in the room and is a fitting end to this chapter:

No adequate explanation of the purpose which the gigantic size of the Sheringham axe could serve has been offered.[48]

DATE	LOCATION	SIZE	No.	FEATURES	PAGE
		LONDON			
1808	The Guildhall	26in	1	Giant femur	97
1761	Aldeuman Buri	32in	1	Thigh bone	97
1666	St Mary's Church, Woolchurch	Giant	1	Female thigh bone	98
1830	King John's Palace, Old Ford	Giant	3	Cartload of bones	98
1590	St Paul's Cathedral	Giant	1	Thigh bone	99
1813	Bethnal Green	7ft	1	Found in stone coffin	99
1717	Rotherhide	10ft	1	Found in stone coffin	100
1895	Brentford End	7ft	1	Trepanning on skull	100
1770	Canonbury Tower	Giant	2	Possibly two burials	100
		SURREY & KENT			
1914	Morden Road, Mitcham, Surrey	7ft	1	Well-preserved teeth	101
1823	Evelyn Chapel, Wotton, Surrey	9ft 3in	1	Crumbled to dust	101
1763	Charterhouse, Surrey	8ft	1	From femur measurement	101
1932	Guildown, Guildford, Surrey	Giant	2	Giant jaw	102
1800s	Hythe Church, Kent	7ft	1	Danish warriors	102
1933	Deal Mines, Kent	Giant	1	Seven feet below surface	103
1901	Dover, Kent	Giant	1	Roman finds with burial	103
		HERTFORDSHIRE & BEDFORDSHIRE			
1600s	Weston Church	Giant	1	Femur found in churchyard	107
1712	Roman Verulamium, St Albans	8ft	1	Roman warrior	108
1800s	Dunstable Monastery, Beds	7ft	2	Perfectly preserved	108
1861	Brent Pelham	Giant	1	Legend of Piers Shonks	111
		CAMBRIDGESHIRE & ESSEX			
1854	War Ditches, Cherry Hinton	7ft	3	Iron Age Ring Fort	115
1950s	Heydon Ditch, Cambridge	7ft	4	Ghostly warriors reported	117
1922	Orton Waterville, Peterborough	7ft	1	Found in pub	118
1953	Nazeing, Essex	7ft	5	Family of giants	118
1701	Wrabness, Essex	7ft	1	Found in fields	119
1701	Walton, Essex	Giant	1	Large teeth and bones	119
1905	Finchingfield, Essex	7ft	1	Perfect set of teeth	119
1937	Thundersley, Essex	7ft	1	Burn marks and monolith	119
		NORFOLK & LINCOLNSHIRE			
1977	Burgh Castle, Norfolk	7ft 4in	1	Saxon skeleton	120
1903	St Peter's, Holbeach, Lincs	7ft 2in	1	Found with curious key	121

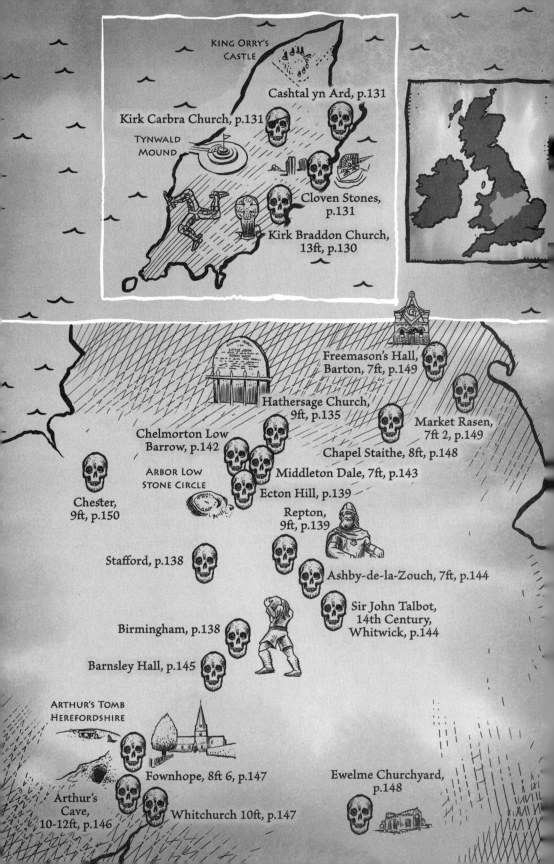

KING ORRY'S CASTLE

Cashtal yn Ard, p.131

Kirk Carbra Church, p.131

TYNWALD MOUND

Cloven Stones, p.131

Kirk Braddon Church, 13ft, p.130

Freemason's Hall, Barton, 7ft, p.149

Hathersage Church, 9ft, p.135

Market Rasen, 7ft 2, p.149

Chelmorton Low Barrow, p.142

Chapel Staithe, 8ft, p.148

ARBOR LOW STONE CIRCLE

Middleton Dale, 7ft, p.143

Chester, 9ft, p.150

Ecton Hill, p.139

Repton, 9ft, p.139

Stafford, p.138

Ashby-de-la-Zouch, 7ft, p.144

Sir John Talbot, 14th Century, Whitwick, p.144

Birmingham, p.138

Barnsley Hall, p.145

ARTHUR'S TOMB HEREFORDSHIRE

Fownhope, 8ft 6, p.147

Ewelme Churchyard, p.148

Arthur's Cave, 10-12ft, p.146

Whitchurch 10ft, p.147

5

MIDDLE ENGLAND AND THE CENTRE OF BRITAIN

The Midlands are home to landscape features and megalithic temples that reveal an underworld of giant lore with some notable bones and skeletons unearthed deep within the British heartland. However, we begin our survey at the original saced centre of Albion, the Isle of Man.

THE ISLE OF MAN

The Isle of Man is located at the precise centre of the British Isles. Scotland, Ireland, Wales and England are neatly contained within a great circle centred on this isle, and all these countries are within a 50 mile radius of the island. It is also aligned on the primary axis between Land's End in Cornwall and John o' Groats in Northern Scotland. Traditions state that it was a Druidic centre of learning and the chief Druid had his college and home at Kirk Michael.[1] It was where Celtic rulers and kings would send their offspring to be taught in the arts of astronomy, astrology, natural philosophy and geomancy.[2]

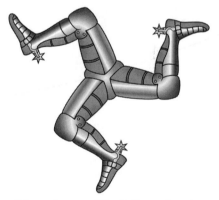

"Three Legs of Man", the national symbol of the Isle of Man.

The name of the island is thought to have come from the ancient Celtic sea god *Manannan* who was also an adept of the 'sowers of thunder' skill set, reputed to be able to control the forces of nature. According to the *Book of Fermoy*, a manuscript from the 14th - 15th century, "*he was a pagan, a lawgiver among the Tuatha Dé Danann, and a necromancer*

125

Map showing the British Isles encompassed by a great circle centred on the Isle of Man. The alignment between John o' Groats and Land's End perfectly bisects the centre of the island. The smaller 100 mile wide circle touches all four countries, England, Wales, Ireland and Scotland. Finding the sacred centre was an important tradition in ancient cultures worldwide and mounds or stones that laws and proclamations were given from marked the omphalos. On the Isle of Man the Tynwald Mound has this honour, but an earlier 'centre' was once used near St Luke's Church. Image by John Michell, 1994.

possessed of power to envelope himself and others in a mist, so that they could not be seen by their enemies." [3]

These next few legends come from the *Third Manx Scrapbook* by W W Gill (1938), a private publication that eventually saw the light of day in 1963.[4] In the section on giants it gives some fascinating stories that have now mostly been forgotten, including references to the *Fenoderee* (Fenodyree), a hairy supernatural creature that is said to be a large elemental being. His descriptions are reminiscent of Bigfoot, a tradition rarely mentioned in the British Isles, although legends of 'hairy men' are known in Wales and a few other areas.

The account was originally from a book called *The Undersea Giants* and gives a story about a miner from Laxey who got lost in mining tunnels:

> *While trying to find his way out he came to a great room hollowed in the rock, with tables and chairs in it all of stone. The light in the room was yellowish, and seemed to come from above. Six great powerful men in queer rough clothes were sitting there and staring at him. He asked them in Manx to direct him to the upper world. They spoke among themselves in a language he could not understand, and then one turned to him and told him he was under Laxey Bay, and a good way out too. 'This is the castle of the giants that used to be living in Laxey, and you are the first man that ever found his way down here.' They said they were just putting in the time till the Island would be fit for heroes to live in again, and they had been waiting there hundreds of years. Then one of them gave him a cled on the head, and he didn't know anything till he found himself outside the big door.*

He tried again to find the area he had seen the giants, but failed. In later years other variations of the story emerged, this one from a Miss Douglas:

> *...and I remember that in a mine disaster that occurred when I was about eight years old one or two of the bodies were not found, and some people thought the giants had taken those men, and they weren't dead at all.*

A shadowy giant figure known as 'The Big Man' was witnessed by Willy Carberry around the time of Samhain (1st November) near Ballure Glen. Willy described him as being *"half the height of a telegraph pole....He was fully dressed, and carried a long staff in his hand."* Other stories of giants have them throwing rocks to one another over Dhoon Glen. It is said that three of these stones landed close together and are called *Meir-ny-Foawyr* (Fingers of the Giants). A large rock on the highest point of Shellag brows, close to the Jackdaw's Church, bears the likeness of the fingerprints of a giant. The boulder is named *Claghvedn* or *White Stone*.

What are locally described as *"the tracks of a giant who lived on North Barrule"* are said to be visible near the river below Thallooqueen, Maughold. In tradition a giant threw a big boulder from the top of the mountain into the river, and then he came down and scrambled across. He slipped and his cheek hit the rock in one place, his foot also left an imprint. In an area called *Track-ny-Foawr*, Ballafayle, also in Maughold, overlooking a small bay there appears to be a rough likeness of a very large human right foot. It is about 3 inches deep in the rock with a prominent big toe. Now lost, the local children often used to gather there to place their feet in it. It is said that the giant originally swam over from Cumberland, climbed the cliff by a path and landed there on his journey. Later incomers to the Isle of Man were said to be of giant stature:

> *From the 8th Century AD the island gradually received an influx of Norsemen or Vikings who apparently intermingled peacefully with the local Celts. Two centuries after the Norse first settled on Man a King Arthur like figure arrived on the island in the form of Godred Crovon, known as Gorree. He was said to be a giant of a man who was already King of the Scottish Isles. He arrived at Ramsey on the NE coast in 1079 by Viking longship and sailed down the Sulby river to Milntown which nestles below the northern flank of the 300m high Sky Hill which butts up to the glacial plain. Here a battle was fought and won against a Manx army. After sparing the lives of those he had conquered, Godred Crovan followed the Via Regia or Manx Royal Road south over the northern uplands to the original Tynwald assembly site at the sacred centre of the isle where he proclaimed himself King Orry. He is referred to affectionately by the Manx people as their first and best king.[5]*

King Orry had several sites associated with him, including the two Neolithic chambered tombs of *King Orry's Grave*, and *Cashtal yn Ard*. There was also a Bronze Age burial mound found near the present Tynwald site at St John's known as the *Giant's Grave*, with a chambered tomb of the same name less than a mile away. At Peel Castle, outside its boundary were some mounds known locally as the *Giant's Graves*, being 90 feet long and 5 feet wide.[6]

The Tynwald Mound on the Isle of Man.

Left: Giant's Grave close to the modern Tynwald Mound. Right: King Orry's Grave.

A standing stone sits beside Jurby Church that was said to have been thrown by a giant, "*...from either Snafield or some of the adjoining mountains, after a companion who had insulted him, but who contrived to escape his rage by wading or swimming from Jurby to the coast of Scotland.*" [7] There are numerous legends of these colossi including one that may have been linked with the Druids.

THE GIANT'S LABYRINTH

A ruined castle situated at Castletown and known as Rushen was reputed to have an underground maze inhabited by a giant guardian. The story states that long ago some determined townsmen decided to thread this maze, seeking out the giant to learn his timeless magic secrets. Armed with staves and torches they descended into the old labyrinth and upon reaching its centre encountered a blind giant with a long beard sitting in contemplation upon a single pillar stone. The giant was wise and peaceful and he asked the men how things were in the island as he had not ventured out of his maze for hundreds of years. He also politely requested that one of the party should shake hands with him before leaving. One of the frightened men proffered an iron bar which the giant squeezed, remarking happily that 'there were still men left in the Isle of Man.' [8]

Anthony Roberts focussed on the symbolism of this elaborate myth, pertaining to the sacred labyrinth and how these were found all over the country mainly in turf form. Also, a central pillar could represent the *omphalos* of the island, and of Britain.

GIANT BONES AT KIRK BRADDON CHURCH

The discovery of a huge thigh bone and skull was reported in 1731. It is also featured in the *Historical and Statistical Account of the Isle of Man*, 1845. [9]

However strange their tradition may seem of the Island being once inhabited by giants, my own eyes were witness to something that does not a little keep it in countenance: As they were digging a vault in Kirk Braddon church-yard, there was found the leg-bone of a man very near four feet in length from the ankle to the knee. Nothing but ocular demonstration could have convinced me of the truth of it ; but the natives seemed little to regard it, having, as they said, frequently dug up bones of the same size. [10]

One of the Celtic Crosses at Kirk Braddon Church.

A femur four feet in length would indicate a skeleton over 13 feet in height. Hugh visited the church in July 2019 and although no evidence of giants was visible, some beautifully carved Celtic crosses were on display inside the entrance, and as the author stated, more oversized bones were found nearby, suggesting it was simply normal for the islanders to find such things!

GIANT SKULL AT KIRK CARBRA CHURCH

The same manuscript followed up with a stunning find at another Church on the island:

> *They told me that but a few months before my arrival (about 1710), there was found in Kirk Carbra church-yard, a human head of that monstrous circumference that a bushel would hardly cover it, and that nothing was more common, when they were digging, than to throw up ribs and hands conforming to the leg I had seen.*[11]

THE CLOVEN STONES GIANT

Among the many megalithic sites on the Isle of Man is the Cloven Stones, a burial chamber located in the village of Baldrine. The site consists of two large uprights and some smaller stones forming a forecourt to a tomb. The rest is under a mound that forms the front garden of a private house. Perhaps the owners did not know what was buried beneath it because in the *Swarbreck Manuscript*, written in 1815, is this report of an extraordinary discovery:

Illustration of the Cloven Stones.

> *Mr Millburne informed us that about seven years since, he with two or three miners opened the mount to a depth of five feet and discovered a human skull and some thigh bones, which from their uncommon size, must have belonged to a person of gigantic stature.*[12]

GIANT'S BONES IN KING ORRY'S CASTLE

Cashtal yn Ard was originally a megalithic cairn with five chambers which extended over 130 feet, located a few miles south of Ramsey. It is also known as

King Orry's Castle or *Castle of the Heights*. It is strangely similar to the Bronze Age Giants' Tombs of Sardinia, with a curved forecourt and long chamber. It was originally excavated in the 1930s:

> *They are all too large to be human and their condition suggests they are probably relatively modern. Neither the pieces of skull, nor the human teeth found in 1933 can be found in the museum.*[13]

We asked at the museum where the bones and teeth were now, but frustratingly they were nowhere to be found. The disappearance of important human remains is an issue we often encounter in the search for evidence of giants.

Cashtal yn Ard with Hugh, Matt Chapman, Jenn Bolm & David H.Childress.

MIDDLE ENGLAND

The central counties of England were in Anglo Saxon times called the *Kingdom of Mercia* along with *Northumbria* to the north and *Wessex* to the south and west. The area roughly corresponds to what is today called the 'Midlands,' and has numerous tales of giants.

In the Malvern Hills of Gloucestershire, the discoverer of leys Alfred Watkins, visited 'The Giant's Cave' that had strange carvings on its back wall, once thought to be the haunt of a local titan. Just below the cave entrance is a partly smoothed rock called the sacrificial stone, once called the *Shew Stone*. On the morning of the summer solstice the sun rises over the cave and beams onto the polished side of the monolith, the moment of the 'the giant's sacrifice.'[14] This alignment extends between these two 'giant sites' into a classic ley visiting Gospel Oak, through three churches, terminating at Aconbury Camp, an ancient beacon hilltop site, all lining up with the summer solstice sunrise. It may be a lingering memory of Pagan fire rituals upon beacon hills lighting up lengths of the country in a unified celebration of the turning of the seasons.

The Giant's Cave in the Malvern Hills.

At nearby Uleybury, also in Gloucestershire, there is a remarkable place of primeval sepulchre called the *Giant's Chamber*.[15] The legends of this area do not end there, as another stone throwing colossus was also busy in the same location:

> To destroy the spire of Hereford Cathedral was the goal of a giant
> who lived in a cave near Adam's Rocks (Hereford and Worcester).
> But the rock he threw landed in a meadow near Longworth Mill,
> Bartestree, and has now been broken up.[19]

The Wrekin, a remarkably large earthwork on top of a hill has a similar theme attached to it much like the Gorm mythos, with a persistent legend that some local giants were involved in its construction. The story has a Welsh giant wanting to destroy the town of Shrewsbury but he got stopped by a cobbler who was carrying a sack of worn shoes, who convinced the giant that he had worn out all the shoes he was transporting because the town was so far away. The giant dropped his spadeful of earth at that spot, creating the Wrekin and returned home.[20] This is a recurring theme in giant lore.

Pyon Hill and Butthouse Knapp near Canon Pyon (Hereford and Worcester) has a similar legend:

> Robin Hood and Little John were each carrying a spadeful of earth
> with the intention of burying the monks at Wormsley, but a cobbler
> used the 'worn shoes trick' on them and they fell for it.[21]

Little John was a giant, so let's begin our survey of skeletal remains in the county of the legendary outlaw Robin Hood and his Merry Men.

NOTTINGHAMSHIRE

The Robin Hood stories are dated to around the same time as *The History of the Kings of Britain*, (12th Century) and take place in Sherwood Forest. Heroic tales of stealing from the rich to give to the poor, echoes through history. "*Robin Hood and his Merry Men have been variously explained as fairies, a coven of witches or even as lingering survivals of the Druidic cultus, and many of their exploits have a heavily symbolic flavour lurking behind the more dramatically overt sociology.*"[22]

Robin Hood meets Little John from
an 1883 book by Howard Pyle.

Some of this symbology is found in the stories of Little John, who seemed to be a relic of an earlier culture and only joined Robin's gang after a long fought battle with the hero. He is prominent in all the tales of Robin Hood. The sobriquet "Little" is a form of irony, as he is usually depicted as a gigantic, seven-foot-tall warrior of the British forests, skilled with bow and quarterstaff. Near Sherwood Forest, skeletal reminders of Little John were said to have been unearthed. In Hathersage, Derbyshire a small church contains one of the largest graves in the county. A headstone reads *"Little John, the Friend and Lieutenant of Robin Hood,"* marking the location of a ten-foot-long tomb. Originally it was reported to be: .."*marked by two stones, thirteen feet four inches apart.*"[23]

Postcard of Little John's Grave in Hathersage Churchyard.

This is not just a legend, because in the 19th century a 29½ inch long thigh bone was excavated from the grave,[24] followed by a series of unfortunate accidents to the discoverer:

> *In 1784 the local church vicar, Charles Spencer-Stanhope (d.1874) wrote that the squires brother, William Shuttleworth hung a thigh bone, reputedly from Little John's grave in his room. However as it was thought to be bringing poor fortune to its owner, it was ordered to be reburied by his clerk. But the clerk kept the labelled bone in his window as a curio. When the father of Charles Spencer-Stanhope (Walter Spencer-Stanhope of Cannon Hall and Horsforth Hall 1749-1821) and Sir George Strickland were visiting Hathersage, Strickland is reported to have 'run away with it' and it has never been recovered. It was William or James Shuttleworth who in 1784 had the grave body exhumed, the thigh bone was measured at 29½ inches.*[25]

A 29½ inch femur is unequivocally that of a 9-foot-tall giant. His famous longbow was six feet long and was said to have been on display inside the church for a long period of time before it ended up in the stately confines of Cannon Hall, near Barnsley:

> *Little John's cuirass of chain mail, his bow and arrows hung for*

many years in the chancel of Hathersage church... I have examined the bow which is made of spliced yew, about six feet in length, though the ends where the horn tips were attached are broken off. It required a power of 160 pounds to draw the bow to its full extent. Only 60 pounds is the power used by men now at archery meetings.[26]

The bow is now thought to be in Cawthorne Museum in South Yorkshire, but it is not labelled or confirmed as being the original bow.

'Little John's bow' held by archaeologist H.C. Haldane.

WEST MIDLANDS

Guy of Warwick was an English giant-killing hero whose romantic stories were well known from the 13th to the 17th century. One tale has him defeating the Danish giant Colbrand, while another has him battling an African giant called Amorant when visiting the holy land. His tales were told in aristocratic courts and were still being shared in ballads in the 19th century.

Guy of Warwick fighting the giant Colbrand in single combat, from The history of the famous exploits of Guy Earl of Warwick (1680).

The Gilbertstone, a volcanic erratic, is connected in local folklore to a giant named Gilbert. This gave its name to the area of Gilbertstone on the border of Yardley and Bickenhill. It also once marked the point of a sharp turn in the boundary between Warwickshire and Worcestershire. Gilbert was said to have carried the block and laid it down at a mysterious angle to change the boundary to his own advantage, suggesting its location had geomantic significance. It is currently displayed in the grounds of Blakesley Hall, a Tudor building on Blakesley Road, Yardley, Birmingham. It dates to 10,000 BC.

Hugh's nephew Joe sitting upon the Gilbertstone.

Also in Birmingham, along Warstones Lane, a giant spectre is often seen in and around a famous cemetery that has two tiers of catacombs, whose unhealthy vapours led to the *Birmingham Cemeteries Act* which required that non-interred coffins should be sealed with lead or pitch. Folklore tells us this:

Postcard of the War Stone.

A giant who lived at Birmingham castle was killed when the Giant of Dudley threw a stone at him (launched from Dudley). The rock also demolished the castle. The stone was erected as a memento, and the lane named after the War Stone used as a weapon. The rock is now on a plinth in the cemetery.[27]

Originally it was known as the Hoar Stone, which was likely derived from the Old English *har stan feld*, meaning 'boundary stone field'. The name slowly changed to the similar sounding 'War Stone', hence the current name. It was used as a parish boundary where the manors of Aston, Birmingham and Handsworth met. The throwing of the stone is another indication of possible geomantic knowledge.

BONES OF GIANT UNEARTHED

At the centre of Birmingham an interesting discovery was made in 1925, although no exact measurements were given. Could this be the giant who resided at Birmingham Castle mentioned in the previous legend?

Bones of a giant were discovered on Monday during excavations...
according to Dr. Bunting, being undoubtedly a man of great stature
and strength.[28]

> **BONES OF A GIANT UNEARTHED.**
>
> Bones of a giant were discovered on Monday during excavations in the centre of Birmingham on the site of an old priory of mediæval times. The skeleton was not complete. In the opinion of the Police Surgeon, to whom the bones were submitted, they had been in the earth for hundreds of years. They were lying 14 feet below the surface, and with them were other human bones belonging to a smaller frame. They appear to be part of the skeltons of two men; one of them, according to Dr. Bunting, being undoubtedly a man of great stature and strength.
>
> It seems probable that the site of the discovery was part of the burial ground of the Priory which, in early days, formed the centre of Birmingham.

THE STAFFORD GIANT

This next account comes from the writings of Dr. Robert Plot in his book *The Natural History of Staffordshire* (1686). He was an English naturalist, first Professor of Chemistry at the University of Oxford, the first curator of the Ashmolean Museum and the secretary of the Royal Society of London. Here he describes a huge jawbone and teeth found in the centre of Stafford:

I received a certain proof from Mr. William Feak, alderman of Stafford, who gave me the jaw-bone of a man or woman, with a tooth yet remaining in it, near double the magnitude of those men ordinarily have, which was found in the south chancel of the collegiate church of St Marie, in Stafford, where now lyes the gravestone of Ann, the wife of Humphry Perry; which is enough to shew that mankind is no more abated in stature than it is in age, the world still affording us a Goliath now and then as well as of old.[29]

DERBYSHIRE

BONES OF EXTRAORDINARY SIZE
Near Arbor Low Stone Circle

This matter of fact account, again from Dr. Robert Plot, is just a few miles west of the massive Arbor Low stone circle in the Peak District in Derbyshire. Arbor Low was called *Eordburh-hlaw* by the Saxons, which means 'built by giants':

> *In the digging open a Low on Ecton Hill, near Warslow, in this county, there were found men's bones, as I was told, of an extraordinary size, which were preserved for some time by one Mr. Hamilton, vicar of Alstonfield; and I was informed of the like dugg up at Mare, in the foundation of the tower; but these being buried again, or otherwise disposed of before I came there, I can say little to them.*[30]

Arbor Low is a Neolithic henge monument set amid high moor land. The site was called 'Eordburh-hlaw' by the Saxons, which translates to 'built by giants'. Photo H. Newman.

THE REPTON GIANT

We first heard about this 9-foot-tall giant thanks to a 2016 BBC documentary called *The Vikings Uncovered*.[31] The presenter was as shocked as we were when the story was told to him on camera by Professor Martin Biddle, telling him that a 9ft giant was discovered where he was standing a few hundred years previously. As we dug into the story we found that Dr. Robert Bigsby, in his

History of Repton in Derbyshire, 1854, discusses a discovery that took place in 1687. An extraordinary grave was unearthed in Allen's Close, Repton, which contained a stone coffin with a 9ft skeleton, with about one hundred other skeletons in the immediate area. In 1729 Dr. Simon Degge collected as many details as he could about the find and subsequently communicated them to The Royal Society. In *Philosophical Transactions for 1734*, Degge says that Thomas Walker, a local labourer who was in old-age, gave him this account:

Artefacts found in the giant's burial at Repton.

About forty years since, cutting hillocks near the surface, he met with an old stone wall; when, clearing further, he found it to be a square enclosure of fifteen feet. It had been covered; but the top was decayed and fallen in, being only supported by wooden joists. In this he found a stone coffin; and, with difficulty removing the cover, saw the skeleton of a human body nine feet long, and round it one hundred skeletons of the ordinary size, laid with the feet pointing to the stone coffin. The head of the great skeleton he gave to Mr. Bowes, master of the free school. I enquired of his son, one of the present masters, concerning it; but it is lost; yet he says he remembers the skull in his father's closet, and that he had often heard his father mention this gigantic corpse, and thinks that the skull was in proportion to a body of that stature. The bottom of this dormitory was covered with broad flat stones, and in the wall was a door-case, with steps to go down to it, whose entrance was forty yards nearer the church and river. The steps and stone were much worn. It is in a close on the north side of the church; and over this repository grows a sycamore tree, planted by the old man when he filled in the earth. The present owner will not suffer it to be opened, the lady of the manor having forbidden it. This was attested to us by several old persons who had seen and measured the skeleton.

Left: Statue of Ivar the Boneless, son of Ragnar Lodbrok. Right: *Birmingham Daily Gazette*, from 19 Dec 1927, revealed further documents and discoveries confirming the existence of the 9ft skeleton.

Dr. Bigsby added to the account:

> *This ancient sepulchre was again opened in 1787, when bones of a very gigantic size, appertaining to numerous skeletons, were discovered together with some remains of warlike instruments.*

Edward J. Wood, author of *Giants and Dwarfs*, commented on the discoveries:

> *There is a tradition connecting this giant's grave with a legendary King Askew, of whom, however, nothing is known. A place called Askew Hill, near Repton, is associated with his name; and he figures as a nine-feet high hero of a romance which Dr. Bigsby introduces into his Visions of the Times of Old. An altar-tomb, supporting a figure in armour, now deposited in the crypt beneath the chancel of Repton Church, is popularly called the tomb of King Askew; but with evident misapprehension, as the armour is of a comparatively modern date.*[32]

The TV show that documented this find believed it could be a legendary Viking leader called Ivar the Boneless who first invaded England in 865 AD. Stories of him being a great warrior may be correct, but nowhere do they mention he was 9ft tall. The 100 skeletons buried next to him has increased to 300 after further excavations were carried out. This 1976 dig revealed further burials that were clearly of Viking origin with the tallest skeleton unearthed being around 6ft tall; a good 3ft shorter than his prehistoric counterpart. In 2018 it was announced that this mass grave could be that of the 'Great Viking Army', but again, no mention of our 9ft friend was made. Perhaps we might consider that the Vikings chose this spot because it was the legendary grave of an ancient Pagan giant who they may have seen as an archaic god. Moreover, in 1829 it was reported that in Derby some workers uncovered a stone coffin containing a remarkable skeleton. However, it seems this is simply a retelling of the famous 9-footer, but it has been included here to complete the story of the Repton Giant:

> *The priory stood nearly upon the site of the King's Head Inn, and in digging the foundation of the buildings in that neighbourhood, a stone coffin was discovered several years ago, containing a body of prodigious size.*[33]

The Viking Sword found in the burials at Repton, perhaps owned by the 9ft tall giant.

UNCOMMONLY LARGE BONES

In opening a barrow near Chelmorton, in 1782, "uncommonly large bones" were found and reported on by James Pilkington. It was imagined that the persons to whom they had belonged must have been at least seven feet high.[34]

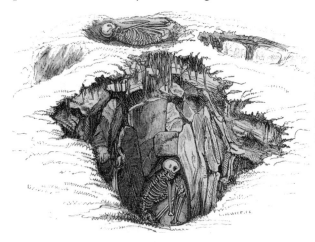

Burial showing skeletons in-situ from a barrow called Hitter Hill (opened in 1862), close to Chelmorton Low Round Barrow in Derbyshire.

SKELETON OF GIANT IN LIMESTONE
Nottingham Evening Post, 14 November 1934

In this account from 1934, a 7ft giant was discovered standing upright in a limestone quarry:

While some men were quarrying at Middleton Dale, yesterday, they discovered the skeleton of a man which I believed to have been buried for thousands of years in the limestone. It measures about 7 feet long, was in an upright position, and is in a perfect state of preservation. The skull is apparently no thicker than an eggshell and only two teeth are missing. The discovery was made in the quarry of the Cupola Mining and Milling Company. The skeleton was removed for expert examination. It is thought that the skeleton may be a valuable link in tracing the evolution of man. On Eyam Edge, nearby, there is a barrow over 100 ft. in diameter, and on Eyam Moor there is a Druidical circle.

SKELETON OF GIANT IN LIMESTONE.

DERBYSHIRE QUARRY DISCOVERY.

GREAT AGE.

While some men were quarrying at Middleton Dale, yesterday, they discovered the skeleton of a man which is believed to have been buried for thousands of years in the limestone.

The Eyam stone circles and mounds are located up on the nearby hills and remind us of the connection between the giants and megalithic sites.

LEICESTERSHIRE

A giant named Bel once took three prodigious leaps, beginning in Mountsorrel on a *sorrel mare* (horse), to a place named *Oneleap*, now corrupted to Wanlip. He then lept another mile, to a village called *Birstall*, a name said to come from the 'bursting' of both himself and his horse, when the violence of the exertion and shock killed him, and there he was buried. The site has ever since been denominated *Bell's Grave* (or Belgrave). "*He leaps like a Bell giant or devil of Mountsorrel,*" became a local proverb.[35] Matt Sibson, of 'Ancient Architects' commented on this in an article he wrote:

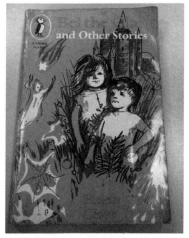

1956 book that tells the story of the giant Bel from Leicestershire.

> *Mountsorrel, Rothley, Wanlip and Birstall form roughly a straight line between Loughborough and Leicester. Belgrave was once a village in its own right but now is part of the city of Leicester... Other versions of the legend states that he started his journey from Belton near Loughborough (Bel's tun) – another fitting place name for this legendary character!*[36]

Further giant discoveries include one recounted by T. R. Potter in 1842, who wrote that in Charnwood Forest, near Charley in Leicestershire, several stone coffins and a lead one near a Giant's Grave were unearthed.[37] It was also in this area, in a town called Whitwick, that a 14th century giant named Sir John Talbot once lived, who was revered as a local hero. His grave is still in the local church, but although it is about seven feet long, it is thought he was much taller. He also had a brother who lived about one mile away from him, who he could speak to even at that distance. Whether they simply had remarkably booming voices, or if it was a metaphor for psychic communication is not clear.[38]

GIANT'S SKELETON FOUND
Deadwood Pioneer Times, 26 April 1942

This obscure account was featured in various newspapers in April 1942 announcing the unearthing of a 7ft skeleton.

> **GIANT'S SKELETON FOUND**
> LONDON—(P)—A skeleton of a man seven feet tall and apparently hundreds of years old was found during quarrying operations near Ashby-de-la-Zouch, Leicestershire.

WORCESTERSHIRE

This strange discovery was on the site of an old Manor House, but in 1907 it was turned into The Barnsley Hall Hospital, also known as the Bromsgrove Lunatic Asylum. The problem was, it was haunted by a tall Edwardian man whose footsteps and figure were often witnessed by patients. It was also here that a giant was once buried in the distant past:

> In the seventeenth century there was at Barnsley Hall, Worcestershire, the reputed thigh-bone of a giant. It was clasped with iron, and locked on to the staircase. It was one foot two inches in circumference at the smallest part.[39]

HEREFORDSHIRE

One of the most famous giant legends in Herefordshire is that of Jack o' Kent who was said to have lived on the Herefordshire/Monmouthshire borders at an unspecified medieval date. He is often portrayed as a giant who jumped off Sugar Loaf Mountain onto the Skirrid, and left his heel mark at the site. He had a throwing contest with the Devil and won, accounting for the eponymous standing stones at Trelleck.[40] Another megalithic site called *Arthur's Stone* marks a spot where the heroic king slew a giant, who fell onto the stones and left marks on one of them. The megalithic chamber dates to 3,700 – 2,700 BC and is situated overlooking the Golden Valley.

Arthur's Stone, Herefordshire. Photo H. Newman.

Herefordshire has several oversized gargantuan skeletal discoveries, including this next example also linked to King Arthur.

THE GIANT OF KING ARTHUR'S CAVE

A skeleton of a *"giant human"* and a *"brass-headed spear"* were discovered in a cave in the Wye Valley between the hills of Little Doward and Great Doward sometime between 1695 and 1700. There is also a connection to King Arthur's Hall, an ancient encampment on the same hill, however this saga turned out to have an unfortunate ending. A book from 1799 entitled *The Excavation Down the Wye from Ross to Monmouth*, by a certain Mr. Heath, first revealed that something of importance had been discovered there in *"a natural tomb, under an arch."*

King Arthur's Cave interior.

The next few quotes come from a report written by Flavell Edwards in the *Woolhope Club Transactions, 1874 - 1876*. It features the collective texts and research of Mr. Heath in regard to the lost giant skeleton. In the text it describes a letter written by 'Mr. White' which was collected by Heath:

> *Something more surprising presented itself to their view, which was the body of a man of very large stature upon the ledge of the rock, and covered over by a natural tomb, an arch of the same rock. He lay at his length, I think, upon his back with his spear by his side. One of them ventured to touch the body of this once mighty man, and all sunk down in dust.*

The report continued:

> *The common account that passeth, of the length of the longest bone of the middle finger and the bones of the leg and thigh... reported to be twice the length of the same bones of a common man, that is 5 feet 8 or 10 inches, which was about the stature of Mr. George White.*

The conclusion is quite startling:

> *Gibson, in his third and last edition of Camden's Britannia, has recorded it that the length of all the joints was twice the length of others of this age. If so, the man must be 11 or 12 feet.*

In another version it claims that hidden treasure was thought to be buried in King Arthur's Hall, saying that's how one of the other discoverers stumbled across the giant skeleton:

In the position of the body, its crumbling to dust, and the gigantic size of the bones, Llewellin fully corroborates Mr. White's account.

However, the skeleton was lost when a local surgeon named Mr. Pye took it to sea on a voyage to Jamaica and his ship sank. The 'Curse of the Giant Hunters' got to Mr. Pye and his giant it seems. More skeletons may still be buried deep in the cave:

Heath mentions a tradition as still current in his time (1799) that King Arthur's Hall extends underground from thence to New Wear, a distance of more than a mile.

THE WHITCHURCH GIANT

In an account recorded nearby, we have a skeleton that was said to be twice the size of a human, although this could be the same skeleton as in the previous account:

At Doward Hill, in Whitchurch, Herefordshire, some men who were digging found a cavity which seemed to have been arched over, and in it a human skeleton, which appeared to have been more than double the stature of the tallest man now known.[41]

8 FEET 6 INCH SKELETON
The Daily Bulletin, 29 September 1882

The following article reveals more giant discoveries in Herefordshire:

An interesting discovery has been made at Fownhope, near Hereford. Mr. Stone, builder, has been engaged for some time in the restoration of St. Mary's Church there. On Thursday morning, whilst his men were excavating beneath the church, they came upon a brick vault with an arched roof, and in this vault was found a handsome oak coffin of extraordinary length and breadth. The coffin crumbled to pieces when touched, disclosing a human skeleton of gigantic proportions, which, when the air struck it, dissolved into dust. The length of the body from head to feet was nearly 8 feet 6 inches, and the breadth 3 feet 6 inches.

OXFORDSHIRE

A remarkable discovery was made in Ewelme Churchyard in 1763:

> *Under the date of January, that several human bones of a very gigantic size had then lately been dug up in the chancel of the church of Ewelm, near the Duchess of Suffolk's tomb.* [42]

The giant bones were found very close to the famous Tomb of the Duchess of Suffolk in Ewelme Church.

The burial in question is the tomb of Alice de la Pole, Duchess of Suffolk (1404 - 1475). However, there is scant record of the nearby discovery except a handful of news accounts published at the same time. What happened to the bones remains a mystery.

LINCOLNSHIRE

In *South Wales Daily News*, 23 October 1900, it describes giant skeletons being unearthed, assumed to have been Danish Warriors:

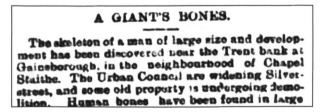

A GIANT'S BONES.

The skeleton of a man of large size and development has been discovered near the Trent bank at Gainsborough, in the neighbourhood of Chapel Staithe. The Urban Council are widening Silver-street, and some old property is undergoing demolition. Human bones have been found in large

> *The skeleton of a man of large size....Human bones have been found in large quantities in the neighbourhood of Chapel Staithe.... the condition of some of the bones that have been discovered from a time pointing to death in savage warfare. One of the thigh bones must have belonged to a man eight feet high.*

The same story reappeared in March 1902 although they were now thought to be 'Vikings'. The incident seemed to have affected the locals in a negative way, as another grave was destroyed in a rage! This from the *Taunton Courier and Western Advertiser,* 19 March 1902, gives us both stories:

> Nine skeletons of men of enormous stature have been unearthed near Gainsborough. The bones are supposed to be those of Vikings who were killed in some long-forgotten piratical raid in the Trent Valley.
>
> At Southborough on Wednesday night an outrage was committed on the grave of the little girl O'Rourke, recently murdered at Tonbridge by Harold Apted. The tombstone was pulled down and an artificial wreath torn to pieces,

GIANTS OF OLD
Yorkshire Post and Leeds Intelligencer, 23 November 1953

A newspaper on the same day as the revelation of the Piltdown Man hoax, revealed something not as old, but much larger:

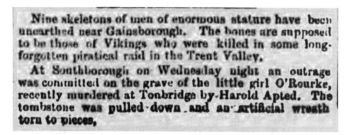

Hoaxer who faked a bone of contention

IF the perpetrator of the Piltdown hoax is still alive his pride at having kept the jest going for 40 years must be mingled with sadness that modern science has now exposed part of his "Piltdown skull" as a deliberate fake. I wonder if, now that he has attained maturity and possibly a position of distinction in the scientific world, he looks back with horror on those madcap days in 1911

Giants of old

THERE is no fake about the giant skeleton of a man measuring 7ft. 2in., and probably at least 1,000 years old, which was found buried 10 feet deep near the village church at Swallow, near Market Rasen, and was reinterred on Saturday. It is the second giant skeleton which the

Rector (the Rev. Cyril H. Jacoby) has seen. The first one was dug up at the near-by village of Irby-on-Humber when Mr. Jacoby was the incumbent there. "The one at Irby was bigger still," Mr. Jacoby told me, "and plaits of hair were twined round the skull."

He thinks that both skeletons are those of ancient Norsemen who may have fallen in battle in the district, in which there are many old battlefields.

There is no fake about the giant skeleton of a man measuring 7ft 2in, and probably at least 1,000 years old, which was found buried 10 feet deep near the village church at Swallow, near Market Rasen. and was reinterred on Saturday. It is the second giant skeleton which the Rector (the Rev. Cyril H. Jacoby) has seen. The first one was dug up at the nearby village of Irby-on-Humber when Mr. Jacoby was the incumbent there. 'The one at Irby was bigger still,' Mr. Jacoby told me,' and plaits of hair were twinned around the skull.

GIANT'S SKELETON UNEARTHED
Dundee Courier, 27 December 1933

Finally, the last report from Lincolnshire was discovered by a secret society with revelations that more were found previously:

While some workmen were digging in a field near the Freemasons Hall, Barton, Lincolnshire, they suddenly came upon the skeleton man over 7ft. high...Several skeletons were found in the same vicinity some time ago.

GIANT'S SKELETON UNEARTHED

While some workmen were digging in a field at the rear of the Freemasons' Hall at Barton, Lincolnshire, they suddenly came upon the skeleton of a man, over 7ft. high, who had been buried about 8ft. deep.

Beside it was the skeleton of a dog. The spot where the skeleton was found is close to the place where the sheep and cattle market used to be held. Several skeletons were found in the same vicinity some time ago.

CHESHIRE

THE CHESTER 9FT GIANT

A Short History and Description of Chester, by W. C. Jones, 1807

This next report describes a 9ft skeleton unearthed in Chester, and according to Sir Thomas Elyot, this was very much connected to the biblical giants of old:

According to Sir Thomas Elliott [Elyot], the original name given to this city, was Neomagus, so called from Magus, son of Samothes, son of Japhet, its founder, 240 years after the flood....a skeleton of prodigious size (some say 9 feet in length) being dug up in Pepper Street.

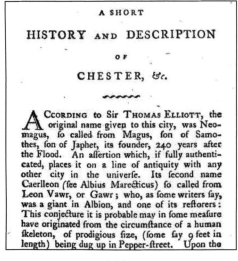

A SHORT

HISTORY AND DESCRIPTION

OF

CHESTER, &c.

ACCORDING to Sir THOMAS ELLIOTT, the original name given to this city, was Neomagus, fo called from Magus, fon of Samothes, fon of Japhet, its founder, 240 years after the Flood. An affertion which, if fully authenticated, places it on a line of antiquity with any other city in the univerfe. Its fecond name Caerlleon (fee Albius Marecticus) fo called from Leon Vawr, or Gawr; who, as fome writers fay, was a giant in Albion, and one of its reftorers: This conjecture it is probable may in fome meafure have originated from the circumftance of a human fkeleton, of prodigious fize, (fome fay 9 feet in length) being dug up in Pepper-ftreet. Upon the

We complete our survey of the counties of England by heading into the northern regions, once called *Northumberland* by the Saxons, and reveal that the giants prevailed there too.

Gazetteer of Giants - Isle of Man & the Midlands

	LOCATION	SIZE		FEATURES	PAGE
	Isle of Man				
1731	Kirk Braddon Church	13ft	1	4ft femur	130
1710	Kirk Carbra Church	Giant	1	Giant skull	131
1815	The Cloven Stones	Giant	1	Megalithic burial site	131
1933	Cashtal yn Ard, King Orry's Castle	Giant	1	Museum lost the skeleton	131
	The Midlands				
1784	Hathersage Church, Notts	9ft	1	Little John's Grave	135
1925	City Centre, Birmingham	Giant	1	14ft below surface	138
1686	St Mary's Church, Staffordshire	Giant	1	Huge jaw and teeth	138
1686	Ecton Hill, Derbyshire	Giant	1	Near Arbor Low stone circle	139
1687	Repton, Derbyshire	9ft	2	On BBC documentary	139
1782	Chelmorton Barrow, Derbyshire	7ft	1	Uncommonly large bones	142
1934	Middleton Dale, Derbyshire	7ft	1	Found upright in limestone	143
1942	Ashby-de-la-Zouch, Leicestershire	7ft	1	Found in quarry	144
1700	Barnsley Hall, Worcestershire	Giant	1	Huge femur & ghost giant	145
1695	King Arthur's Cave, Herefordshire	11-12ft	1	Giant curse as ship sunk	146
1700	Whitchurch, Herefordshire	11ft	1	Could be same as above	147
1882	Fownhope, Herefordshire	8ft 6in	1	Crumbled to dust	147
1763	Ewelme Churchyard, Oxfordshire	Giant	1	Duchess of Suffolk's tomb	148
1900	Chapel Staithe, Lincolnshire	8ft	1	Possibly Danish warriors	148
1853	Market Rasen, Lincolnshire	7ft 2in	1	Plaits of hair on skull	149
1933	Barton, Lincolnshire	7ft	1	Near Freemason's Hall	149
1807	Chester, Cheshire	9ft	1	Biblical connections	150

Aspatria, 7ft, p.162

LONG MEG AND HER DAUGHTERS STONE CIRCLE

Alnwick Priory, p.166

Newcastle, p.171

St Bee's, 13ft 6, p.160

Keswick, 7ft-30ft, p.163

Penrith, p.158

Corbridge, 21ft, p.164

River Cor, p.165

Furness, p.170

Westmorland, 7ft, p.166

GIANT'S GRAVE, PENRITH

Sunderland, 9ft, p.172

WADE'S CAUSEWAY

Settle Caves, 7ft, p.169

Thirsk, 7ft, p.169

Clitheroe, 7ft, p.171

Ripon Cathedral, 7ft 3, p.167

RUDSTON MONOLITH

Humbleton, 7ft, p.168

6

THE NORTHERN GIANTS

Northern England features the five counties of the North containing a plethora of local giants and traditions all over this mysterious landscape. The Danish giant *Wade* is present in certain counties, as well as quoit throwing and dropping stones from aprons such as on Pock Stones Moor in North Yorkshire near Wharfedale. Another giant called *Rombold* dropped great apronfuls of stones on Ilkley Moor in West Yorkshire. This was a story so profound that a 10ft effigy of the giant was cast in bronze and is situated in an open café in the Airedale Shopping Centre in Keighley:

> *Folklore has it that Rombold originally resided with his wife just outside Keighley on Ilkley Moor, an ancient moor containing many signs of stone and bronze age occupation. A domestic disagreement found Rombold racing from Ilkley to Almscliff hotly pursued by his wife, who was armed with an apronful of stones which she dropped as she chased him. To avoid her, he is reputed to have leapt from one side of one of the valleys clear to the other. Presumably the huge stone he is shouldering in the statue is one of the boulders he*

The giant Rombold, now on display as an art piece in the local shopping centre.

hefted to hurl back at his raging wife. But another tale that I have heard is that Rombold was a great giant, a god-like character like the Norse god Woden, and resided on Ilkley Moor. When angered, Rombold generated thunder to register his displeasure by hurling enormous boulders across the valleys.[1]

Once again, we hear references to weather fluctuations, specifically thunder, caused by giants.

Binsey Fell in Cumbria was said to have been created when a giant accidentally dropped an apronful of rocks that he was supposed to be using to build a church, in order to redeem himself from his life of sin and violence. A large ghostly figure wearing an apron is said to appear when the mist grows thick on certain nights.[2]

Binsey Fell Cairn, dropped there by a giant. Photo Andrew Locking.

LONG MEG AND HER DAUGHTERS STONE CIRCLE

Long Meg and her Daughters is one of Britain's largest stone circles. Unlike the beautifully placed *Castlerigg* in Cumbria, Long Meg has a modern road carved through it in a relatively bland landscape. It is 359ft wide and has a Flattened B type geometry as discovered by Alexander Thom. Long Meg is actually the outlier stone standing around 12 feet high, oriented to the midwinter sunset (like at Stonehenge) and has beautiful spiral carvings on it. Its relationship to giants is briefly mentioned in folklore, although stories of petrification due to unlawful love and the inability to count the stones are more well-known legendary associations. In the 17th century the largest monolith was thought to have got its name from the saying *"As long as Meg of Westminster"* of which Fuller in his *Worthies* (1662) tells us:

This is applied to persons very tall... That such a gyant woman was in Westminster, cannot be proved by any good witness, (I pass not for a late lying pamphlet) though some in proof thereof her Gravestone on the south-side of the Cloisteres, which (I confess) is as long and large and entire Marble, as ever I beheld.

Long Meg and her Daughters Stone Circle. Long Meg is the tall stone at the bottom of the image. Photo H. Newman.

The pamphlet mentioned was in fact *The Life of Long Meg of Westminster* (1635), which describes a giantess being buried under a large stone at Westminster. The large block of blue-black marble in the south cloister of Westminster Abbey, over the grave of Gervasius de Blois, is referred to as Long Meg, but its connection to Cumbria's second most impressive stone circle is questionable.

Two other stories hint at archaic traditions of giants. One of them is that if a piece of Long Meg were broken off, the stone would bleed.[3] The other recounts when Colonel Lacy expanded Lacy's Caves on the River Eden attempting to blast the circle away, a huge storm broke out and the workmen fled in terror.

THE GIANT WADE

A single standing stone remains just south of a road between Whitby and Loftus in Yorkshire. In the early nineteenth century a second menhir stood alongside it. The monoliths were said to mark the grave of the giant Wade. Camden mentions them in his *Britannia* (1637):

Here within the hill between two entire and solid stones about seven foot high lieth entombed: which stones because they stand eleven foot asunder, the people doubt not to affirm, that he was a mighty giant.

The giant, who was said to have been as tall as the distance between the stones (11ft), was also thought to have been the resident of the nearby ruined Mulgrave Castle. The structure once bore his name, and was said to have been built by Wade and his giantess wife called *Bell* (or Bel). The couple had one hammer to help with construction, so would throw it in a straight line to one another. Another local story tells of two other stones near Goldsborough that had the same reputation of marking the length of an interred giant, but this time they were 100ft apart from one another. One of the stones still survives.[4]

WADE'S CAUSEWAY, NORTH YORKSHIRE

Wade's Causeway on the North York Moors is a 6000 year old sinuous linear monument nearing 25 miles in length. It is constructed of flagstones and although once deemed to be a Roman Road, it is now thought to be of Neolithic origin. The traditions of Wade became mixed with British lore and this legend correlates with others we have uncovered:

Section of Wade's Causeway, c.1918.

[Wade] is represented as having been of gigantic stature... His wife...was also of enormous size, and, according to the legend, carried in her apron the stones with which her husband made the causeway that still bears his name.[5]

The stories relate that he created it to make it easy for his wife Bel to take herself or her cow to market. The name Bel became associated with the structure, and although modern academics believe this is not important, it could relate to Belinus, the ancient British King who laid down sanctified roads across Britain as written about by Geoffrey of Monmouth in the 12th century.

THE GIANT'S CAVE OF EDENHALL

Based near Honeypot Farm, Edenhall in Cumbria, the *Three Caves of Isis*

Parlis are often associated with the Penrith Giant Sir Hugh Cesario. In 1875, a schoolteacher claimed at that time that the giant was still living in the caves.[6] However, the stories go much further back in time:

> *There is a holy well dedicated to St. Ninian and the caves are enlarged out of Lower Permian sandstones and their associated breccias and purple shales. Near this is a chasm in a rock, called the Maiden's Step, from a traditional account of the escape of a beautiful virgin from Torquin, the giant, who, after exercising upon all occasions every kind of brutality and depredation within his reach, retired to the cave, his stronghold.*[7]

Torquin is in fact the legendary *Tarquin* who featured in the annals of Arthur. A carving of the giant was witnessed next to the cave entrance in 1914 (with no head) by Revd Arthur Heelis.[8] This was supposed to record the tradition of a maiden who, surprised by the giant after she strayed too near the caves, escaped with a long leap across the chasm. Another version has a giant called Isis who, capturing cattle and men, dragged them to his cave to eat them. The word Isis is obviously recognized as an ancient Egyptian goddess so it is unclear

Isis Parlis Caves.

how it reached the remote hills of Cumbria. Perhaps it was an interpretation of *Idris*, a name given to a legendary Welsh giant just over the border, a prolific character we will investigate in the next chapter. At nearby Eamont Bridge in Cumbria another cave by a river was once the dwelling place of a cannibalistic giant called *Isir*. However, this could be based on the Edenhall giant myth.[9]

GIANT'S STONES, GRAVES AND CAVES

Numerous Giant's Graves exist all over Britain with some notable examples in the North, including *Samson the Giant's Chamber* at Lazonby in Cumberland.[10] Barrasford, in Northumberland once had some standing stones that were thought to have been thrown into place by local giants.[11] This example featured in *Giants and Dwarfs*:

> *At Rutchester, Northumberland, a trough-like excavation has been made in the solid rock. Its use is not known; but it was once popularly called the Giant's Grave.*[12]

The London Magazine (1791) discusses customs that once took place in a *Giant's Cave* in the area:

> *In some parts of the North of England it has been the custom, from time immemorial, for the lads and lasses of the neighbouring villages to collect together at springs or rivers, on some Sunday in May, to drink sugar and water, where the lasses give the treat. This is called Sugar and Water Sunday. They afterwards adjourn to the public-house, and the lads return the compliment in cakes, ale, punch, etc., and a vast concourse of both sexes always assemble in the Giant's Cave on the third Sunday in May for this purpose.*

There are local stories of a giant and dwarf going out to battle and sharing the victory.[13] Another tale relates that a dwarf was choked in the fraternal embrace of a giant, with the consolation that it was the giant's nature to squeeze too hard. An old proverb says, "*A giant will starve on what will surfeit a dwarf,*" and Erasmus refers to the adage of "*Drawing a pigmy's frock over the shoulders of a giant.*"[14]

CUMBRIA

Penrith in Cumbria is an important location in the history of England. The name Penrith is thought to mean either 'red hill' (after the red sandstone of Beacon Hill) or 'chief ford' (due to the river crossing at Eamont Bridge). The battles fought by Urien and his son Owain against the Angles in the 6th century became part of Arthurian legend, and widespread Neolithic and Bronze Age cultures flourished in this area and throughout the Lake District.

THE GIANT'S GRAVE AND THUMB OF PENRITH

Evidence of giants are abundant in Cumbria and one striking example may be found at the church of St Andrew's in Penrith. The eleven-foot-tall weathered cross shafts (that may have originally been Neolithic menhirs) are spaced fifteen feet apart and have numerous 'hog-back' stones between them. The *Giant's Grave* shows curious symbolism of traditional Anglican, Celtic and Norse art. A certain Dr. Todd is quoted as saying in Walker's *History of Penrith* (1857):

> *The common vulgar report is that one Ewen or Owen Cæsarius, a very extraordinary person, famous in these parts for hunting and fighting, about fourteen hundred years ago, whom no hand but that of death could overcome, lies buried in this place. His stature,*

as the story says, was prodigious beyond that of the Patagonians, in South America, seventeen feet high, that the pillars at his head and feet denote it, and the four rough unpolished stones, betwixt, represent so many wild boars, which had the honour to be killed by this wonderful giant.

The Giants Grave, in the Church Yard, Penrith engraved by R.Sands
after a picture by Thomas Allom, published in 1833.

Owen Caesarius was the king of Cumbria between 920 and 937 AD. Further local tradition states it is the burial place of the mythical giant Sir Ewan, who lived in the Giant's Caves on the banks of the river Eamont nearby. Whoever was buried here, a magnificent femur bone and mighty sword were once discovered in the grave:

> *This place was opened by William Turner, who there found the great long shank bones and a broad Sword.*[15]

How long this bone was is unclear, but the old British name of Giant's Grave and the traditions of giants in this area do suggest this was an important personage buried here. Prehistoric cup and ring marks are also found on the stones suggesting they may re-used megaliths. In the same churchyard a wheel-head cross called the *Giant's Thumb* dates to around 920 AD and was erected by Owen, King of Cumbria, in honour of his father.

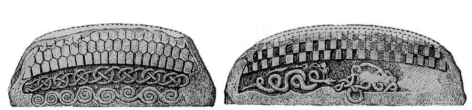

Details of two of the hog-back stones at Penrith Churchyard.

A similar *Giant's Grave* with two large standing stones sits at the foot of Black Combe Hill in Cumbria, near the Irish Sea. The smaller menhir is over eight feet tall and has three cup and ring marks, whilst the taller stone at ten feet tall has just one.[16] They are spaced about fifteen feet apart indicating the height of the supposed giant who was buried there.

Giant's Grave at Black Combe.

THE ST BEE'S GIANT - 13.5 FT

This is the first recorded account of a discovery that has become somewhat of a legend in the world of giantology. A nearly fourteen-foot-skeleton, as well as some remarkable weaponry were unearthed:

> *A true report of Hugh Hodson, of Thorneway, in Cumberland, to Sr. Rob Cewell of a giant found at St Bees, in Cumberland. The said giant was buried 4 yards deep in the ground, which is now a cornfield. It was 4 yards and a half long, and was in complete armour: his sword and his battle-axe lying by him. His sword was two spans broad, and more than two yards long. The head of his battle-axe a yard long, the shaft of it all of iron, as thick as a man's thigh, and more than two yards long. His teeth were 6 inches long, and 2 inches broad; his forehead was more than two spans and a half broad. His chine bone could contain 3 pecks of oatmeal. His armour, sword, and battle-axe are at Mr. Sand's, of Redington (Rottington), and at Jilr. Wyber's, of St Bees.*[17]

This mighty giant would have been about thirteen-and-a-half feet tall and his sword about six feet long if the standard yard is used. We first saw this account in *Giants and Dwarfs* where the author extracted it from an unknown manuscript in the library of the *Dean and Chapter of Carlisle*.

The St Bee's giant reconstructed with armour described in the account. By E. Martin.

THE ASPATRIA GIANT - 7FT

Curiously, another giant was unearthed from a 90-foot-wide burial mound near Aspatria, 25 miles north of St Bees. In 1789, Mr Rigg, local surgeon and antiquarian, excavated a barrow on Beacon Hill. Inside the mound was a stone-lined grave, with a very tall, but poorly-preserved human skeleton accompanied by various grave goods. This update in 1872 summarises the full story:

> *Mr Rooke...gives an account of an excavation at a place called Aspatria, a little farther westward, and near St Bees. They cleared away a barrow about 90 feet in diameter, and at 3 feet below the original surface of the ground found a cyst in which lay the skeleton of a man of gigantic stature. As he lay extended, he measured 7 feet from the head to the ankle. His feet were decayed and rotted off. At his side, near the shoulder-blade, was an iron sword 4 feet in length, the handle elegantly ornamented with inlaid silver flowers; a gold fibula or buckle was also found, with portions of the shield and his battle-axe. One of the most curious things found was the bit of snaffle-bridle, which is so modern looking that it would not excite interest if seen on a stall in Seven Dials. The main interest resides in its similarity to that which Stukeley found at Silbury Hill.*[18]

Also inside the burial chamber were:

> *Two large cobble stones which inclosed the west side of the kistvaen... On these stones are various emblematical figures in rude sculpture, though some of the circles are exactly formed, and the rims and crosses within them are cut in relief.*[19]

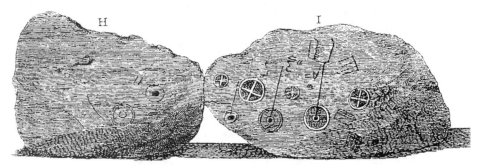

Antiquarian illustrations of the two carved stones of Aspatria, 1792.

Further discoveries were made and given to the British Museum, only to be lost. The stones are presently thought to be of Viking or earlier Celtic origin. The mound is now completely obliterated but its location is known. The site was excavated again in October 1997 due to Beacon Hill being a perfect place for a massive phone mast. One thing that caught our attention in the archaeological report was discussion about one of the tibia. It mentions that it was unusually long and thin:

> It is unlikely that the fragments form part of the same bone as this would indicate a shaft length of >36 cm... this would from an extremely long thin bone which although possible is not probable.[20]

THE KESWICK GIANT - 7FT OR 30FT TALL?

The Aspatria Giant is thought to later have become the *Keswick Giant* as some of his teeth and bones, and possibly his skull, found their way into the possession of antiquities collector Peter Crosthwaite, who, in 1781, established his museum (also known as *Cabinet of Curiosities*), on Keswick's main street. Estimations of the skeleton's height ranged from 7ft, 10ft, 13.5ft, 21ft or even 30ft tall. The numerous accounts make it clear that is was extremely oversized.

In 1825 C. Hulbert describes the discovery in detail, including the sword of "*extraordinary size*" found with the enormous skeleton that was, "*so rusted, that on being moved it easily dropped to pieces.*" It also recounts that the local neighbourhood were involved and each took a part of the giant skeleton, with "*several of his teeth, with some other relics, are still to be seen at Mr. Croshwaite's Museum in Keswick.*"[21] We contacted Keswick Museum about this and they said they knew nothing about it from their history, but this newspaper account and illustration describes a skull of gigantic proportions that may have been of a skeleton about ten feet tall.[22]

> THE GIANT SKULL—From the measure-ments of the giant skull at the Museum of Kes-wick, Cumberland county, England, the stature of the giant must have been in excess of 10 feet, or about the height of the Philistine Goliath.
> TUESDAY—"A U. S. PORT OF ENTRY NOT ON THE SEA."

However, in *Origins of English History* by Charles Isaac Elton, it confirms that the skeleton was seven feet tall.[23] Either way, this cartoon-like illustration from the news report is all we have to support this case.

PRODIGIOUS SKULL OF A GIANT
Keswick Museum, England
Photo by McNamara

THE SKELETON OF THE GIANT COR

Although often thought to be the previously mentioned discovery, the so-called *Cor Giant* appears to be quite different from the Keswick/Aspatria Giant and was found much further to the east. It clearly got confused with this other discovery, due the connection of both of them to the Keswick Museum. The original account featured in *The History of Northumberland* by Robert Morden, published in the 18th century, about a discovery made in 1660. Excavations at Corbridge in Northumberland revealed a shocking find. This is the most northerly town of the Roman Empire with a section of Hadrian's Wall nearby:

> *...there was found out accidentally, about forty years ago, a thing remarkable. The bank of a small torrent which comes from the wall side of this town being worn away by some impetuous land flood, the skeleton of a man appeared of a very extraordinary and prodigious size. The length of his thigh bone was within a very little of two yards, and the skull, teeth and other parts proportionably monstrous, so that by fair computation the true length of the whole body may be well reckoned at seven yards. Some parts of it were in the possession of the Right Honourable the late Earl of Derwentwater at Dilston in 1695.*

Seven yards is around 21 feet tall. The next account detailed the original find and what happened to the bones. In 1881 Robert Foster wrote:

> *A singularly large bone found here was hung up in the kitchen of the Old George Inn, in the flesh market, Newcastle for many years. This bone was purchased by the proprietor of the Keswick Museum*

where it is shown as the rib of the giant Cor found at Corbridge.[24]

Forster also mentioned a huge skull had been found nearby in the 1800s:

Mr Adam Harle states that when a youth he found in the field a little to the west of this place a human skull of immense size and wonderfully perfect.[25]

This next report was published by Peter Crosthwaite's son, Daniel, in 1826. It lists most of the curiosities the owner had accumulated, including the legendary Cor Giant's rib, but this time included a remarkably large skull that could either be that of the original Cor Giant, the discovery made by Adam Harle, or the Aspatria giant previously mentioned:

Among many oddities was a rib from the skeleton of the Corbridge Giant, dated about 1700. The skull was large enough to hold three gallons of water; the thigh bone was just under six feet long. It was estimated that this 'prodigious monster' was 21 feet tall.[26]

Two giant skeletons (or the remains of them) ended up in the curious collection of Peter Crosthwaite at the Keswick Museum from two different places. The mystery of these giant discoveries has left a trail of confusion through the centuries, so Hugh decided to go there for himself and talk to the curator. However, she had only heard about the giant's rib that was on display during the nineteenth century, but nothing else, no skulls, teeth or other evidence. The only reference that Hugh found in the museum was this sign (No. 20) mentioning the rib (pictured below).

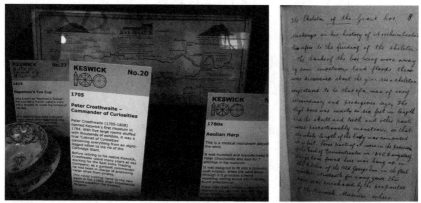

Left: Sign at the museum mentioning the giant rib.
Right: Hand-written account from 1881 by Forster that describes the Cor Giant.

Also, once on display at the Keswick Museum was a pair of shoes belonging to Mr O'Brien, an Irish giant, who was eight feet three inches tall, as well as a clog fourteen inches long worn by local giant Jacob Barker, of Hutton Roof, near Greystoke.[27]

Furthermore, in 1881 George Neasham discusses legends associated with the Cor Giant. This may signify, once again, a coded reference to surveying the land by giants throwing hammers between different locations:

> *Tradition says that Corbridge, Benfieldside, and Consett, were founded by three Brother Giants, Cor, Ben, and Con, who are stated to have had a huge hammer in common, which each could throw nine miles.*[28]

HUGH THE GIANT AND THE 7FT SKELETON

During the seventeenth century a giant named Hugh Hird was said to have lived in Troutbeck, Westmoreland. It is recorded in the parish church that he was buried there. *The History of Westmoreland* contains accounts of his prodigious strength; and amongst other things it stated that he went on a mission for the king, being sent to London for that purpose by Lord Dacre.[29] Also, in this area the following was discovered, although clearly from a much earlier time:

> *At Rassett Pike there was found, 11 or 12 feet below the surface, the skeleton of a man who must have been at least seven feet in height.*[30]

NORTHUMBERLAND
THE GIANTS OF ALNMOUTH

In a weekly column of local folklore called 'The Antiquities Column' in *The Welsh Worker,* from 1886, we find this account about a discovery in Alnmouth:

> *I understand that bones of a huge size have raised themselves from time to time in Alnmouth, Northumberland which have caused the belief that a race of giants at one time had been living there.*[31]

We looked further into this account and found the original report in *The Berwick Advertiser* of a seven-foot skeleton discovered at nearby Alnwick Abbey in 1884.[32]

ALNWICK.

ALNWICK ABBEY.—During excavations at Alnwick Abbey, a large skeleton has been found, about seven feet in height. It is supposed by some to have been that of Henry, the second Lord Percy of Alnwick, who was buried in the abbey in the year 1368, but unfortunately for this idea, he is described in the Chronicle of Alnwick Abbey as being " a man of little stature, but brave, faithful, and grateful; and content with the lordship left by his father, he wished to obtain the lands or possessions of no one.".—Dickson's Translation. With regard to the excavations, it is understood that when completed the remains will be covered with concrete, so that the foundations of the Abbey will in all nature time be visible to the antiquary.

THE ABBEY. -- The excavations at this place are being continued with vigour, and have shown that the Abbey was one of the finest in the kingdom. The great church was 237 feet long, not much less than Westminster Hall. The chapter-house was remarkable in being circular, a form of which there appears to be only one other example in England—at Worcester. The chapter-house is also most interesting in containing the tombs of the first of the De Vescys, lords of Alnwick—namely, William De Vescy, son of Eustace Fitz-John. William was interred there in 1184, by the side of his wife Burga ; and the stone coffins with their contents remain where they were deposited 700 years ago.

YORKSHIRE

It is always surprising to us that where giant legends persist, giant bones are often unearthed somewhere nearby. This first account comes from *Giants and Dwarfs* and like some other churches we have investigated, giants were things of wonder and long bones were often found on display there.

RIPON CATHEDRAL GIANT THIGH BONE

At the bone-house at Ripon Cathedral is a brightly-polished femur or thigh-bone which could have belonged only to a man not less than seven feet two or three inches in height.[33]

The *bone-house* mentioned was a crypt full of hundreds of bones and skulls that was open to the public between around 1815 and 1865. Doctor Augustus Bozzi Granville described his visit in 1841:

The massive door was open, and I descended into the crypt, the three vaulted arches of which are filled with skulls and detached jawbones, with leg and thigh bones, in some parts piled twelve feet deep... when suddenly the eyeless sockets of some unknown genius, destined to perish in obscurity in life as well as in death, stares you in the face; or the gaping jaws of some decrepit lawyer yawn like the portal of death.[34]

THE OSSUARY or BONE HOUSE at RIPON CATHEDRAL.

The space now houses the 'Song School' and the 'Chapel of the Resurrection' and no bones are any longer on display.

THE GIANT OF HUMBLETON

An Historical, Topographical, and Descriptive View of the County of Northumberland by Eneas Mackenzie, 1825

A stone-lined burial was discovered in 1811, with a 7ft skeleton inside:

> *At the bottom of the hill, where stands Humbleton Burn House, and close to the barn, the plough in 1811 struck against a large stone. On removing this impediment, a human skeleton was exposed to view, lying in a kistvaen, formed of six large flags [stones]. The bones were in a high state of preservation, of a close texture, and remarkably large. From the specimens sent by the late Mr. Alexander Kerr, of Wooler, to the publishers, the skeleton must have been seven feet long.*

In 1870-72, John Marius Wilson's *Imperial Gazetteer of England and Wales* confirms the find 60 years later:

> *An urn and a stone coffin, inclosing a gigantic skeleton, were discovered here in 1811.*

168

GIANT'S BONES IN BRITAIN
Western Gazette, 5 August 1938, p.12

Settle is a small town in North Yorkshire. The discovery of a seven-foot-tall skeleton was made in some caves in the nearby hills by an early caving organisation. Sir Arthur Keith was a prominent anatomist and anthropologist and was involved in the classification of the skeleton of the 'Irish Giant' Charles Byrne. It seems his interest in giantology did not end there, as he presented a paper at a conference detailing the finds.

> GIANT'S BONES IN BRITAIN.
> Sir Arthur Keith, the anthropologist, who is one of the oustanding personalities of the British Speleological Association, now holding its annual conference at Settle, Yorkshire, has classified bones of a man and woman found in a Settle cave as having belonged to the pre-iron age. Although the man was at the most 17-years-old, he was at least seven feet in height.

THE THIRSK GIANT

Thirsk Museum in Yorkshire is one of the only museums in Britain that has giant human bones on display. The skeleton was discovered at Castle Garth in 1994 and is known as the *Saxon Giant* as he lived in the area around 600 AD. Hugh visited the museum to measure the foot and jaw, but the curator did not have the key so we could not acquire the exact measurements. The skeleton is estimated to be seven feet tall.

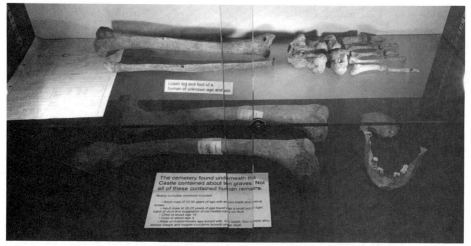

The remains of the 7ft skeleton found at Castle Garth in 1994.

Several other burials were also discovered but little remains of the 12th century castle except for the few finds that are on display in this tiny museum.

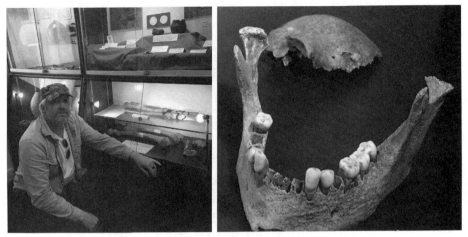

Left: Hugh at Thirsk Museum. Right: The jaw bone of the 7ft giant from Castle Garth.

THE YORKSHIRE CYCLOPS

There is an old legend of a one-eyed-giant in the neighbouring village of Dalton. There was once an elongated mound that existed outside Dalton Mill and according to author Joseph Jacobs in the late 1800s, you could have once seen the long blade of iron that was described as a 'scythe'. This was called *the giant's knife*, and the story related to it outlines a kidnapped boy called Jack, who after seven years of imprisonment, grabbed the knife and stabbed the 'cyclops' in his eye, before skinning the giant's dog and wearing its fur as a disguise. He then escaped through the legs of the giant.[35]

LANCASHIRE

This account appears to be talking about a discovery near Broughton-in-Furness, that is just east of one of England's most impressive stone circles, *Swinside*, in Cumbria. The two stone circles mentioned in the report are most likely to be *Heathwaite Fell* and the *Giant's Grave*, which are located adjacent to one another nearby. The Giant's Grave has only three or four stones remaining but the accounts give a tantalising glimpse into the lore of the area and how it connects the stone sites with giants:

> *In Furness, North Lancashire, near two barrows on Heathwaite, are two small stone circles, probably of the Druid era, which are*

170

*called giants' graves, and which on being excavated, about 1842,
were found to contain the bones of men, covered with a flat stone.
There is a tradition among the old inhabitants that giants formerly
lived at the place, and were buried there. The last of the race
is said to have been shot by an arrow upon the adjacent hill of
Blawithknott.*[36]

GIANT SKELETON FOUND AT CLITHEROE
Northern Daily Telegraph, 23 September 1904

This account details the discovery of a seven-foot-skeleton found whilst mining
lime near Clitheroe in the early 1900s. The disused lime quarry is still in
existence.

GIANT SKELETON FOUND AT CLITHEROE.

The quarrymen employed by Messrs Carter and Co., Bellman Park Limeworks, Clitheroe, yesterday exhumed the skeleton of a man much above the normal stature, whilst "feying" or baring the rock. In a declivity two feet below the sod lay the bones in a natural position, with head to the west and feet to the east. Unfortunately the workmen disturbed it with a pick, and some of the bones were smashed. A second skull, apparently buried with the body of the man, was found, but no trace of trunk bones could be found. The teeth and bones are in perfect preservation. Local antiquaries cannot name the period to which the man may have belonged, but placed in position, the bones show that his height must have been quite seven feet.

TYNE & WEAR
EIGHT FOOT TALL GIANT FOUND IN CLAY
The Gentleman's Magazine, November 1757, p.529

This account is from just south of the current city of Newcastle upon Tyne, but
we could find no further information in any other news outlets:

*During the course of this month, as some colliers were sinking a new
pit on Gateshead Moor, near Newcastle, they found the entire skeleton
of a man of a gigantic size in a bed of stiff clay, about seven feet from
the surface. Near the skeleton were found three small pieces of very
ancient coin. The person, when living, must have been near eight feet
high; the bones laid compact, measuring seven feet eight inches, and
must have lain there many 100s years.*[37]

RUINED RESIDENCE OF A GIANT

Slightly further northeast along the River Tyne is a strange report of a ruined building that, in legend, was the abode of an archaic giant:

> *The ruined residence of a giant was pointed out not many years ago at West Charlton, on the North Tyne. It encompassed an acre of ground, with strong walls built of stones four feet thick. Its size, strength, and antiquity, in the absence of any exact knowledge of its history, induced the credulous people living in its vicinity to believe that it had been the habitation of some mighty giant in the old days.*[38]

NINE FEET AND UPWARDS
The Gentleman's Magazine, April 1758, p.191

This next report gives mention of a remarkable skeleton unearthed just north of the centre of Sunderland:

> *At a quarry near Fullwell-hills, near Sunderland, the skeleton of a man was found, measured nine feet and upwards.*

The Annual Register for the same year corroborates this story, and a further letter that was printed in *The Gentleman's Magazine*, October 1763 gives more accurate details with a slight increase in height:

> *A few weeks ago a gentleman from Durham showed me some large teeth and two Roman coins. The teeth, he said, he took out of the jaw of a gigantic skeleton of a man, and the coins were found in the grave near it... In the middle of this bank was found the skeleton of a human body, which measured nine feet six inches in length; the shin bone measuring two feet three inches from the knee to the ankle; the head lay to the West, and was defended from the superincumbent earth by four large flat stones, which the relater, a man of great probity, who was present when the skeleton was measured, and who himself took the teeth out of the jaw, saw removed. The coins were found on the South side of the skeleton, near the right hand. Signed, P. Collinson.*[39]

Throughout the north of England we have encountered the remains of a powerful culture of giants and although they may be from different eras, their presence suggests that this phenomenon is not confined to one particular region and may be part of a legacy going back many thousands of years. A good deal of them were buried with care, often with curious coins, artefacts and armour, some with stone graves or mounds built over them, suggesting these were tribal leaders, Druids, or some other people of valour.

The next part of the book leaves England, crossing the border into the mountains of Wales, the true homeland of the giants.

Gazetteer of Giants - The North

DATE	LOCATION	SIZE	No.	FEATURES	PAGE
	CUMBRIA & NORTHUMBERLAND				
1808	Penrith Churchyard, Cumbria	17ft?	1	Celtic 'Giant's Grave'	158
920	St. Bee's, Cumbria	13.5ft	1	Huge sword and armour	160
1789	Aspatria, Cumbria	7ft	1	Intricate armour	162
1781	Keswick Giant, Cumbria	7ft-30ft	1	Carved inscription	163
1600s	Corbridge, Nortumberland	21ft	1	Giant rib on display	164
1800s	Near Corbridge	Giant	1	Perfect giant skull	165
1885	Westmorland, Cumbria	7ft	1	11-12 feet below surface	166
1884	Alnwick, Northumberland	7ft	1	Possible race of giants	166
	YORKSHIRE				
1885	Ripon Cathedral, Yorkshire	7ft 3in	1	Found in 'Bone House'	167
1811	Humbleton, Yorkshire	Giant	1	Found in stone coffin	168
1938	Settle, North Yorkshire	7ft +	1	Young adult 17 years old	169
1994	Castle Garth, Thirsk, Yorks	7ft	1	On display in museum	169
	LANCASHIRE				
1842	Giant's Grave, Furness, Lancs	Giant	1	Rumours of giants in area	170
1904	Clitheroe, Lancashire	7ft	1	Discovered in quarry	171
	TYNE & WEAR				
1757	Newcastle, Tyne & Wear	8ft	1	Found in clay quarry	171
1758	Sunderland, Tyne & Wear	9ft 6in	1	Giant teeth and jaw	172

7

GIANT LORE OF WALES

The mountains of Wales have hidden an array of legendary giants who were said to be able to shape the landscape and were bestowed with magical powers. The annals of King Arthur, the cult of Bran (see *London and the Southeast* Chapter), as well as the tales of Jack-the-Giant Killer, all feature in arguably the richest area in Britain for giant lore. Dozens of giants are specifically named, including *Idris Gawr* (see illustration on opposite page) and it is thanks to a manuscript written in the late sixteenth century that the records of these archaic colossi became rooted in the historical record.

Scholar and doctor Sion Dafydd Rhys (John Davies Rhys, 1534 - c.1609) of Brecon, was responding to the harsh attacks on the earlier works of Geoffrey of Monmouth, and in doing so created one of the most important documents in giantology (that was thankfully reprinted, most notably as part of *The Giants of Wales* by Chris Grooms in 1993).[1] Davies compiled numerous traditions often related to ancient sites and provided a survey of Britain as a whole, but with a focus on Cymru (Wales).

Naming over fifty giants in the process, Grooms brought to light a lost tradition that was part of a resurgence of giantology sweeping the planet during the 1500s. Navigators such as Ferdinand Magellan and Sir Francis Drake had witnessed live giants in Patagonia and North America, and this may have triggered Davies Rhys to compile his manuscript. Although only about ten pages of the much larger volume focussed on giants, it featured important details of giant's bones and skeletons unearthed across Britain. In 1993, Chris Grooms took this to another level with the publication of *The Giants of Wales*.[2] This uncovered many more skeletal discoveries and obscure local folklore in the ancient records, that we will look at in detail in the next chapter.

The giants of Welsh mythology are in every way similar, both physically and culturally, to their counterparts found in the rest of the world's myths. Allowing for a certain distortion of distance and local colouring, they conform to the general historical pattern of once superior beings of enormous size, wielding magical powers, who were remnants of the highly civilised antediluvian world. Beings who had slowly degenerated after its destruction, declining into brutal savagery and hatred of men.[3]

THE CHILDREN OF DON

The 'Fairy Races' of Cymru (Wales) are as widespread as they are in Ireland and Scotland. Many of the elaborate tales connecting fairies, giants, heroes and immortal gods and goddesses found their way into the 'Four Ancient Books of Wales' which included *The Black Book of Carmarthen, The Book of Haneirin, The Book of Taliesin* and *The Red Book of Hergest*, all of which date from the twelfth to the fifteenth centuries AD.[4]

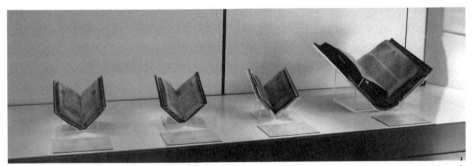

The 'Four Ancient Books of Wales' were displayed together at the National Library of Wales in 2014.

The poems themselves are from a much earlier epoch and may go back to the very origins of Britain. Numerous references to giants are found in these texts and the bards of old maintained their oral tradition by accurately passing down these songs and stories from generation to generation, preserving the esoteric and mostly lost traditions of the Druids. One of the most profound stories describes a strange race of half men, half gods, known as the *Children of Don*.[5]

The Children Of Don were powerful figures who came into the land of Wales (or Prydain as it was sometimes called) from a lost, sunken country.[6]

The Children of Don have many similarities to the Tuatha de Danann of

Ireland. They were superior beings who were masters of magical forces, and like their Irish counterparts, were wanderers from a lost, sunken land. They were described as immortal, superhuman, strong, vital, were of large stature and *"wielded magic as normally as later warriors wielded swords."*[7] They brought with them the arts and sciences of their former home, including building with Cyclopean masonry, natural science and sorcery. They settled in the mountains of Gwynedd, where they established citadels of stone and earth, ruling over the land with mortals by their side. The later Bards named these four areas of habitation as Caer Dathyl, Dol Pebin, Caer Arllechwedd and Caer Seon.

By the fourth millennium BC they migrated to other areas when the invading Britons made their way west. Some stayed up in the mountains and are remembered in many of the stories we share in this chapter. They eventually became the stuff of legend and their original prowess devolved into a diluted form of nature elementals. However, if we look carefully at the Children of Don, some interesting traits emerge. We begin with the brother of the Goddess Don who was called *Math*:

> *The patriarchal leader of these mysterious people was called Math, son of Mathonwy, and he was remembered as a white-bearded giant of a man, noble and wise, bearing his long years with strength and dignity. Math was credited (along with magical powers over life and nature) as having the ability to hear whatever word was spoken anywhere in his kingdom.*[8]

This particular skill is what we would call telepathy and was described by the bards as "*wherever a word was spoken, if the wind caught it it was carried to Math.*" He was the lord of Gwynedd in North Wales and his main seat of power was in Caer Dathyl.

Next in the hierarchy, a master magician called Gwydion ap Don (son of the Goddess Don), was also of extraordinary stature with pale features and long golden hair. He was said to be a shape-shifter, able to transform into animals, birds and other creatures. He was referred to as a high master of the occult sciences and was a powerful psychic, whose voice was described by bards as being: "*Bell-like and musical, with the power to charm humans, birds and animals into a state of happiness and ecstasy.*"[9]

Govannon was the master smith and god of skills, who created magical fire-weapons and other items that sound like descriptions of high technology. Although he was a master artificer and engineer, he also had human attributes and would get angered, being responsible for the death of Dylan, 'Son of the Waves' (which translates as great navigator).

The goddess figure Aranrhod was also a member of the Children of Don. Her name translates as 'silver wheel' and the early stories relate something akin to a flying chariot that later became simplified into the silvery moon. Her scheming, sorcery, bouts of anger, revenge and the cursing of her son were some of her more negative traits, although she was remembered as being the most beautiful woman in the world.[10] Her relentless magic eventually caused a flood that has Atlantean and biblical connotations, although on a much more local scale. With her two sisters, she cast a spell over a drinking well and gained power over the clouds, storms and tides causing a destructive and decisive event that wiped out all in the domain of Caer Aranrhod.[11]

In other parts of the old texts, Taliesin mentions one more god of interest to our studies. The giant Lieu was of divine birth. His mother Aranrhod bore him (and Dylan) by stepping over the magic wand of her uncle Math. Comparisons are made to the Celtic god Lugh, who was also a mighty man of magic.

The epoch of the Children of Don got absorbed into history and legend and few mentions of them remained until the four books were published six centuries ago. Their legacy lasts in the prevailing stories throughout the sacred landscape of Wales in different forms, but their dramatic sagas became diluted fairy tales and fragmented bedtime stories over time.

THE IRISH ORIGINS OF THE WELSH GIANTS

The Mabinogion[12] is formed from one of the 'Four Ancient Books of Wales', *The Red Book of Hergest*, which was compiled from earlier oral traditions. In the second branch of *The Mabinogion* it tells the story of Llasar Llaes Gyfnewid and his wife Cymidei Cymeinfoll. Llasar is described as *"a large man with yellow red hair... huge, monstrous... with an evil look about him."* Cymidei is said to be *"twice his size"* in height. They emerge from a lake in Ireland with Llasar carrying a magic cauldron on his back. Their offspring cause all sorts of problems to the Irish nobles. After continued oppression, the Irish retaliated by trying to burn them in an iron house. They escaped and ended up in Wales, where the giant King Bran gave permission for them to move to "high places" and build fortresses. They were said to have populated the country with giants. In *The Mabinogion*, Bendigeidfran says: *"I quartered them everywhere in my domain, and they are numerous and prosper everywhere, and fortify whatever*

178

place they happen to be in with men and arms, the best that anyone has seen."

The Irish origins of Wales was not a popular idea, threatening the unity of Britain at the time. It appears that Irish giants, perhaps the Fomorians when they were driven out of Ireland by the Tuatha de Danann, migrated to Wales. The cauldron also appeared to have magical qualities and was later used as a weapon against the Irish. Other stories describe the first giants as benevolent and these elaborate tales persisted for thousands of years thanks to poets and storytellers in the courts of the kings and also through the oral traditions of the Druids. However, it seems as though the mythical unity of Britain was being subverted by some of the tales.

The Destruction of the Cauldron of Rebirth, by T. Prytherch.

HU GADARN - THE SERPENT GOD

Hu Gadarn of ancient Wales.

One of the early 'culture heroes' or giant gods of ancient Wales was *Hu Gadarn* ('Hu the Mighty' or 'good/strong'). He was often associated with 'serpent power', an alchemical form of earth and weather magic, who invented song to strengthen memory and record. He also taught laws and sacred teachings to his followers.[13] He was renowned for taking over territories with 'peace and justice', rather than war and terror. His origins related to a land called Gwlad Yr Haf (also Deffrobani), or 'Land of Summer,' where he is said to have led the Cymry (or Celts) out from to settle in Wales.

Anthony Roberts relates this to the lost continent of Atlantis, but this is also the name for Somerset, derived from the Old English Somersæte.[14] It has also been connected to the Bosporus area of Turkey and even Sri Lanka! He also points out that Hu Gadarn taught the Cymry to remove raw materials such as stone and metal from the earth to construct into megalithic sites, temples, ornaments and weapons. His association with the plough, an agricultural instrument, represents fertility, suggesting he may have introduced early farming techniques into Britain. This also echoes the stories of the Dagda from Ireland that we will explore in the next chapter.

IDRIS THE GIANT - THE HOLY ASTRONOMER

The name Idris may have designated a line of ruling chieftains over many years.[15]

Idris is one of the most well-known names in Wales relating to giants and the landscape. In some accounts, such as the *Annals of Ulster* and *Tigernach*, he is recorded in 623 AD as being a chieftain and even 'King of the Britons'. Four giants including Idris, as well as Yscydion, Ysbryn and Offrwm proclaimed their territory over a series of hills and mountains in the Dogelly area.[16] Idris, the group's leader was skilled in poetry, astronomy and philosophy. He was so large that he could sit upon this mountain and survey his kingdom and the heavens:

Landmarks associated with giants can be found across Britain particularly in hilly and mountainous locations. Cadair Idris is named after the giant Idris who holds his seat on this haunting and formidable mountaintop.[17]

Cadair Idris is a lofty, rocky outcrop in the shape of a huge legendary seat, made of its three peaks. It was said to be imbued with magical powers and those who sat in it either went mad or had revelations concerning the occult teachings of the giants. The summit of the mountain is known as *Penygader* (top of the chair/stronghold). In 2010, local artist Nick Bullen built a 'giant's chair' and took it up the mountain in honour of Idris, but fell afoul of countryside rules, so it had to be removed.[18] It is now on display at the Eden Project in Cornwall.

Druids and occult students still conduct their rituals on this spot, sensing the power that emanates from the mountain. Strange lights are said to appear around the summit for the first few days of a new year.[19] Others claim that Cadair Idris is the hunting ground of Gwyn ap Nudd, Lord of the Underworld 'Annwn', who is escorted by a pack of supernatural red-eared hounds that herd a person's soul into the underworld.

Early painting of the peak of Cadair Idris.

Llyn Cau, the beautiful glacial lake on Cadair Idris, is not only said to be bottomless, but also the home of a 'water dragon', which once terrorised the locals. King Arthur is said to have captured it, tied it behind his horse and dragged it up the mountain to release it in Llyn Cau.[20] Idris was said to have used the nearby lake of Llyn y Gader as his bathing pool.[21]

Idris is known as one of the 'Holy Astronomers of the Island of Britain,'

Lyn Cau lake on Cadair Idris.

181

and "*So great was their knowledge of the stars, and of their natures and influences, that they could foretell whatever anyone might wish to know till the Day of Judgement.*"[22] Megalithic sites were attributed to Idris all over Wales, many with notable names and legends connected to geomancy and astronomy. Idris was also described as a "*giant astronomer-magician.*"[23]

Hugh Evans, author of *The Origin of the Zodiac*,[24] has discovered a gigantic 'star map' around this area of Gwynedd ('Heaven' in old Welsh) centred on Cadair Idris, that shows all the northern celestial hemisphere constellations listed by Ptolemy. There are similarities to the 'Glastonbury Zodiac' as discovered by Katharine Maltwood in the 1920s, although Evans has found constellations mapped on to the landscape, rather than astrological signs. Being an amateur astronomer for most of his life, Evans' findings suggest that the astronomical stories of Idris may contain elements of reality, recorded in myth and the landscape.

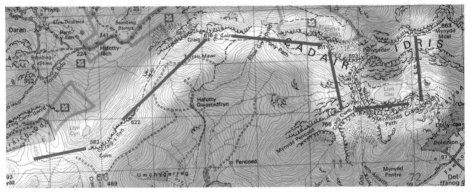

Ursa Major (or the Plough constellation) centred on Cadair Idris, by Hugh Evans.

Lolo Morganwg, a controversial Welsh antiquarian, bard and collector of ancient literature, noted that an area around the mountain called 'The Great Triangle' was known as 'Idris' Triangle' (Tryfal Idris), that has three standing stones on each corner. One of these is called Llech Idris and stands ten feet tall.[25] It was said that the giant threw or kicked it there from the summit of Cadair Idris.[26] Further landscape triangles have been found connecting major megalithic sites in Wales and Britain, a geomantic tradition we will look at in the *Giant Geomancers* chapter.

Caer Idris Hillfort is a small promontory fort on Anglesey, with three stone banks on the edge of a limestone scarp. A lost stone circle just north of this area was called *The Astronomers Stones*, another tantalising testament to the talents of Idris. It is also the alleged location of the battle that took place between the Romans and the Druids in 60 AD.[27] It is called 'The Hill of Graves.'[28]

GIANTESSES DROPPING STONES

The giantess-and-the-apron mythos is repeated all over Wales. The Welsh folk tales narrate a persistent mother goddess figure dropping stones from her apron to create mounds, castles, bridges, megalithic sites, cairns or hills in an ordered, linear fashion, over many miles. Numerous stories were collected by Chris Grooms in *The Giants of Wales* which we will investigate further in the *Giant Geomancers* chapter. The stories maintain a coherent strand, with minor variations occurring in their themes. In some cases she is startled by someone and at other times she is getting revenge, such as in the case of the *Town of the Giants* (Tre'r Ceiri) where she got stopped by a noble knight before throwing fired stones on the local's crops.[29] Sometimes it's a male giant dropping the stones, but there are hundreds of legends of this kind in Wales:

> *At Bwlch Y Ddufaen two giants were said to have created a stone bridge across the pass, and its remains are to be seen even now as massive square blocks situated on both sides of the valley.*[30]

In the parish of Caerhun in northern Caernarvonshire, two cairns lie near the top of *Bwlch y Ddeufaen* (Pass of the Two Stones). William Williams, in his *Observations of the Snowdon Mountains* (1802), reveals a story of a huge giant travelling to the Island of Mona with his giantess wife with the intention of settling there with the first inhabitants. He took large stones under his arms, while she took smaller stones in her apron. However, as we have seen in similar stories around England, a man with a large box of worn shoes met them on their journey, and told them he'd worn them all out by trying to reach Mona, tricking them into believing it wasn't worth going there because of the distance. The giant dropped the two long stones 100 yards apart and the space between them is said to be where he was eventually buried. The two stones today are known as *The Giant's Walking Stick*.[31]

On the west coast of Anglesey, *Barclodiad y Gawres* (Apronful of the Giantess) is a decorated cruciform passage grave. It is now partially reconstructed and has stunning carved stones featuring spirals, zig-zags, lozenges and chevrons that have been compared to the Boyne Valley sites of Ireland:

> *The local tradition says that the giantess (Ceridwen) and druidical goddess, in her sacred service of preparing a holy altar, took a huge stone in her 'barclog' or 'arffedog' [Apron], from a special place,*

and proceeded to a special and worthy glade, for putting down her 'Maen Bethel' [Bethel Stone], but unfortunately, the string of her apron broke and the blessed stone fell here. [32]

In some traditions Ceridwen was seen as a heroine from a time before the deluge and was associated with the 'Cauldron of Knowledge' as well as being connected to Druidical wisdom. Whatever the truth of this, it is known that sacred rituals took place in this chamber looking over the sea towards Ireland.

Barclodiad y Gawres (Apronful of the Giantess), Anglesey. Photo H.Newman.

Another name given to many single standing stones is *Baich y Cawr*, which translates as 'The Giant's Burden'. Anthony Roberts saw this as the local giants being forced to carry the monoliths due to their anger at the new religion of Christianity taking over their lands. *Crud Y Gawres*, or 'The Giantess' Cradle' is a huge erratic boulder near Ardudwy with a small area hollowed out in the middle. Legend states that the giantess would rock her giant baby within the stone. More erratics are found on an isolated conical crag near Corwen:

> *The top of the hill was known to the ancients as Caer Drewyn, the giant who shaped it being wrongly called Drewyn by the later Celts. The word is a distortion of 'Trewyn' or 'Tref Wyn', meaning the castle of Gwynn ap Nudd, the king of the Fairies who sometimes appeared as a giant.* [33]

The giant of Caer Drewyn built a large stone hillfort north of the town of Corwen, Denbighshire, as a stronghold for his wife who herded cows into it for milking. He could make the entire mountaintop invisible, usually during thunderstorms. This magical art that some of these giants were adepts of is found in countless myths across Britain. Not only could they manipulate the weather, but they could leave a spell on their burial place, so that when it was disturbed by later generations great tempests would arise, often scaring off the

The massive Furze Platt Hand-Axe from Kent. Photo by James Zicik

Above: Maeshowe Tomb in Orkney, where a 10ft skeleton and mummies were found. Below: Gia
bronze dirks with the famous Rudham Dirk at the bottom, on display at the British Museum.

Saint Patrick discovering a 12-foot giant skeleton in Ireland. By Dan Lish.

The skeleton of Charles Byrne (1761–1783) the Irish Giant at the Hunterian Museum at the Royal College of Surgeons of England.

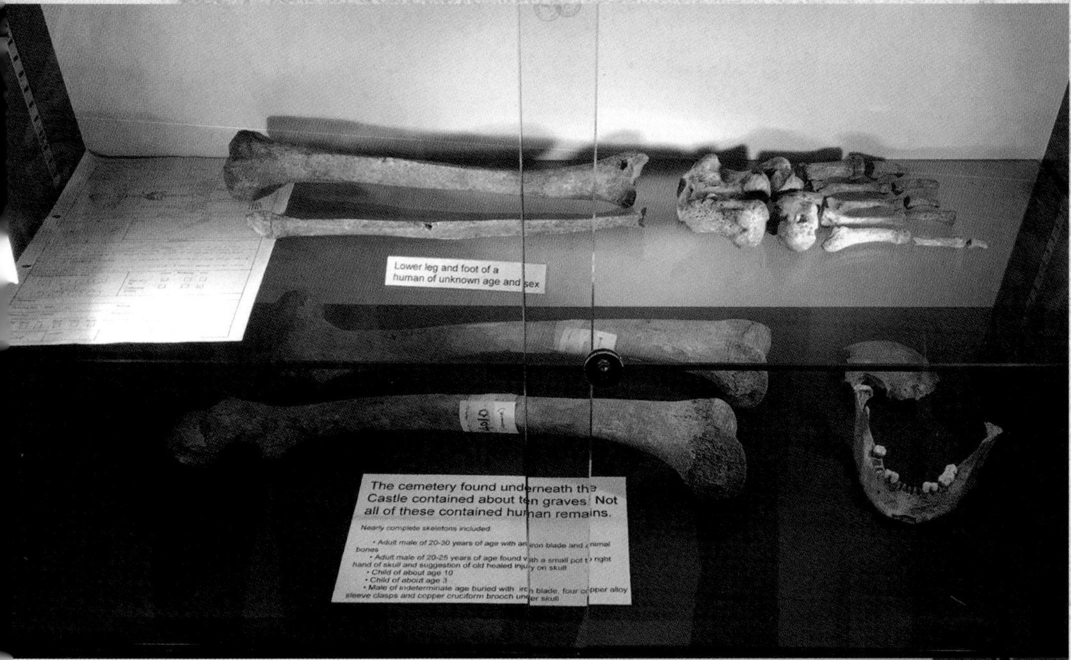
Lower leg and foot of a
human of unknown age and sex
The cemetery found underneath the
Castle contained about ten graves. Not
all of these contained human remains.
Nearly complete skeletons included
• Adult male of 20-30 years of age with an iron blade and animal bones
• Adult male of 20-25 years of age found with a small pot to right hand of skull and suggestion of old healed injury on skull
• Child of about age 10
• Child of about age 3
• Male of indeterminate age buried with iron blade, four copper alloy sleeve clasps and copper cruciform brooch under skull

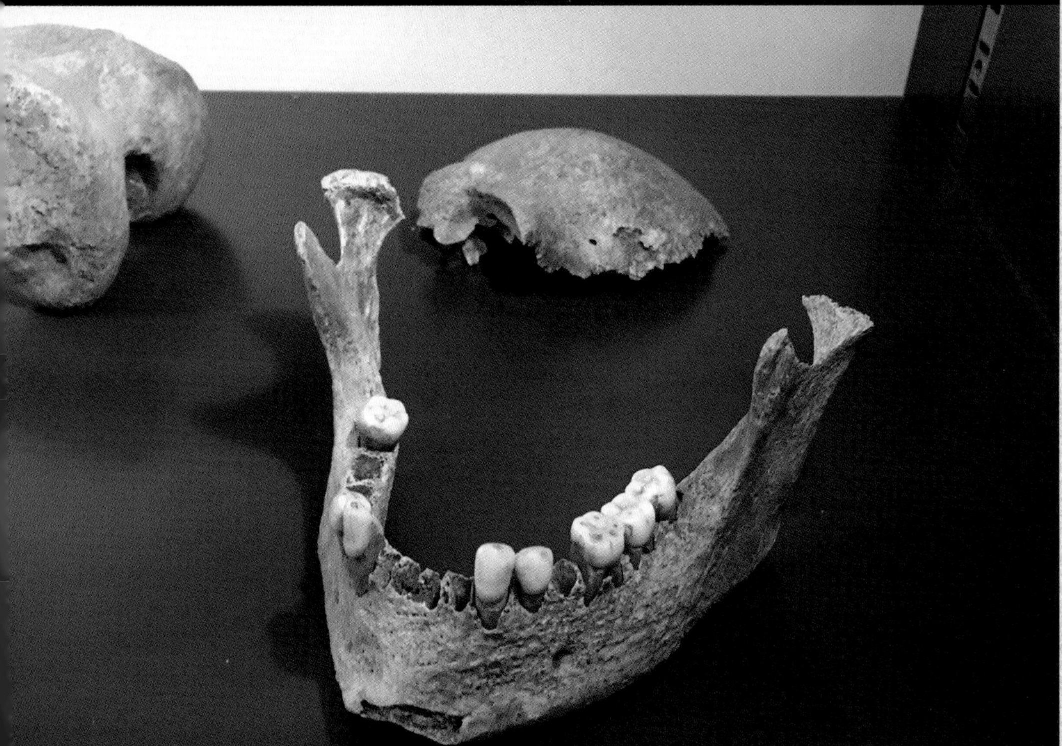

The Thirsk Museum Giant is a skeleton of a 7-foot individual found at Castle Garth, Yorkshire.

Giant Idris, the Holy Astronomer of Wales by Dan Lish.

Tom Hickathrift's grave is 8-feet long and lies in the graveyard at Tilney All Saints, England.

TOP TEN GIANT DISCOVERIES IN BRITAIN

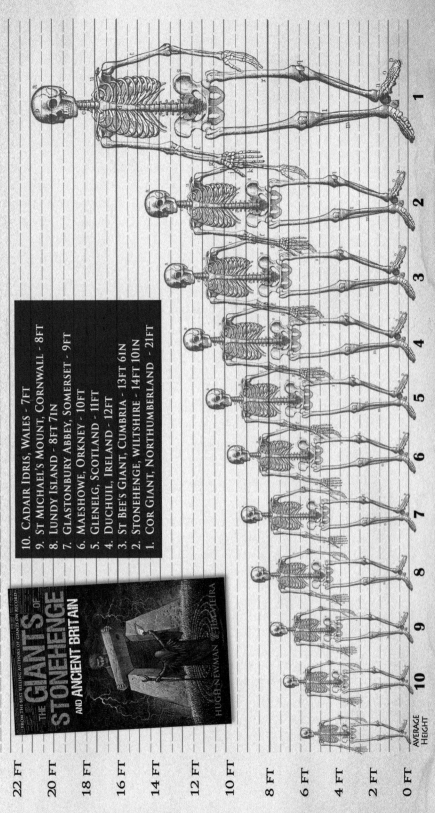

10. CADAIR IDRIS, WALES - 7FT
9. ST MICHAEL'S MOUNT, CORNWALL - 8FT
8. LUNDY ISLAND - 8FT 7IN
7. GLASTONBURY ABBEY, SOMERSET - 9FT
6. MAESHOWE, ORKNEY - 10FT
5. GLENELG, SCOTLAND - 11FT
4. DUCHUIL, IRELAND - 12FT
3. ST BEE'S GIANT, CUMBRIA - 13FT 6IN
2. STONEHENGE, WILTSHIRE - 14FT 10IN
1. COR GIANT, NORTHUMBERLAND - 21FT

FROM THE BEST-SELLING AUTHORS OF *GIANTS ON RECORD*

THE GIANTS OF STONEHENGE AND ANCIENT BRITAIN

HUGH NEWMAN & JIM VIEIRA

22 FT
20 FT
18 FT
16 FT
14 FT
12 FT
10 FT
8 FT
6 FT
4 FT
2 FT
0 FT

1 2 3 4 5 6 7 8 9 10 AVERAGE HEIGHT

grave robbers. A full examination of weather wielding powers of the giants will be revealed in the *Sowers of Thunder* chapter.

The massive stone walls of Caer Drewyn Hillfort, dated to around 500 BC.

MOLL THE GIANTESS

The giantess *Moll Walbee*, also known as Maud de Breos, built Hay Castle in Hay-on-Wye (Powys) in a single night. This was told in a legend that features apronfuls of stones and pebbles in her shoes (see *Giant Geomancers* chapter). Another site attributed to Moll is the great castle of Paincastle, a few miles from Llowes which is about 3 miles from the Wales-England border. It was once a thriving area in the 12th century, when great battles were said to have taken place between different factions up to the 1400s when Owain Glyndŵr, the first Prince of Wales led the rebellion against the English.[34] Legend states that Moll may have once had a giant husband who resided at Painscastle. He carried off young women and was killed when one was being rescued from his castle.[35]

Giantess Moll Walbee carrying an aapronful of stones.

THE BLACK GIANTS

There are numerous 'black giant' traditions that were known about before the time of Geoffrey's *Histories*, some of them featuring a one-eyed cyclops. In other accounts they were 'gorms' or oppressors who could control animals with their minds, as in the tale of Owein.[36] Other stories saw them as guardians of castles carrying large clubs, such as the mysterious *Black Porter* (Gwr du Mawr)

185

mentioned in the story of 'Culhwch and Olwen'.[37] Although the 'black' comes from the dark garments they wore, their dark beards and sometimes 'black' eyes, it is likely also to refer to their skills of sorcery. They were recorded as ferocious warriors who the English named the 'Black Army', *"because they had suffered under their victorious power in many a battle."*[38] In the Welsh romance *Peredur Son of Efrawg*, a black giant destroys kingdoms with merciless zest. Alltgawrddu in Glamorgan was thought to be a stronghold of the black giant warriors and Morgan Rhys in his unpublished *Traditions in Glamorganshire* (from *The Cambrian Journal* 1855) heard from a local landowner that the giants had: *"…black hair, beard, and eyes, and that a rim of caerulean hue encircled the black part of their eyes."* Caerulean is a shade of blue-green.

A site in the parish of Llantrisant, Glamorgan called Ysgawrddu translates as 'Below the Black Giant.'[39] Nearby are ancient quarries and iron-ore excavation sites. *The Giant's Glade* is next to a beautiful waterfall near Caban Coch resovoir in Breknochshire. The folk tale *Qwein* (Chwedyl Larlles y Ffynnawn) gives another legend of the black giants:

> *And journey along that until thou come to a great clearing as of a level field, and a mound in the middle of the clearing, and a big black man thou shalt see on the middle of the mound who is not smaller than two of the men of this world.* [40]

About ¾ of a mile away from this mound are remnants of Bronze Age activity that includes a stone row and an arrangement called *Saith Maen* (Seven Stones),

Left: Cerrig Duon Stone Circle with Maen Mawr behind it. Painting by Paul Blades.
Right: Llangar Church is located near Caer Drewyn and has this unusual effigy that may represent the local giant.

northwest of the site. The alignment of this stone row is pointing to the nearby stone circle and avenues of *Cerrig Duon* ('the black stones') four miles to the northeast.[41] The stone avenue leads to a massive block called *Maen Mawr* (Big Stone), the reputed grave of a giant.

THE SLEEPING GIANT

The Sleeping Giant of Cribarth is a nearby hill in the Brecon Beacons that gets its nickname from the shape its contours form against the sky, of a giant man appearing to be laying down having an afternoon nap. It can be seen from the 'Black Stones' of Cerrig Duon.

The Sleeping Giant Mountain in Brecon Beacons National Park, Powys.

THE GIANT RHITTA OF SNOWDON

The mountainous region of Snowdonia was the abode of the giant *Rhitta* (Rhudda or Ricca) who held court in his 'high place'. He was said to be 'twice as large as any other person ever seen'. In some traditions, Snowdonia is thought to be the main mystical stronghold of the Welsh giants and immortal gods. "Ar y drum oer dramaur, Yno gorwedd Ricca Gawr", translates as *"On the ridge cold and vast, There the Giant Ricca lies,"* written by the poet Rhys Goch Eryri, a local resident of the parish of Beddgelert during the early fifteenth century.[42]

Aerial view of some of the peaks of Snowdonia.

It was called Dinas Afferaon or 'The Fortress of the Higher Powers'. This city was remembered by many of the early historians and chroniclers such as Nennius, the renowned Geoffrey of Monmouth, and the even earlier bardic writers and poets. It was allegedly built of Cyclopean blocks of stone which were said to have been transported and fixed by the 'magic of the giants when the world was young.' [43]

Rhitta was a fearsome warrior who took on a confederate army put together by the 28 kings and tribal leaders of Britain and defeated them with ease, cutting off their beards and making a robe from them to protect him from the cold. His arrogance contributed to his downfall when he demanded Arthur give him his beard. He was eventually defeated by the legendary king who, after winning the battle, commanded that a cairn be built over his remains. This later became known as *Gwyddfa Rhudda* (Rhitta's Cairn). On the opposite page is an artist's impression of Rhitta facing King Arthur wearing his fashionable 'beard cloak'.

Photograph from 1880 of Rhitta's Cairn by F. Bedford
from National Media Museum Collection.

Over the proceeding centuries his original name dissolved into history and the burial cairn Gwyddfa Rhudda became known as *Yr Wyddfa* and was later demolished to build a hotel, now with a modern cairn at the summit. Another version states he marched southwards to claim Arthur's beard, but the king stood up to him and made him shave his own beard off and retreated, "...*much humbled in stature but much wiser in knowledge.*" [44] Another account of a giant catching men and tearing off their beards was recorded in *The Giants of Cynllwd*

Valley (Cawri Cynllwyd). The said giant was called *Each*.[45] There is an obscure reference in a bardic poem from the mid-fifteenth century that is somewhat cryptic, and may encode geomantic secrets about his domain:

> *Rhitta the giant gave a house of timber, giving it a white fortress against the frost of January; underneath, nine degrees (of strength) ahead of us, lifting stones that giants could not lift.*[46]

ARTHURIAN GIANTS

Arthur battling a giant. G.H.Thomas, 1862.

The Arthurian mythos prevailed in Wales after the era of the giants, presenting him both as a giant, and a giant slayer, with over 200 ancient sites attached to his name.[47] In general, the earlier legends of Arthur have him defeating giants in ways that correspond to the stories of Jack-the-Giant-Killer. It is much later in Welsh lore that he actually becomes a giant, perhaps in the sixteenth century when many prehistoric sites were given new names. However, stories of heroic and royal giants do appear to go back much earlier, perhaps to the early middle ages circa 600 AD. For example, Arthur the Giant is said to have thrown a stone from Carn Fadrun to Trefgwm, several miles away, whilst his wife took three stones in her apron and placed them as uprights for the dolmen. It became known as one of many sites with this name *Coetan Arthur*. Here, we also see the 'apron' motif, usually associated with the giantesses, converted to the Arthurian mythos. In his *Itinerary* (1543), John Leland confirmed the role Arthur had during the sixteenth century in the Elan Valley in Radnorshire, remembered as being a prolific giant killer, "*The dwellers say also that the giant was buried thereby, and saw the place.*"[48] This is very close to where a giant skeleton was unearthed at Rhayader Church (see next chapter).

One example linked to the stories of giants and Arthur includes Llyphan, Pysgog and Howell, who were three giants married to three sorceresses. The witches wove numerous spells over the landscape while their

190

partners built megalithic sites marking spots to honour their wives' magic. The witches were later slain by the nephew of King Arthur, Walwen (also Walwyn, Gwalchmai or Gawain) who was one of the knights of the Round Table. He had to trick them, as their magic was powerful, but it was the only way to get the deed done. In local Pembrokeshire legend, a giant called Walwen was buried at Walwyn's Castle near Haverfordwest and the legend relates, "*Here resides Walwyn or Gwalchmai, a prince of gigantic size*" in a grave "*fourteen feet in length.*" Walwyn's Castle was once visited by William the Conqueror. Richard Fenton in 1811 commented "*...probably the bank had been the tumulus that covered the skeleton of the gigantic hero Walwyn.*"[49]

Bedd Arthur (Arthur's Grave) is a megalithic oval overlooking the Bluestone outcrop of Carn Meini, Pembrokeshire. Its shape and design has led to speculation that it influenced the original horseshoe/oval of Bluestones at Stonehenge. Nearby *Carn Arthur* was originally attributed to a giant hero buried in prehistoric times.

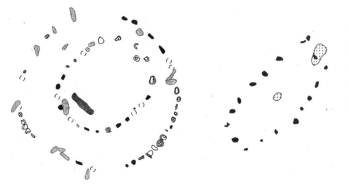

Left: Stonehenge. Right: Bedd Arthur Stone Oval. Both contain Bluestones from Preseli and are of similar shape and size.

MERLIN THE GIANT

Merlin (Myrddin) is popular in British legendary history, first making an appearance in Geoffrey's *Histories* and being inextricably linked to King Arthur. Although often associated with other giants, he only makes one appearance in Welsh folklore as a giant himself. In Carmarthen, Jonathan Ceredig Davies was informed that, "*Merlin was such a giant that he could jump over the Vale of Towy.*" [50] He was also said to have been fathered by one of the 'Incubi,' who he inherited his supernatural powers, prophetic abilities and shapeshifting talents from, indicating a link to the story of Albina and her offspring becoming giant in size. His mother was a human female who may also have been a witch.

It is said in parts of Wales that the first name of Britain was 'Myrddin's Precinct' (mentioned in the Welsh text, *The White Book of Rhydderch*). This was from a time before the land was even populated. Numerous ancient sites are attributed to Merlin, including a ruined burial chamber, now with only two stones, residing at a local farm called *Merlin's Quoit* (Coetan Myrddin). Also nearby is *Merlin's Stone* (Carreg Myrddin). This was said to be the place where Merlin had hidden a priceless treasure. The stone once fell on one of the many treasure hunters and killed him, fulfilling an old prophecy that a raven would drink a man's blood off Merlin's Stone.

Merlin being as cool as a cucumber somewhere really awesome a long time ago.

Across the road is the hillfort of Bryn Myrddin. This is one of the supposed areas of the imprisonment of the great Magician, who was captured by the enigmatic *Lady of the Lake*. Having taught the Lady everything he knew about the mystic arts, the wicked water-nymph turned on him. She used her new-found powers of sorcery to enclose him in a cave beneath the hillside. According to tradition, he still dwells there, and some say you can still hear his subterranean groans.[51]

CANNIBALISTIC GIANTS

Cannibalism is widespread in giant lore. Canthrig Bwt was a notorious giantess in the area of Gwynedd. She was said to have lived under a great boulder in Nant Peris and had a culinary taste for local children.[52] Another elaborate tale talks about the *Giantess of the Pass* (Cawres y Bwlch) just north of Llangollen in Denbighshire. Saint Collen was walking through the pass and encountered the giantess, who had a habit of capturing people and eating them. The saint prepared his sword and cut off one of the giantess' arms. The wounded giantess then used her own arm as a weapon to attack the saint, so Collen cut off her other arm and killed her, washing her blood off in a local sacred spring.[53] A very similar story, but without lost limbs, is repeated in Merionshire.[54]

Christian saints became the slayers of cannibalistic giants. *Gwedros the Giant* would kidnap locals from their houses to eat them. Saint Cynog hid inside the giant's cave amongst the corpses where he was caught by the colossi and had a chunk taken out of his thigh, believing he was going to be eaten, "*Then came, chosen soul, a weapon for you to overcome a devil; a torque from heaven.*"[55] With this magical torque he slayed the giant instantly. Thunderdel was a two headed cannibalistic giant from North Wales, who attacked Jack-the-Giant-Killer, but was defeated by him.[56] There are few recorded accounts of real cannibals in ancient Britain. One of these talks of the barbarous people called the 'Attacotti' who inhabited the most northern mountains of Scotland during Roman times. Hugh Thomas, when he was a youth, saw them eat human flesh at banquets as a delicacy.[57]

GIANTS AND STONE

Near the Church of Saint Tyfrydog at Clorach on Anglesey, is a standing stone resembling a man bearing a burden. According to legend he was stealing gold (or a Bible) from the local church and was turned to stone by the Saint:[58]

> *And in the land of Caerfyrddin in Cynwil Gayo was a giant called Cynwil Gawr, and that is the reason, perhaps, why the place is still called Cynwil, and he was a godly man.*[59]

St Tyfrydog's Stone.

Mynydd Myfyr (Myfyr's Mountain) in Denbighshire is a series of earthworks including a 'Giant's Grave' with a nearby standing stone at the point at which four parishes met: Llanelidan, Bryneglwys, Gwyddelwern and Llansantffraid Glyndyfrdwyand. It also stood on the border between Denbighshire and Merionethshire and is very close to the border between England and Wales. To the west is a lake, the only one in Welsh tradition that has the name of a giant, *Rhuddwyn Gawr*. There is a local story about three giant kings or lords who would meet at a cube-shaped stone named *Table of the Three Lords* (Bwrdd y Tri Arglwydd). The triad of giants is a theme remembered in many stories in Wales, but the fact this was a 'moot' or 'ting' site of the giants is compelling as

The cuboid stone of Bwrdd y Tri Arglwydd is 3ft by 3ft wide.

it is on the boundary of multiple parishes, two counties and two countries, as well as within a sacred area of earthworks, menhirs, castles and lakes all linked with local giants. Could the location have been a prehistoric survey point, and could The Table of the Three Lords have been an early kingship stone?

Although giants were usually associated with mounds, cairns and hill forts, a handful of stone circles became attached to giant lore. A stone ring attributed to giants just over the border in Shropshire is *Mitchell's Fold* which has some elaborate folklore attached to it:

> *There is a traditional folk story that a giant whose marvellous cow gave unlimited amounts of milk used the circle until a malicious witch milked the cow, using a sieve until it was drained dry, as a result of which it fled to Warwickshire where it became the Dun cow. As a punishment, the witch was turned into stone and surrounded by other stones to prevent her escaping. What became of the giant is unknown. Local folklore also suggests this is the actual place where King Arthur withdrew Excalibur from one of the stones in the circle and then became king of the Britains.*[60]

Mitchell's Fold also provides a connection to Stonehenge, as it was reported in the 18th century that it originally had a 'trilithon' structure within it, but this has now since been destroyed.

The Giant's Stone in Denbighshire is an 11ft 6in monolith south of Eriviat Hall. A giant was said to have perfectly carved a thin slab from the stone and left it leaning against it.[61] Also in Denbighshire, the *Giant's Footprints* is a large stone with footprints carved into it, but although now lost, it is said to have rested on Mynydd Hiraethog.[62] *The Big Red of the Birches* (Coch Mawr y Fedw) is a giant featured in several tales recorded from the Liverpool and Birkenhead Eisteddfods between 1869-72. He was said to have lived above Dolgellau in the birch woods on the high slopes north-west of Cadair Idris. One tradition stood out that related to the peculiar movement of a large stone:

> *Another time, when they were building a barn...four of the stone masons failed to raise some stone...they came back and the stone had been put in its place; they figured that the giant had raised it himself. Further stories have him stealing pigs, carrying heavy loads and throwing stones over vast distances.*[63]

Tre'r Ceiri is located just south of Bachwen Dolmen and is one of the most impressive ancient monuments in Wales, dating to around 200 BC. The stone settlement's walls are largely intact, and are as high at 13 feet in some places.

Bachwen Dolmen is located to the northwest of Cadair Idris and close to Tre'r Ceiri, created by watercolourist Moses Griffith during the eighteenth century.

The Cairn that sits upon its summit belongs to the early Bronze Age and is sometimes called 'The Crown of the Giant's Island'. Ifor Williams connects the site with the Caer Dathyl featured in *The Mabinogion*.[64] A local folk story recounts that people in the local village offended a giantess who lived in the hillfort. As revenge she collected an apronful of hot stones, that had been heated in the sulpherous fire of the 'Phantoms' (ellyll), who lived on the mountain. She intended to throw them at their fertile fields to destroy the crops, but a dignified knight came to meet her and she dropped them on that spot, where they still exist today. Due to its long term use the hillfort has been speculated over the centuries to have been built by the Tuatha de Danann of Ireland.[65]

Tre'r Ceiri hillfort. Artist's reconstruction, viewed from the north-west (A. Smith, GAT).

THE WELSH BIGFOOT

In a tale collected from Owen Jones, it describes a wild giant who may be one of the only references to a Bigfoot type creature in the British Isles.[66] Nant Gwynant is a valley in North Wales in Caernarfon, and between here and Beddgelert is a location that locals call *The Hairy Man's Cave* (Ogof y Gwr Blewog):

> *One day...one of the shepherds looked from the mountain when he saw a clamp o ddyn mawr [lump of a large man] covered in long red hair sitting on a hill near the pass.*[67]

The same story was elaborated upon by John Rhys in *Celtic Folklore: Welsh and Manx*, 1901:

> *I may mention a cave near a small stream not far from Llyn Gwynain, about a mile and a half above Dinas Emrys... called Ogo'r Gwr Blew, 'the Hairy Man's Cave'; and the story relates how the Gwr Blew who lived in it was fatally wounded by a woman who happened to be at home, alone, in one of the nearest farm houses when the Gwr Blew came to plunder it.* [68]

Another hairy giant creature called *Brenin Llwyd* apparently lives in the Snowdonia mountains in Wales. The beast is also called *The Monarch of the Mists* or *The Grey King* and is said to haunt various mountain tops. Brenin Llwyd is often associated with Snowdonia, particularly Cadair Idris, and is sometimes confused with *Idris Gawr* (Idris the Giant). A very similar creature, the *Big Grey Man,* inhabits the summit and passes of Ben Macdhui in the Cairngorm mountains of Scotland (see next chapter). Both examples are described as gigantic, furry creatures with the general shape of humans. They are often referred to as the Celtic equivalent of the Sasquatch (Bigfoot) and reports of the creatures go back to the middle ages.[69]

GIANTS AND THE FAIRY REALM

We complete this chapter by exploring the elemental beings of Wales with a story collected by school children in the Blaenrhymni region of Glamorganshire in 1911. The book reveals a tale called *Giant of Gilfach-fargoed* by D. Glyn Williams. It discusses the fairies of the Rhymney Valley who for aeons had led peaceful and joyous lives until a cruel giant with a *"strange staff with a living snake coiled around it"* arrived in the area. He was able to catch the fairies with it and eat them. With the giant transgressing the laws of the fairies, one of them, whose parents had been eaten by the giant, decided to kill him.[70] The nasty odour coming from the decomposing giant urged the locals to cremate him, but a piece of strange black rock caught fire and smouldered in the pit. The wisest of the fairies realised this was a great tool for heating and cooking and the fairies were able to survive the winters. This was, of course, coal.

The Fairy Frolic at the Cromlech, from *British Goblins,* by Wirt Sikesby, 1880.

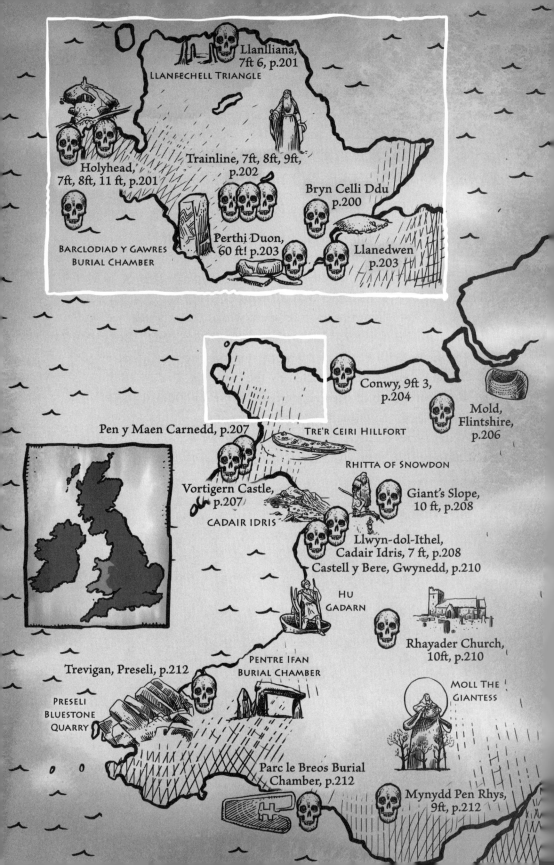

Llanlliana,
7ft 6, p.201

LLANFECHELL TRIANGLE

Holyhead,
7ft, 8ft, 11 ft, p.201

Trainline, 7ft, 8ft, 9ft,
p.202

Bryn Celli Ddu
p.200

Perthi Duon,
60 ft! p.203

Llanedwen
p.203

BARCLODIAD Y GAWRES
BURIAL CHAMBER

Conwy, 9ft 3,
p.204

Mold,
Flintshire,
p.206

Pen y Maen Carnedd, p.207

TRE'R CEIRI HILLFORT

RHITTA OF SNOWDON

Vortigern Castle,
p.207

Giant's Slope,
10 ft, p.208

CADAIR IDRIS

Llwyn-dol-Ithel,
Cadair Idris, 7 ft, p.208
Castell y Bere, Gwynedd, p.210

HU
GADARN

Rhayader Church,
10ft, p.210

Trevigan, Preseli, p.212

PENTRE IFAN
BURIAL CHAMBER

MOLL THE
GIANTESS

PRESELI
BLUESTONE
QUARRY

Parc le Breos Burial
Chamber, p.212

Mynydd Pen Rhys,
9ft, p.212

8

THE GIANTS OF WALES 'CEWRI CYMRU'

THE ANGLESEY GIANTS

She was called many different names, that is the Ynys Dywell (Dark Island) for the woods that covered her; Ynys y Cederyn (Island of the Mighty) because of her distinction as a rearing-place for heroes and giants, and the original home of the powerful Druids.[1]

'The Great Temple and Grove of the Druids at Treedrew in Anglesey,' now thought to be lost, was recorded by William Stukeley in 1725.

We begin our survey of Wales on the small island of Anglesey, the last stronghold of the Druids during the Roman invasion in 60 AD. They defended their honour with dignity, but the Romans defeated the robed elite whose oral history mostly died with them. Long before this, stone circles, dolmens, burial tombs and standing stones were erected all over the island.

THE MOUND IN THE DARK GROVE

The most famous site on Anglesey is *Bryn Celli Ddu* (The Mound in the Dark Grove). It slowly transformed into the mound with a megalithic chamber we see today. It is oriented to the summer solstice sunrise and was originally an oval megalithic ring, with about 17 stones. An early discovery from a farmer in 1780 hints at a powerful ruling elite being buried at the site. He found "...*large human bones lying near the pillar.*"[2] How 'large' these were was never expressed.

Aerial view of Bryn Celli Ddu, with the Gorsedd in the background in the top right.

The site contains a carved pillar at the back end of the chamber, still in-situ, but the finest one is now in the National Museum with a concrete replica in its place. Between Bryn Celli Ddu Chamber and the Llandaniel Fab Menhir is the rocky outcrop called *The Gorsedd*. Bryn Celli Ddu may have been preceded by another much earlier ritual structure, as a short avenue of postholes next to the henge was uncovered during excavations in 1928. Charcoal found in the bottom of these was carbon dated to the Late Mesolithic period c.4000 BC. This is 4000 years after the great wooden post holes of Stonehenge, but show similar traits, as they were made of pine and on an east-west axis.

At exactly eleven miles to the west on the same latitude is the aforementioned *Barclodiad y Gawres* burial chamber, also known as the *Apronful of the Giantess*. One further example of traditions of giants relating to surveying is on the north west tip of the island, at a site once called the *Giants' Stones* (Meini-y-cawri). It is a set of three stones marked as a triangle with another menhir about 600 yards away. Locally the site is called *The Druids' Stones*,[3] but it is also called *Llanfechell* and *Mein Hirion*.

The Three Stones of Llanfechell, also known as Mein Hirion, that have another standing stone 600 yards to the south east.

SEVEN FEET SIX INCHES TALL

In the *John o' Groat Journal*, 30 July, 1841, we find this account of a 7ft 6in skeleton being unearthed on the most northerly village of Wales in Anglesey that now has the name Llanlliana:

> *1840 - November, at Llanlinan, near Alnwick, a human skeleton was found in a kind of stone coffin, in a state of preservation, seven feet six inches in length.*

8FT, 9FT AND 11FT GIANTS ON ANGLESEY

In 1898, a lecture was given by Welsh folklorist Rev. Ellas Owen of Llanyblodwel in Bala describing various giant myths of Wales, as well as some fascinating accounts of bones and skeletons being found on the island.[4] It was also outlined on 21st October 1899 when *The Rhyl Journal* ran an article titled *Welsh Folk Lore, Giants and Other Shadowy Matters* written by Rev. Elias Owen:

> *Angharad Llwyd in her 'History of Anglesey', states that many graves were found in Holyhead containing skeletons of enormous size. The graves were walled, the bottoms paved and so closely covered with flags that no soil could get in. Some of these graves measured 8ft, some 9ft, and one in particular measured 11ft in length. The skeletons of a corresponding size, were all entire within these graves until handled and exposed to the air, when they fell into dust, except the teeth.*[5]

More information came forth when we found an old copy of the book the original reports came from *A History of the Island of Mona, or Anglesey, by* Angharad Llwyd, 1835:

> *In removing the ruins of this chapel some years ago, a stone coffin, containing human bones of prodigious size, was found in the north side of the chapel… I was assured by several masons, and others, that in digging the foundations of houses, now building to the south of the church, they found many graves, containing skeletons of enormous size.* [6]

It then goes on to give the accounts mentioned in the lecture and relayed one interesting thing about the teeth:

> *[The teeth] were so hard, as to require some pains to be broken with their hammers.*

7FT AND 8FT GIANT SKELETONS UNEARTHED

Further accounts mentioned in the lecture (and text) include numerous discoveries in Anglesey:

> *Mr. Owen Jones, of Manchester, told me that the navvies engaged on the railway between Bangor and Holyhead came upon skeletons in Anglesey 7ft and 8ft in length, and the author of the 'History of Aberconwy' states that many bones of very large dimensions were found.*

We looked further into who might be the author of the *History of Aber Conwy* and found that the full title was *The History And Antiquities Of The Town Of Aberconwy* and was authored by Robert Williams in 1835. Indeed, it does mention an account of giant skeletons and how the town was almost depopulated by the plague in 1607:

> *About forty years ago the lower half of the high-street was repaved, and considerably lowered, when a vast number of skeletons were discovered lying side by side the whole way; many of the bones were of very large dimensions and some of the jaw-bones are said to have exceeded a span in the distance of the extremities.* [7]

A span is the distance of an outstretched hand, so indeed this giant gentleman had a rather powerful jaw that was well over the average size of the time. If you stretch out your hand in front of your face you will see what we mean.

THE GIANTESS AND THE PETTICOAT

At Llanedwen in Anglesey, Edward Lloyd relates an unusual account of a giantess from the 1780s:

> *It is certain that there lived in Anglesey men and women of a gigantic stature; and such a woman was buried in the Church of Llanedwen, who was commonly called Y Ferch a'r Bais las (The Girl with the Blue Petticoat). About 80 years ago some jolly young men were digging her grave, and found her bones wholly answerable to the tradition of her stature.*[8]

The blue petticoat is often associated with faeries. In his *Cambrian Superstitions*, William Howells states that the fairies *"danced in blue petticoats, paying the strictest regard to each other, and never deviating from a perfect circle."*[9]

A TOLERABLY BIG FELLOW - 20 YARDS LONG!

In *Miscellaneous Antiquarian Notices* in the 1846 *Archaeologia Cambrensis*, a contributor says this about a local megalithic site:

> *It is called the Perthi Cromlech; near Llanidan or Brynshenkin; nine feet long, seven feet broad, and two feet eight inches deep. There is a tradition that a giant was buried near this Cromlech with his head towards the east, measuring, from my striding, twenty yards at least; a tolerably big fellow.*[10]

Left: Perthi Duon in 1802, sketched by the Reverend John Skinner.
Right: a mid-19th century sketch of the dolmen.

This is referring to a dolmen called *Perthi Duon*, now badly ruined with only of a fallen capstone and two collapsed supporting stones. It dates to 3500 BC, one of the oldest sites on Anglesey.[11] Where the exact burial was is unclear, but 20 yards is outrageously tall for any human being, equating to around 60 feet.

NORTH WALES

A 9ft 3in skeleton was unearthed in the mid 1800s near Conwy in North Wales, reported in the *John o' Groat Journal*, 30th July, 1841:

> *...many bones of very large dimensions were found in or near Conwy. An old friend. Hugh Jones, Garlonau. Llandinorwig, opened on his farm a 'cist-faen,' and within was a skeleton of man 9ft 3in in length.*

A few miles away from Conwy are the *Great Orme Copper Mines* of Llandudno. Large-scale mining began here around 4000 years ago. A few hundred yards from the entrance is an ancient dolmen, dating back to the Bronze Age. At the site on display is a huge stone that has two massive axe head hollows carved out of the rock, suggesting that copper was cast at the site. Again, they are incredibly oversized and leaves one wondering if indeed giants were the miners in this part of ancient Wales.

Over 2500 dolerite and diorite hammers have been found ranging from 2 to 29kg (64lb) in weight. The largest of these would have been far too heavy for a standard human to wield. If this was an axe or hammer head the length of the handle would need to be 9 feet long (based on a modern 20lb sledgehammer having 3ft long handle) and someone of immense strength and height would have been required to wield it.

Aerial photo of the Orme Mines, with an image of the nearby dolmen and one of the axe casts that are of giant size.

BENLLI THE GIANT AND THE MOLD CAPE

Benlli is one of the oldest recorded giants in Welsh lore and is often associated with being an oppressor or tyrant king. He was written about in *Historia Brittonum* by Nennius in the 9th century.[12] The text claims he was an early king of Powys, who was later burned to death after threatening Germanus of Auxerre, a holy man. Folklore across Wales mentions his presence in elaborate stories, and sites associated with him include two lost monoliths that were said to once mark his grave in nearby Foel Fenlli in Denbighshire. This area was also where he was said to have resided and ruled from. Benlli's hill fort at Foel Fenlli represents the oldest personal and place name lore concerning giant legends in Wales, and has connections with Balor of Ireland.

One of the most important artefacts in the British Museum in London is linked with strange accounts of Benlli and a massive skeleton unearthed in a haunted tumulus near the River Alyn in Flintshire. R Thomas Williams first wrote about the discovery in the early 19th century:

> *In the year 1823 (in the month of October), a human skeleton of gigantic size was discovered by accident in a field near this town (Yr Wyddgrug) in a place by the name of 'Bryn-yr-ellyllon' (Hill of the Spirits) by a man plowing...On the chest of the skeleton was found a mantle of gold.*[13]

The full story is related in *Giants and Dwarfs* p.85:

> *In the year 500 lived Benlli Qiiwr, or Benlli the Giant, at Yr Weyddgrug, now called Mold, in Flintshire. The hill upon the summit of which he collected his warriors is still called Moel Benlli, In 1833 the overseer of the highways in those parts caused a tumulus to be removed in order to obtain materials with which to mend the roads, it being supposed that this tumulus from its depth was a gravel pit. At the lower part of it were found some large bones, a skull of greater than the usual size, a bright corslet, and two or three hundred amber beads. The late Dr. Owen Pughe, the celebrated antiquary and historian, ascertained from ancient Welsh manuscripts and the Triads that the person here buried was Benlli. The corslet, which is now in the British Museum, and is called the Lorica or Golden Vest, is of leather, cased with thin fine gold of most beautiful workmanship. The field in which it was*

found, near to the town of Mold, is known by the name of the Cae Ellyllion, or the Field of the Goblin; and a story is current that a man of gigantic stature, with a breastplate or vest of gold, may be seen standing upon the site of the tumulus at night, and that many persons have been much frightened at his appearance.[14]

One of the main reasons the tumulus was excavated was because of a local woman who had a recurring dream about a giant in full armour standing over an earthen mound in the area. The tumulus has since been dated to about 1900–1600 BC. The gold cape may have been ceremonial dress or the royal dress of a local king or chieftain. Between two-and three-hundred amber beads in rows were on the cape when it was created, but only a single bead survives.

The Mold Cape on display at the British Museum, London.

Even though the corslet has been on display for decades in London, there is no mention of the giant skeletal discoveries from the original report. However there could be a connection to the Great Orme Mines located nearby:

The individual who once wore the cape would have possessed great power and wealth, and it is believed this wealth may have been generated by the nearby Great Orme – the largest copper mine in north-west Europe. This would have been a major trading centre for prehistoric communities.[15]

LARGE BONES DISCOVERED IN CAERNARVONSHIRE

A Description of Caernarvonshire is a topographical work written by the antiquary Edmund Hyde Hall (c.1770-1824). In it he described one report and one piece of local folklore of interest to giantologists. The problem is, he leaves no clue as to the actual height of the interred giant. This first account is just south of Baschel Dolmen and Tre'r Ceiri Hillfort that was featured in the previous chapter:

> *About a mile and a half from the church, close upon the road to Nefyn, stands a small tumulus surmounted with a stone of memorial or stele, called Pen y Maen Carnedd. The evidence, coupled with tradition, in favour of this place is supposed to rest upon the discovery of the bones of a large sized man in a grave said to be still dimly visible.*

Upon Mynnyd yr Ystym, a nearby hill, a circular henge or hillfort had a stone cairn nearby called the *Giant's Grave* that legend states was the burial place of a local titan, whose bones were once unearthed there.

VORTIGERN THE GIANT - 13 FEET TALL

Vortigern Castle (Castell Gwrtheyrn) in Pistyll, Caernarvonshire was probably originally a burial site, or at least contained a burial site called *Bedd Gwrtheyrn*. Vortigern was an ancient king of Britain said to have ruled during the fifth century AD. A tumulus made of stone and turf stood on the hill now called Castell Gwrtheyrn. It was opened in 1964, exposing a stone coffin, containing the bones of a very tall man.[16]

In his *Parochial Enquiries* (1695), bishop and antiquarian White Kennett records another giant excavated in the early seventeenth century when the locals made an effort to find the truth about the giants of the area:

> *The inhabitants of the parish of Llanaelhaern where his grave was, for better proof of the truth itself, assembled themselves together, digged down that heap of earth, and removed the great heap of stones, and in a stone chest found the body of a very tall man, for his shin bone was an ell long... Mr Hugh Roberts, who is a landed man in the parish, a certain man of his word, and a careful preserver and searcher of antiquities.*[17]

In England, the ell was usually 45in (1.143m), or a yard and a quarter. The average length of a male tibia (shin bone) today is 17 inches (43cm) so he would have been around 2.5 times the height of an average human today indicating a total height of thirteen feet. Thomas Pennant in 1893 also wrote about what may be the same discovery of a *"stone coffin, containing the bones of a tall man."*[18] It was further corroborated by William Williams of Llandegai (c.1809) believing it may have been connected to Vortigern, although this could have been an entirely different discovery:

> *It is said that in digging in a bank close to this place a stone coffin was found, and in it the skeleton of a man of gigantic stature.*[19]

BONES HUGE IN SIZE

Rhiw'r Cawr (Giant's Slope) in Merioneth was written about in 1827 as having gigantic bones discovered at the site by Owen Ewynne Jones:

> *Rhiw'r Cawr is a little way down from [the farm] Cilwern in Llanymawddwy. At the head of the slope were found bones huge in size - twice the size of the bones of common men. There, according to tradition, is where rests Samson, the great giant of Mawddwy.*[20]

The local tradition has it that children would be sent to the *Giant's Slope* if they had been bad. Also, Samson is a very old giant's name that is found associated with megalithic sites all over Wales.

GIANT SKELETONS OF CADAIR IDRIS

The legend of the giant Idris that we explored in the previous chapter seemed to have come to life in these skeletal discoveries from below Cadair Idris unearthed in 1685 (published in 1795):

> *Lhuyd notes two pairs of extra-ordinarily large skeletons found in the middle of the sixteenth century by peat diggers in a bog at the base of the southern ascent of the mountain at the farm named Llwyn-dol-Ithel, each buried in coffins with hazel rods.*[21]

A more detailed account was given in a passage from 1921 (see next page). The farm where the discovery was made was described as being at the base of Cadair Idris and fellow author, Andrew Collins confirmed the farm does still exist at the GPS coordinates given in the text from 1921. Hugh subsequently visited the farm and the current owners confirmed the finds were genuine.

DIVISION VII (FINDS).

528. *Medieval Interments.*

The following interesting incident is recorded in Lhuyd's *Parochialia* (i, 6) :—

About two- or three- and twenty years agoe certain people digging Turfs at Lhwyn Dol Ithel, after they had digged about 3 yards deep they found a coffin of ab't 7 foot long, made of fir wood, and carved at both ends thereof, which were also guilt, and when ye same was open'd they found two sculls therein, and two skeletons, one of a man—the other of a woman, the bones being something moist and tuff, the same were of great length, viz.—the thigh bones between each knuckle or joynt were 27 inches long ; and within a yard of the place where the Coffin was found, they digged up two other skeletons, one of man, the other of woman, much of ye same length with the former, wet also, being laid on clay and within two roods of them they found another grave in which they found also man's bones, something as they imagined . . . of a smaller stature than ye rest and moist and tuff, also ye corps were so laid in ye coffin yt ye feet of ye one was towards ye head of ye other, and likewise those bones which were double and in the same grave. It was observed that there was laid white Hazel rods ab't 2 iards and a half long with ye bark on along ye sides of ye graves and coffin w'ch were so tuff, that when wrung, it made a writh.

In the additions supplied by Lhuyd to Camden's *Britannia* (ed. 1695) the date of the discovery is given as 1685, and the site of the discovery indicated as a turbary called " Mawnog ystradgwyn near Maes y pandy." This may be placed somewhere on the summit of what the modern Ordnance sheet calls Mynydd Dolffanog (42 N.E., lat. 52° 40' 47", long. 3° 52' 0"), where is also shown a *mwdwl eithin* [a gorse stack] which Lhuyd mentions as " in Ystratgwyn." ' Maes y pandy ' is probably represented by the present house called Ty'n y maes, and Llwyn dol Ithel will be found at lat. 52° 40' 58", long. 3° 53' 6" (42 N.E.).

*An Inventory of the Ancient Monuments in Wales and Monmouthshire, VI County of Merioneth,*1921. Royal Commission on the Ancient and Historical Monuments of Wales, p.528.

Left: Llwyn dol-Ithel Farm, where the giants bones were unearthed. Photo by Andrew Collins. Right: Burial of skeleton in wooden coffin with two hazel wands placed over the legs, with another burial next to it. This is what the Cadair Idris discovery may have looked like. This is in fact from a burial in Norway from an excavation in 1971.

GIANT BONES UNEARTHED NEAR CASTELL Y BERE

Castell y Bere is a castle near Llanfihangel-y-pennant in Gwynedd, built by Llywelyn the Great in the 1220s. It sits at the base of Cadair Idris and may have been an important location in prehistoric times. Some monstrous bones were found near the castle in a farm called *Maes-y-llan* (Field of the Church):

> *When a stone was moved near the cliff a few years ago, on this farm, human bones were found buried underneath the large stone; they were in good condition, and of a huge size; but whose they were and how they came there, no tradition survives to throw light on them.*[22]

Castell y Bere, next to Cadair Idris in Gwynedd.

CENTRAL AND SOUTH WALES

THE GIANT OF RHAYADER

There was once a huge castle dominating the area of Rhayader in central Wales. It's location is the geodetic centre of wales, the omphalos of Cymru, and was much fought over throughout history. This account was sent to us by researcher Rob Underhill, a writer who lives in the village. He discovered this account of a giant warrior from Brian Lawrence, a local historian, who revealed the following to Mr Underhill in a personal letter:

> *Whilst digging the foundations for the tower in 1783 a great number of skeletons were discovered about a foot below the surface*

of the ground, arranged side by side, in a most orderly and regular manner. All their bodies were placed in the same direction except one whose skeleton was of immense size, the thigh bone measuring more than a yard in length. Whose skeleton was placed in the opposite direction to the rest. It was thought that the skeletons were the remains of the garrison soldiers from Rhayader castle who had been put to the sword by Llewellyn ap Iorwerth, Prince of North Wales in 1231. The giant skeleton was thought to have been that of the commander of the castle. The inhabitants of Rhayader are believed to have buried the skeletons in a methodical manner under the belfry of the ancient church. All the bones were gathered up and deposited in one large grave in the churchyard by David Williams, father of the county historian Jonathan Williams. Rees Evans was paid 3/6 for collecting and reburying the bones. The grave can still be seen surrounded by iron railings near the entrance to the tower. When the tower was rebuilt again in 1888, a large quantity of fragments of human bones were found under the foundations, so we may assume that all the bones had not been cleared away originally. (Letter signed by Brian Lawrence).

Hugh at St Clement's Churchyard. The giant's grave is surrounded by iron railings.

Further sources reveal the femur was over one metre in length.[23] Either way a yard or a metre would equate to a remarkably large human being, considering

211

the average thighbone is just under half that size. A yard is 36 inches and a metre is just over 39 inches, so it is estimated to have belonged to someone over ten feet tall.

THE GIANTS OF THE PRESELI MOUNTAINS

A few miles northeast of the Preseli Mountains, in the heart of bluestone country, a remarkable discovery was made and recorded by Richard Fenton:

> *Near the place of Trevigan, in ploughing up a field, it was found at a certain depth to be covered with graves marked out by stone coffins, formed of coarse purple flags from the quarries in the neighbouring cliffs; one of which contained a skeleton much above the ordinary size, with a sword with him, commensurate with the dimensions of the warrior who bore it, of such a length as not to admit of being sheathed by the tallest man of those parts.*[24]

Craig Rhos-y-felin, rhyolite bluestone outcrop in Preseli, Wales, where some of the stones of Stonehenge were quarried. Photo H. Newman.

CAN HIR THE GIANT - NINE FEET IN LENGTH

East of Neath near in Glamorgan, an extraordinary discovery brought a legend to life in the 1800s. Griffith Thomas, in *Hanes Capel y Gyfylchi* writes:

> *In the Pont-rhyd-y-fen district there is a tradition that a giant named Can Hir once roamed the neighbouring mountains. His burial place was at the top of Mynydd Pen Rhys, near gyfylchi Chapel, and when, about a hundred years ago, the mound of*

212

stones was removed, a human skeleton, nine feet in length, was discovered.[25]

PARC LE BREOS - GRAVE REVEALS GIANT SKELETONS

Parc le Breos Burial Chamber (Parc Cwm long cairn) is a partly restored Neolithic Severn-Cotswold type of chambered long barrow. Its traditional name has long been the *Giant's Grave*. It is situated on the Gower Peninsular near Swansea, South Wales and dates to 3850 BC, making it older than West Kennet Long Barrow in Wiltshire. In 1869, an excavation report outlined an important discovery:

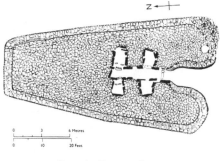

Parc le Breos plan.

> *There are the remains of two remarkable skeletons. One must have been of gigantic proportions... Case 1, the remains of a male of very considerable proportions. Case 3, enclosed separately are the condyloid ends of two femurs representing a skeleton of gigantic size.*[26]

Archaeologist Joshua Pollard noted that males analysed from here were "*particularly robust*" when compared to females.[27]

Illustration of Parc le Breos from the south, 1887.

213

Very close to the chamber is another site that has pushed the date back of human presence in Wales. A 1984 excavation of *Cathole Cave* revealed two tanged points that may be 28,000 years old. Rock art depicting a reindeer from the Upper Paleolithic era was found on the back wall of the cave in September 2010. The intricate carving, measuring 15 x 11cm, has been dated to 14,000 - 15,000 years ago. George Nash, the archeologist who made the discovery, said it was *"the oldest rock art in the British Isles, if not north-western Europe."* [28]

Another discovery called the *Red Lady of Paviland*, is a partial male skeleton (once thought to be a Roman female) that was discovered in a cave about eight miles west of Cathole Cave. It was painted with red ochre which is a rarity in Britain, especially as it is dated to c. 33,000 years ago, the oldest known human burial in Great Britain and the oldest known ceremonial burial in Western Europe. These extremely early dates combined with the giant skeletons found at Parc le Breos could be important in understanding the origins of the Welsh giants. This is a subject we will explore in the final chapter.

Aerial view of Cadair Idris in Gwynedd, Wales.

Gazetteer of Giants - Wales

DATE	LOCATION	SIZE	No.	FEATURES	PAGE
	Anglesey				
1780	Bryn Celli Ddu, Anglesey	Large	1	Neolithic burial chamber	200
1841	Llanlliana, Anglesey	7ft 6in	1	Found in stone coffin	201
1832	Holyhead, Anglesey	7ft - 11ft	3	Strong teeth & many graves	201
1795	Trainline from Aberconwy	7ft - 8ft	4	Huge jaws and teeth	202
1780	Llanedwen, Anglesey	Giant	1	Bones of giantess	203
1846	Perthi Duon Dolmen, Anglesey	60ft !	1	Found under dolmen	203
	Wales				
1841	Conwy, North Wales	9ft 3in	1	Near ancient copper mines	204
1823	Mold, Flintshire	Giant	1	Famous gold cape & ghosts!	206
1800s	Pen y Maen Carnedd, Nefyn, Caernarvonshire	Giant	1	Under carved monolith	207
1695	Vortigern Castle, Pistyll	13ft	1	Found in Vortigern Castle	207
1827	Giant's Slope, Merioneth	10ft +	1	Samson's resting place	208
1685	Cadair Idris, Gwynedd	7ft	2	Found with hazel rods	208
1800s	Castell y Bere, Gwynedd	Giant	1	Found under huge stone	210
1783	Rhayader Church, Powys	10ft	1	In local churchyard	210
1800s	Trevigan, Preseli Mountains	Giant	1	Huge sword discovered	212
1800s	Mynydd Pen Rhys, Port Talbot	9ft	1	Legend of giant Can Hir	212
1869	Parc Le Breos, Glamorgan	Giant	2	Neolithic burial chamber	212

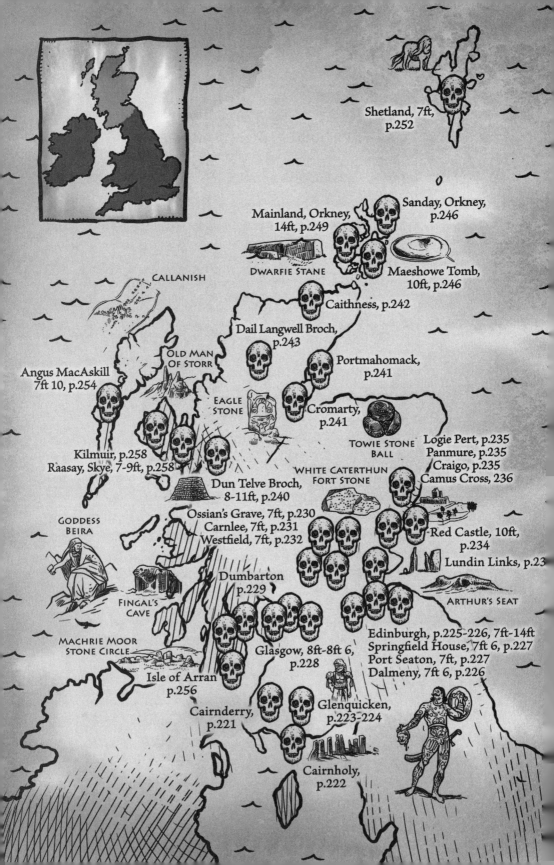

Shetland, 7ft, p.252

Mainland, Orkney, 14ft, p.249

Sanday, Orkney, p.246

DWARFIE STANE

Maeshowe Tomb, 10ft, p.246

CALLANISH

Caithness, p.242

Dail Langwell Broch, p.243

Portmahomack, p.241

OLD MAN OF STORR

Angus MacAskill 7ft 10, p.254

EAGLE STONE

Cromarty, p.241

TOWIE STONE BALL

Logie Pert, p.235
Panmure, p.235
Craigo, p.235
Camus Cross, 236

Kilmuir, p.258
Raasay, Skye, 7-9ft, p.258

WHITE CATERTHUN FORT STONE

Dun Telve Broch, 8-11ft, p.240

Ossian's Grave, 7ft, p.230
Carnlee, 7ft, p.231
Westfield, 7ft, p.232

GODDESS BEIRA

Red Castle, 10ft, p.234

Lundin Links, p.23

FINGAL'S CAVE

Dumbarton p.229

ARTHUR'S SEAT

MACHRIE MOOR STONE CIRCLE

Glasgow, 8ft-8ft 6, p.228

Edinburgh, p.225-226, 7ft-14ft
Springfield House, 7ft 6, p.227
Port Seaton, 7ft, p.227
Dalmeny, 7ft 6, p.226

Isle of Arran p.256

Cairnderry, p.221

Glenquicken, p.223-224

Cairnholy, p.222

9

CALEDONIA - ANCIENT SCOTLAND

Scotland is a land of endless beauty with folk memories of epic proportions, literally. Giant lore is woven into the landscape along with tales of mighty races with supernatural abilities who were said to have been the builders of the ancient stone monuments.

Pictish Warrior with fully-tattooed body.

It is believed the first hunter-gatherers arrived in Scotland around 12,800 years ago, as the ice sheet retreated after the last glaciation. There are also suggestions that Solutreans may have sailed up the western coast of Britain around 15,000 BC and settled in Orkney. These early people began building simple houses around 9500 years ago and the first villages around 6000 years ago, notably the Knap of Howar on Papa Westray in Orkney c.3600 BC, followed by other monuments such as the Ness of Brodgar in Orkney.

These megalith builders may have been the early ancestors of the *Picts* (Painted Ones), who became a confederation of tribal units remembered as fearsome

217

warriors who tattooed their entire bodies. It is thought they originated in Orkney, after arriving from Scythia (Iran and eastern Europe), and then slowly headed south, perhaps instigating the megalithic achievements detailed in this chapter. Historian Stuart McHardy states *"The Picts were in fact the indigenous population of this part of the world."*[1] They carved stunning symbols onto stone and often came together to fight outside enemies. They established themselves in small communities made up of families belonging to a single clan that was presided over by a tribal chief. These clans were known as Caerini, Cornavii, Lugi, Smertae, Decantae, Carnonacae, Caledonii, Selgovae and Votadini.[2]

The Picts, according to Roman historian Tacitus in 83 AD, were known as "Caledonians."[3] However, this was based upon one central Brythonic tribe, the *Caledoni*, but the name Caledonia became the romanticised name for Scotland. Tacitus records their physical characteristics as having "red hair" and "long limbs." Numerous battles with Roman Legions ensued in later times, and got to a point where Hadrian, in 122 AD, ordered the construction of a great wall running from the west to the east coast to keep them out.[4] Even today 10% of Scottish men are directly descended from the Picts/Caledonians.[5] McHardy writes: *"the tribal peoples of Pictish and Scottish origin combined to form the new political entity of Alba which in turn became Scotland."* [6]

THE EGYPTIAN ORIGINS OF SCOTLAND

The foundation of Scotland as a country may be linked to an unlikely source. A 15th century manuscript called the *Scotichronicon* by Walter Bower, details the arrival of an Egyptian Princess around 1350 BC, fleeing her homeland. She landed in the northern regions of Britain and founded an early colony in Scotland, before being pushed out by the local tribes.[7] She fled to Ireland and became part of the lineage of the royal line of Tara, founding the tribe of *Scotti*, who dominated parts of Ireland. A few hundred years later the Scotti tribes returned to Scotland to defeat the Picts and founded the Scottish nation. The very name, Scotland, Bower claimed, derived from these people. The manuscript makes the claim that the Scots predate the arrival of the Picts and Britons in Scotland. This version of Scottish history was accepted unhindered until the 18th century.

Lorraine Evans, in her book *Kingdom of the Ark*,[8] investigated these claims based upon various artefacts, boats and faience beads discovered in the British Isles, that appear to be of Egyptian origin. She also worked out that Scota was a daughter of the heretic Pharoah Akhenaten, who ruled in Egypt at around this time period. Scota was also referenced in the Irish *Book of Leinster* (c.1150 AD) as landing, with her fleet, on the coast of Ireland. Oddly, several artefacts of Egyptian origin appear to be related to some of the giant sites we feature in this book, and may also provide clues about other migrations to Britain in the distant past. This is a subject we will tackle in the final chapter.

The founders of Scotland of medieval legend, Scota with Goídel Glas, voyaging from Egypt, as depicted in a 15th-century manuscript the *Scotichronicon* by Walter Bower.

GIANT LORE OF SCOTLAND

Let us now look at some of the voluminous giant lore found in Scotland. We begin with renowned mythologist Donald Alexander Mackenzie (1873 - 1936) who helps us get a clear vision of this in his *Scottish Folk Lore and Folk Life*:

Donald A. Mackenzie.

> *Throughout Scotland there are a number of hills which, by reason of their situation rather than heights, might be referred to as positions of strategic importance. All of them command wide prospects, and from each other hill, or two hills of like character, may be visible. They have all distinctive names*

with folk-lore associations, and a number have archaeological relics or are remembered as beacon hills. Some are referred to as 'seats' or 'chairs' of giants, or of heroes or saints who supplanted giants, and are connected in folk-tales with megaliths, or ice-carried boulders, supposed to be flung by giants. As stated, these giants are in Gaelic referred to as the Famhairean (Fomorians). In the Gaelic Bible, Genesis, vi, 4 is rendered: Bha famhairean air an talem 'snag latish sin, 'there were giants on the earth in these days.' The Fomorians of Scotland occupy not only hills, but caves among the mountains, while some come from the sea and others have strongholds in islands.[9]

The Fomorian giants are more commonly connected with Ireland, but they also have a strong presence in the mythology of Scotland. The Fomorians were portrayed as wicked giants, warlike masters of black magic and survivors who landed on the shores of the British Isles after a great flood. One of these is a boulder-flinging giant who occupied *Dun-Fhomhair* (Fomorian's Hill) at Kilmorack, overlooking Beauly. Two famous Fomorian giants were said to face each other on eminences on opposite sides of Loch Ness. The Southern Giant, according to a local folk tale, became angry with the Northern Giant on Dun Binniligh, Abriachan, and hurled a huge black boulder at him. The Northern Giant retaliated by flinging a huge white boulder. These boulders can still be seen today. There are other stone-throwing giants in Nairnshire, Morayshire, Banffshire, Aberdeenshire, Kincardineshire, Angus, Fife, Perthshire and the Lothians also along the western coast of Scotland. Local history also speaks of mighty men of renown. This is found in the *History of Galloway* from 1841:

Nature seems to have been peculiarly profuse in its bounties to the Celtic nations. Their persons were large, robust, and well formed; and they excelled in running, wrestling, climbing, and swimming. Both history and tradition assert these facts; and from the writings of Tacitus they receive extensive corroboration. Strabo mentions the Britons as taller in stature than the Gauls and differing a little from them in the colour of their hair. 'For proof of their tallness,' he says, 'I myself saw very youths taller; by half a foot, then the tallest men.' Besides, in some of the sepulchral remains of the earliest inhabitants of North Britain, have been discovered human bones of a large size.[10]

THE SOUTH: DUMFRIES AND GALLOWAY

LEGEND OF THE BIG MAN

The council area of Dumfries and Galloway is located in the Western Southern Uplands of Scotland. This area is a hotbed for ancient supernatural traditions. One wonders what to make of tales of banshees, brownies, fairies, goblins, and kelpies. Is it some kind of strange reality or a superstitious people trying to make sense of the natural world? Legends of giants were recorded in Balmaghie, an ecclesiastical and civil parish in the historical county of Kirkcudbrightshire in Dumfries and Galloway. The story goes that three giants terrorised the inhabitants of Barstobrick but were eventually killed by a man named McGhee, who was granted lands for his good deed. Another tradition states that a family in Carsphairn were crushed to death when a giant on the hill of Dundeuch threw a boulder on their house.

Reminiscent of the giants Gog and Magog being paraded yearly through the streets of London, in 2010, Dumfries and Galloway had the 26-foot-tall giant called the *Big Man* walk down the streets for the local arts festival, controlled by 12 puppeteers. The story is as follows:

Dumfries and Galloway, 'Big Man Walking'.

> A creature of legend, the Big Man arrived many thousands of years ago, when Scotland was a rough country of vast forests and wild beasts… He helped the people, cleared the land, carved inlets for fishing boats, and he chased the nightmare beasts from the forests. The Big Man created a land in which people could live and thrive. Then he was gone.[11]

Gigantic bones found in Cairnderry were reported in the *Kirkcudbrightshire Ordinance Survey name books 1848-1851*:

...about four years ago a great quantity of the tumulus was removed, within which were found human bones which denoted the mighty dead to have been a person of gigantic stature, these relics are now in the possession of Rigby Wason Esquire, Corwar House, Ayrshire.[12]

STONE TOMBS OF GIANTS

Other ancient stone monuments in the area were also the burial places of giant men. This account describes two excavations, one at the Neolithic chambered tomb of Cairnholy:

Remains of Druidical circles are found in different places, and in 1778, while removing some stones from a tumulus, were discovered a coffin containing a skeleton of gigantic size, an urn inclosing ashes, and an earthen vessel for holding water. In 1809 was found a coffin of rude form, containing a skeleton of large size at Cairnholy.[13]

The name Cairnholy comes from the Gaelic *Càrn na h-ulaidhe* (cairn of the stone tomb).

The two cairns of Cairnholy were built in the 4th millennium BC. They are known as *Clyde Cairns*, a type of tomb characteristic to South West Scotland. Both tombs are now open to the sky, their covering stones taken long ago. Cairnholy I is the more elaborate of the two, while Cairnholy II is said to be the tomb of the mythical Scottish king Galdus, the chamber mentioned in the account above.

GLENQUICKEN MOOR STONE CIRCLE

This is one of the rare accounts of giants being directly related to stone circles, reported in 1875:

...in about the year 1778, in removing a quantity of stones for building dikes from a large tumulus in Glenquicken Moor, there was found a stone coffin, containing a human skeleton, which was greatly above the ordinary size. There was also found in this sepulchral monument an urn containing ashes and an earthen pitcher. This tumulus is called Cairnywanie. Thus, we have an account of two skeletons of very large size, found in Glenquicken Moor at different times. These facts seem to confirm the tradition that a battle had taken place here at some very remote period. [14]

Glenquicken stone circle. Two others were reported in the area, but are now lost.

Glenquicken has twenty-nine stones used in the construction with one at its centre and dates to 1000 BC. However, the noted archaeoastronomer Prof. Alexander Thom proposed that the orientation between this and a now destroyed circle nearby was aligned upon the rising of the star Antares in 1860 BC. Archaeologist Aubrey Burl said it is *"the finest of all centre stone circles."* [15]

BONES OF A MAN OF GIGANTIC SIZE

This report from 1853 discusses the demolished chapel of Riddell on the moors of Glenquicken. One of the strategies that the Christians used to convert the Pagans to the Catholic faith was to build churches over the ancient sacred sites:

Tradition carries their antiquity to a point extremely remote, and is in some degree sanctioned by the discovery of two stone coffins,- one containing an earthen pot filled with ashes and arms, bearing a legible date, A.D. 727, the other dated 936, and filled with the bones of a man of gigantic size. These coffins were discovered in the foundation of what was by then, long ceased to be, the chapel of Riddell. [16]

THE GIANT KING OF SCOTLAND

A final account from the area introduces one of the mysterious Scottish stone spheres. Described in the report is a three-inch diameter stone ball interred with the Scottish king Galdus that is described as *"perfectly round...and highly polished."* This may be one of the 425 stone spheres found all over Scotland:

> *Kirkmabreck: On removing a large cairn on the moor of Glenquicken, said to be a tomb of a King of Scotland (Aldus M'Galdus), the workmen came upon a stone coffin containing a skeleton. The coffin also contained a stone ball, about three inches in diameter, perfectly round and highly polished, and an arrowhead. In the same place another cairn contained an extraordinarily large skeleton.*[17]

The size of the ball indicates it is one of the Neolithic stone spheres found in the area of recumbent stone circles in Aberdeenshire and Orkney. Most were geometric but some were polished smooth and two examples like this have been found at Loughcrew in Ireland.

Eight of the geometric stone spheres on display at the Hunterian Museum in Glasgow. Photo Martin Morrison.

EDINBURGH, GLASGOW AND THE LOTHIANS

ARTHUR'S SEAT

Rising 822 feet above Edinburgh is the extinct volcano known as *Arthur's Seat*, providing panoramic views of the city and beyond. Its name is said to have been derived from the legendary King Arthur and the idea that this may have been the location of Camelot. In other parts of the British Isles, and after the Arthurian romances became popular, it was often the case that 'giant' sites

224

became 'Arthur' sites, and this may have also happened here. Arthur's Seat has been the mythical backdrop for many novels such Jules Verne's, *The Underground City* in which Nell, a young girl who is an inhabitant of the subterranean metropolis, is taken to Arthur's Seat to view her first sunrise. Folklorist Donald Mackenzie revealed stories of 'sleeping giants' under various hills in Scotland. Those under Arthur's Seat are said to be Arthur and his knights.

Arthur's Seat, Edinburgh.

STONE COFFINS YIELD GIANTS

Unsurprisingly, we have extracted from the local records numerous accounts of enormous bones that were unearthed and reported in the area. From the *London Magazine or Monthly Gentleman's Intelligencer,* 1838, we have several more finds associated with stone monuments:

> *Edinburgh Sept. 1st, at St. Fort, there are a number of round heaps of stones laid together in a regular and uniform manner, which have always been taken for sepulchral monuments or tombs: upon opening one of them lately, there was an entire skeleton of a human body, enclosed in a coffin of slate stones, the bottom was composed of a large, smooth slate... The bones measured seven feet in length and are certainly the remains of a very large man. A physician from Edinburgh, was coming there accidentally after the skeleton was interred, made open another of these tumuli and after digging about six feet came upon another stone coffin like the former but more regular and larger. The remains of some inscription plainly appeared but by no cleaning could be made legible. When the upper part of the coffin was removed, there appeared a skeleton lying all in order with the head to the east, as the other had been found. All the bones were in proper order, and of an ivory color, firm and*

225

no way porous. The length of the skeleton measured seven feet five inches. It does not appear from any records that the natives used this manner of burying.[18]

10FT TO 14FT TALL SCOTTISH GIANTS

Here is an account from Alexander Low's *History of Scotland,* 1826, that speaks of a royal burial from the 9th century:

Fyn Mac Coul, of Finnanus the son of Caelsus, who was of Scottish descent, and lived in the time of Eugenius II, one of the fabulous race of kings, was supposed to be an ancient giant seven cubits high.[19]

Seven cubits is around ten and a half feet tall. Author Hector Boethius (1465–1536) revealed an even taller discovery reminiscent of the giants of old:

Hector Boethius in his History of Scotland (1526), relates, that the bones of a man, ironically called Little John, are still preserved, who was supposed to have been 14 feet high.[20]

Add to that another claimed historical giant from the *Encyclopedia Britannica,* 1823:

Funman, a Scotsman, who lived in the time of Eugene the second, King of Scotland, measured 11 feet and a half.[21]

TWO GIANTS DISCOVERED IN DALEMY

The information in these next two accounts was provided by the *Ordnance Survey Named Books Collection of Scotland.* The survey was carried out in the mid 1800's across all parts of the country. Massive volumes of material were collected as mapping and information gathering occurred. An officer in the Ordnance Survey, usually a Royal Engineer or Surveyor, would consult local historians and etymologists to gather all relevant subject matter. Within these voluminous records are, not to our surprise, numerous reports of the unearthing of giant skeletons. This information gathering was carried out by professionals well skilled in the art of observation, objectivity and discernment. We highlight several of these enormous exhumations in this chapter.

A 7ft 6in and 7ft 6½in skeleton were recorded in the *West Lothian Ordnance Survey named books, 1855-1859*:

226

This mound or tumulus has long since been removed. At the east side of the tumulus was discovered a human skeleton of gigantic dimensions At 7 ft 6½ Inches and enclosed by stone flags on each side and on top by a large granite Slab.[22]

And:

When improving and levelling the grounds surrounding Springfield House, some years ago a great quantity of human remains were discovered...a complete Skeleton was found, which measured 7ft 6 in length, and the breadth across the shoulders was nearly twice that of an ordinary man of the present age.[23]

In nearby East Lothian we find the following:

A long barrow opened in the neighborhood of Port Seaton, East-Lothian, in 1833, a skeleton was found laid at full length within a rude cist. It indicated the remains of a man nearly seven feet high, but the bones crumbled to dust soon after exposure to the air.[24]

Traprain Law in East Lothian, where hundreds of roundhouses of the Votadini tribe were located.

The Romans identified nine major tribes after arriving in Scotland in 79 AD. Ptolemy did much of the documenting as he was working on his map of the world at the time. The fierce Votadini tribe had a large settlement around Traprain Law in East Lothian, a large volcanic hill, with territory throughout the Lothians. Might some of the skeletons recorded here be giant warriors from their ranks?

From the *Westmorland Gazette*, November, 27th 1819, we have this discovery:

On Thursday, as some weavers who are employed in cultivating the moss of Paisley were cutting a deep drain, they found the bones of a human skeleton of extraordinary size. All of the small bones could not be found; but an idea of the actual stature of the individual to whom they belonged may be formed, from the thigh

227

bone measuring two feet three inches and a half long. At the same time was found a piece of iron very much corroded with rust, about seven inches long, which seems evidently, from its tapering nearly to a point at one end, and being cleft at the other for fitting on a shaft, to have been the head of a spear. The bones are hard, and of a deep black colour, and when rubbed a little assume a very polished appearance. The whole has been presented to the Paisley Philosophical Institution and is at present deposited there.

A 27 ½ inch femur would make the owner easily over eight feet tall.

A GIANT COFFIN AND A GIANT BODY ROBBED

Here we have a case from Glasgow where the finder speculates that the remains of the skeleton found justified the 8½ foot coffin it rested in. This discovery was made in 1792 near the ruined St Bride's Chapel on raised ground called the *Earl's Know*:

Many human bones of an ordinary size were found, and, moreover, fragments of a human skull, and of a lower jaw bone, with the case of teeth, which were perfectly sound, and fragments of thigh bones; these were all of an enormous size, and afforded a convincing proof that the body buried there had required a grave of the dimensions specified.[25]

Also, in Glasgow, we have a bizarre story from the book, *John Kobler: The Reluctant Surgeon*. Kobler and his family were from the area around Glasgow and he went on to prominent fame, with strange associations, as is shown here:

He was a century ahead of his time; one of the most renowned men the English nation has produced. He was the greatest dissector and collector of anatomical specimens in history. His association with the rougher element, with whom he drank bottle for bottle, ensured him a supply of corpses, through body snatching, including the skeleton of a man eight and a half feet tall. The Hunterian Museum, which contained 13,682 of his original specimens, was bombed by Hitler's forces in 1941, and all but 3,000 of these were destroyed.[26]

We never thought we would get to blame the loss of a giant skeleton on the Nazis.

THE MIDDLE COUNTRY: PERTH, ARGYLL AND KINROSS

SAMSON THE GIANT STONE THROWER

A local folk-story tells us that the strongest giant in Scotland lived on the mountain of Ben Ledi, near Callander. He challenged all the Scottish Fomorians to a trial of strength called 'putting the stone' and was the winner in the contest. A large boulder called *Samson's Putting-Stone* lies embedded in the lower eastern slope of the Ben, and is said to have been flung from the summit by Samson. Not far from the stone is a hill fort. Many of the giants are referred to simply as Fomorians, but some have been given personal names or are referred to as 'devils'.

The Ben Ledi giant evidently acquired his name after the introduction of Christianity, having been compared to Samson. The giant of Norman's Law in Fife was known as the 'devil'. He hurled a boulder across the Tay against the giant of Law Hill, Dundee, but it fell short. This boulder is known as the *De'il's (Devil's) Stane*, and is protected by an iron railing.

Left: The Samson Stone. Right: Ben Ledi Mountain.

THE GIANT OF ARGYLL

Donald MacKenzie shared the tale of an Argyll giant who *"was tall as an oak tree and very strong."* This giant was noted to have killed many other giants by throwing boulders on them. After he stole a pot of gold from the castle of the King of Ardnamurchan, he met his demise with an arrow through the forehead shot by one of the king's bravest men.

A report of numerous gigantic skulls and bones from Argyll featured in *The Riddle of Prehistoric Britain,* by Comyns Beaumont:

Some years ago, the late J.J. Bell, the Scots historian and novelist, who lived near Dumbarton, told me that after a violent storm along the Argyllshire coast he was shown a quantity of giants' skulls and bones of considerable size exposed to view by a fall of cliff. They also crumbled to dust.[27]

Giant Legends of Finn and Ossian

Another local legend states that near the village of Kinloch Rannoch in Perth and Kinross is a hill reputed to resemble the head, shoulders, and torso of a man. It has been given the name of *The Sleeping Giant*. Local myth says that the giant will wake up only when he hears the sound of his master's flute. In the same area we have the stories of Ossian (or Oisin) who was the giant son of the legendary giant Fionn mac Cumhaill (anglicised to Finn McCool).

Ossian's Stone, purported to be his grave site, is located in Perthshire and was moved from its original location by a General Wade and his men who were creating military roads in Scotland in the early 1700s. The enormous stone was a cap stone for a cist or burial chamber that it once covered there. This curious giant's grave was lucky to have survived as a drawing from 1834 coupled with several other early photographs and descriptions of the place, which clearly showed other antiquarian remains and a circular ring of stones surrounding it.

Left: Old photograph of Ossian's Grave Stone.
Right: Antiquarian Fred Cole's 1911 plan of the stone and surrounding ring.

Here is a summary of the events surrounding the discovery from Otta Swire's *The Highlands and their Legends*:

Perthshire says that he (Ossian) was buried just like any other man of his rank and time, in a stone coffin in the Sma Glen, and that his bones would be resting still but for General Wade. When Wade's men were preparing the foundations of a road to the North, they came upon an old stone cast or coffin of great age. This was opened and seen to contain a very large skeleton. But the men of the Highland clans who were working on the road became much upset when they saw what had been done, and explained that these were the bones of the great Ossian and should not have been disturbed nor his grave opened. The consequences might be appalling. Indeed, so angry and frightened were they that the young engineer in charge of that section thought it unwise to proceed without new and exact orders, and so sent a message to General Wade.

In the night all the bones and the heavy stone coffin which had held them disappeared and the next day the workmen denied all knowledge of them. It appeared that none of them had ever seen a coffin or bones, nor heard of Ossian. In fact, no one knew anything about anything. But some soldiers in camp nearby had sent sentries out and one of these had reported large numbers of men on the move. The sentry had claimed to have seen a long procession of torches moving through the gloom of the mist and to have heard wild, unearthly chanting sweeping by on the wind. Where Ossian's bones, if his they were, now lie, no one knows. But it does appear from the large skeletons and old songs and stories that there must once have been in the Highlands a tribe or clan of men larger and perhaps more civilized than their neighbors. Traditionally the Feinn were fair and very tall, not quite 'Giants' in the fairy tale sense but larger than men, say 9 to 10 feet in height; 'his fault is the fault of Fionn' was said in Gaelic of a little man, because Fionn's only fault is believed to have been his small size. He was smaller than his men, only 8 feet or so some claim.[28]

STONE CAIRNS AND GIANT STONE COFFINS

Within the confines of the Perth and Kinross area are found many ancient monuments such as the *Croft Moraig Stone Circle* (2000 BC) and the *Dundurn Pictish Fort* (500 AD to 800 AD). In this next account from 1859, James Taylor shares his astonishment at exploring a truly massive stone mound and discovering an amazing skeleton:

On the hill above the moor of Ardoch...are two great heaps of stones: the one called Carn-wichel, the other Cairnlee. The former is the greatest curiosity of this kind that I have ever met with. The quantity of great rough stones lying above one another almost surpasses belief, which made me have the curiosity to measure it; and I found the whole heap to be about one hundred and eighty-two feet in length, thirty in sloping height and forty-five in breadth at the bottom. In this cairn there has been found a stone coffin, containing a skeleton seven feet long.[29]

An earlier version of this account is found in James Browne's, *A History of the Highlands and the Highland Clans*:

In a large oblong cairn, about a mile west from Ardoch, in Perthshire, there was found a stone coffin, containing a human skeleton seven feet long.[30]

Continuing with another stone coffin in a cairn we have this from *Fife and Kinross-shire Ordinance Survey name books, 1853-1855*:

When that large cairn which Stood in Westfield policies, a little to the right of the house was removed about 30 years ago a vast number of Urns full of ashes were found which the workmen called Cans and many of them finely Carved, some more coarsely than others. There was also a Stone Coffin found a little South from the Cairns when digging a ditch for a Stripe of planting on the roadside, containing a very large Skeleton at full length with uncommonly large teeth, but when it was exposed to the atmosphere or touched it Crumbled down into dust.[31]

DRUIDIC TEMPLES: A STONE CIRCLE AND A GIANT

The three massive standing stones described in the next report are the Lundin Links megaliths which stand between fourteen and seventeen feet tall. They are preserved as part of a modern golf course. The stones were raised in roughly 2000 BC, are made of sandstone, and legend holds that the site was used by the Druids for ancient rituals. This report is also from the *Fife and Kinross-shire Ordinance Survey name books*:

Three upright stones situated in a triangular form, they are of red sand stone and said to be from a rock on Largo beach, bearing no trace of any inscription; being rough unhewn blocks. In the days of Sibbald a fourth stone of equal magnitude stood near the present ones. Mr. James Wilson who formerly held the adjoining lands says he frequently raised stone Coffins near them when ploughing, containing urns filled with Calcined bones and one holding the complete Skeleton of a man of enormous size, the skull of whom is at present either in Largo House or in the Museum of the United College of St. Andrews.[32]

Lundin Links Standing Stones. Photo H. Newman.

ANCIENT SEVEN FOOTERS IN ABUNDANCE

We finish in the region with two more reports of seven-foot-tall skeletons discovered in stone coffins from *Perthshire Ordinance Survey Name Location, 1859-1862, Volume 6:*

> *Many stone coffins have been found at different times in digging about the Camps or near them; and the skeletons contained in them have been of an uncommon size. About a mile west from the camps, a stone coffin was found containing a skeleton seven feet long. A mile and a half distant in the Muir of Ochill, another was found of the same length, in Cairn Wochill. These have generally been in cairns or heaps of stones, which may be accounted for, from a practice in former days of throwing a stone upon the respected dead, or upon any place remarkable, in passing by. So, among the Highlanders there is still a saying that of one shall do a favor to another, a 'stone shall be added to his cairn;' - that is, his grave shall be remembered and respected.*[33]

THE MIDDLE COUNTRY: ANGUS
A GIANTESS HELPS BUILD FORTS

Angus was a county known officially as Forfarshire from the 18th century until 1928, bordering Kincardineshire to the north-east, Aberdeenshire to the north and Perthshire to the west. The area that now comprises Angus has been occupied since at least the Neolithic period.

There are several notable giant legends to be found in the area. Situated up the Angus Glens are the Caterthun Forts north of Brechin. The White and Brown Caterthuns are examples of ancient drystone and turf forts respectively. There is a tradition that the walls were built with the assistance of a giantess, who strode up the hill along with the materials gathered in her apron. Her apron strings broke one day and a particularly large stone plummeted to the ground just outside the walls.[34] The boulder can still be viewed just to the north west of the fort and is a fine example of a Bronze Age cup-marked rock.

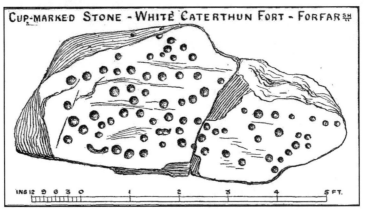

The cup-marked stone from White Caterthun Fort, c.1882.

A TEN-FOOT GIANT AND A THREE-FOOT DWARF

Red Castle is an impressive stronghold that was built by William the Lion and then later granted to Sir Walter de Berkeley in 1194. It is situated on a prominent spot near the sea in Angus, North East Scotland. Legend has it that Sir Walter employed two devoted

Illustration of Red Castle.

234

servants, a ten-foot giant called Daniel Cajanus and a three-foot-tall Swedish dwarf called Linicus Calvus. It is recorded that Daniel died defending the castle against a Viking raid and Linicus died soon after of a broken heart because of the loss of his close friend.

TUMULI, STONE COFFINS AND GIANTS

We have several reports of enormous skeletons found in the historical records of the area:

> In a tumulus opened in the parish of Logie Pert, stone-coffins were found containing skeletons of gigantic dimensions. On some of the uncultivated Grampian Moors are vestiges of the ancient Caledonian dwellings, consisting of large slab stones planted together in a circle without cement.[35]

Next, from the *Forfarshire (Angus) Ordinance Survey name book 1857-1861*:

> About the year 1620 the Tumulus was opened by order of Sir Patrick Maule of Panmure 'Afterwards the first Earl' in the presence of a number of gentlemen, when a skeleton of gigantic dimensions, in good preservation was found, nothing being imperfect.[36]

Also, we have this from the same area:

> The only antiquities in the Parish are the three tumuli, on three laws of Craigo, mentioned in the former Statistical Account, and situated nearly a mile west of the house of Craigo. Two of these tumuli have been opened, and in one of them was found a stone coffin, containing a human skeleton almost entire, the bones of an extraordinary size, of a deep-yellow color, and very brittle. In the other tumulus opened, there were found, about a foot from the surface, four human skeletons of gigantic proportions, and near to these, a beautiful black ring like ebony, of a fine polish, and in perfect preservation. The ring was 12 inches in circumference, and 4 in diameter, flat in the inside, rounded without, and capable of fitting a large wrist. In the same tumulus was found an urn full of ashes. From the discovery of these skeletons of extraordinary size, both here and in other parts of Scotland, some confirmation would seem to be given to the ancient tradition, that at one time there was a race of giants in this country.[37]

Yet again, we have more reports of giants, stone coffins, and this time, a beautifully crafted ebony ring showing evidence of sophisticated workmanship. Who could these people have been? It may have been the *Venicones*, one of nine tribes encountered by the Romans who lived around the River Tay. They were one of the few groups in Northern Britain at this time known to bury their dead in stone graves.

IN THOSE DAYS THERE WERE GIANTS HE SAID

This detailed entry is sure to raise a few eyebrows, sourced from *Historic Scenes in Forfarshire*, by William Marshall, 1875:

> *And certainly, it is most remarkable that about 1620, the then Lord Panmure opened the tumulus at Camus Cross, in the presence of a number of gentlemen, when a skeleton of gigantic dimensions, in good preservation, was discovered, nothing being imperfect but the skull, a part of which was wanting. Buchanan farther says, 'To this day, when the wind raises the sand at Balbride, many bones are uncovered, of larger dimensions that can well agree with the stature of men of those times.'...We may observe that in those days there were giants-some among the native population, and more among their hostile invaders so that references to gigantic skeletons, gigantic tombs etc.. are not to be sneered at as fabulous. The fact is so well authenticated that it requires a great deal more credulity to disbelieve then to believe it.*[38]

THE BIG GREY MAN OF BEN MACDUI

Am Fear Liath Mor is Scottish Gaelic for *Big Grey Man of Macdhui*, the name of a giant ghostly creature (pictured on left) who is said to haunt the summit and passes of Ben Macdui. The mountain is located in Aberdeenshire and is the second highest in the United Kingdom at 4,295ft. There have been numerous sightings of this creature described as being covered with short hair. He had a humanoid appearance and most accounts have him standing over ten feet tall. Nearly all the incidents include the sound of footsteps crunching in the gravel just out of

Ben Macdui as seen from Carn Liath.

sight. Footprints nineteen inches long and fourteen inches wide have also been reported, but the first recorded sighting of the creature was from poet James Hogg in 1791, who fled home in panic leaving the sheep he was tending and described it as 30ft tall:

> It was a giant... at least thirty feet high, and equally proportioned, and very near me. I was actually struck powerless with astonishment and terror.[39]

THE NORTH: INVERNESS-SHIRE/ ROSS AND CROMARTY
MOUNTAIN OF THE GIANTESS

Ben Nevis is the highest mountain in the British Isles. Standing at 4,411ft above sea level, it is at the western end of the Grampian Mountains in the Lochaber area of the Scottish Highlands. It is also the mythological home of a giantess. In Scottish mythology, the giant goddess Beira built the mountains of Scotland with her magic hammer, and is described here by Donald Alexander MacKenzie in 1917:

> Dark Beira was the mother of all the gods and goddesses in Scotland. She was of great height and very old, and everyone feared her. When roused to anger she was as fierce as the biting north wind and harsh as the tempest-stricken sea. Each winter she reigned as Queen of the Four Red Divisions of the world, and none disputed her sway. But when the sweet spring season drew nigh, her subjects began to rebel against her and to long for the coming

237

of the Summer King, Angus of the White Steed, and Bride, his beautiful queen, who were loved by all, for they were the bringers of plenty and of bright and happy days. It enraged Beira greatly to find her power passing away, and she tried her utmost to prolong the winter season by raising spring storms and sending blighting frost to kill early flowers and keep the grass from growing.[40]

Drawing of Beira by John Duncan, 1917.

BUILT BY GIANTS IN A SINGLE NIGHT

Historian Otta Swire shares a few interesting folk-tales of giants in *Highlands and their Legends,* 1963.[41] In this account she talks about the *Cave of Raitts,* a Souterrain (an underground stone-lined tunnel) near Lynchat which is believed to be dated to the later Iron Age (around 100-400 AD). It has a stone built underground chamber in a horseshoe shape, first excavated in 1835:

> *Traditionally, it was built by a giant race in a single night. The giantess dug it out and carried the earth and rubbish to the River Spey in their aprons. There they dumped it. Meanwhile the giantess was quarrying stone in the distant hills. As soon as the excavating was finished, they carried down the stones and began to build. The whole was finished before daylight.*

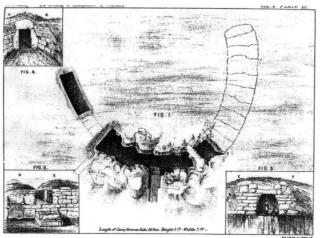

UNDERGROUND BUILDING DISCOVERED AT BELLEVILLE
PARISH OF ALVEY INVERNESS-SHIRE

STONE THROWING GIANTS AND THE EAGLE STONE

Swire shares another giant tale from the same book:

> *Near the old church of Fodderty, is a standing stone known as the Eagle Stone. Once there lived in the hills nearby a Giant of remarkable strength who loved 'putting the stone'; indeed, he is reputed to have been the originator of that sport. In those days some of the Feinn lived on Knockfarrel and the giant challenged Fionn to a stone throwing competition. Fionn refused but offered to match his*

The Eagle Stone.

239

dwarf against the 'Stoneputter' as he was called. The giant agreed to this. When he and his dwarf took station on Knockfarrel the giant picked up the eagle stone, which three men could not lift, and flung it across the strath on to this green knoll, where it stuck fast. In proof of the truth of this story the marks of an enormous finger and thumb are shown still visible on one of the stones.

Donald Mackenzie shares more folk tales of giants throwing big stones:

Two boulders lying on the beach near Cromarty were reputed to have been flung by the giant on the hill of Struie (called Gilltrax in Cromarty) in the parish of Edderton. There are giants on three hills near Inverness, known as Torvean, Dunain and Craig Phadrick. They are said to throw a stone hammer from one to the other each morning.[42]

Again, we have a veiled reference to geomancy and the landscape, with the hammer-throwing giants forming a triangle between different hill-tops.

GIANTS FINGERPRINTS

The Cromarty legends are filled with descriptions of stone throwing giants. One stone was hurled by a local titan across Dornoch Firth, and another, yet larger, landed a few miles from Dingwall which bears the impression of the giant's finger and thumb. The stone still rests today in its landing spot.

The stone thrown across Dornoch Firth with giant's fingerprints.

THE FIELD OF THE BIG MEN

This account brings to life the mythic era of giants and is one of the most fascinating reports of giant skeletons uncovered in Scotland to date. The entire story can be found in *Popular Tales of the West Highlands*, 1893.[43] Here we have the shorter version from Otta Swire:

Kylerhea in Skye and Glenelg on the mainland were much beloved of the Fein and they were believed to have lived in Glenelg. At a place called Imir nam Fear Mora, Field of the Big Men, in Glenelg, there were some large burial mounds, traditionally the graves of the Fienn. Macpherson's Ossian and the controversy to which it gave rise resulted in a revival of interest in such matters and one day in the early nineteenth century two of these grave mounds were opened in the presence of 'a number of local gentlemen of repute' including a parish minister. In them are said to have been found stone coffins enclosing skeletons of two men and a woman, all considerably larger than life size as it is known today. It is believed that a doctor examined their heights when alive as having been, one approximately 8 ½ feet and the other almost 11 feet and said that they appeared strong-boned and well proportioned.[44]

The Dun Telve Broch in Glenelg is over 2000 years old.

The witnesses seem to have been going out of their way to point out the enormity of the finds. They were not looking at skinny tall skeletons but enormously thick bones and monstrous skulls.

A GIANT SKULL AND A CASTLE GIANT

A massive skull is described in the following matter-of-fact account from the *Ross and Cromarty Ordinance survey name books 1848-1852*:

> *Human remains found here A.D. 1811. At the removal of this Cairn human bones of a gigantic size were found here, among the rest, a skull sufficiently capacious, according to the description of one of the laborer's, to contain, 'two leppies of beer.'* [45]

More giants in stone coffins from Cromarty reported in 1851:

> *Near the village of Portmahomack, on an eminence called Chapel Hill, a number of human bones have been found in rude coffins of flagstones, and, in the vicinity, several stone chests, each containing an entire skeleton of unusually large size.*[46]

THE NORTH: CAITHNESS AND SUTHERLAND

Caithness and Sutherland are counties in the far north of Scotland. The landscape is rich with the remains of prehistoric occupation. These include the well-known Grey Cairns of Camster, the Stone Lud, the Hill O Many Stones, a complex of sites around Loch Yarrows and over 100 brochs. It is also rich in giant lore and giant discoveries.

The chambered Cairn of Camster in Caithness.

THE CAITHNESS GIANTS

The many ancient stone monuments of Caithness have yielded some strange finds. From the *Caithness Ordinance Survey name books, 1871-1873* is reported an excavation of a stone cairn:

> *Situated on the very summit of Sithean Dubh it (the stone mound) is about 10 feet high, covered with loose Stones, two skeletons of gigantic size (human) were taken out of this place, on or about the year 1831, and were sent to an Antiquarian Society in London, by the late Captain McDonald of Sandside Reay.*[47]

The 'Stone Mound' on the summit of Sithean Dubh .

Also, from the same publication on pg. 71 it reads:

> *I may state that the skeleton and coffin or cist were of large dimensions, being upwards of 7 feet long.*

Curiously, we also have an account of a living giant recorded at nine feet five inches in the historical record of the immediate area:

> *Then there was the Caithness Giant, William Sutherland, commonly called William More, who was born about the end of the fourteenth century, who measured nine feet five inches.*[48]

A GIGANTIC HUMAN SKULL

Let us finish with a fascinating tale of a giant skull found in an ancient broch reported in 1893. We have also included the original news report overleaf:

> *A gentleman traveling recently through Glencassley in the parish of Croich, Ross-shire had his attention drawn to a large cairn which was said to have belonged to the giants which inhabited this glen at the beginning of the Christian era. He was also informed that about twenty years ago a giant's skull was found in the cairn which the inhabitants of Glencassley used as a tar pot, and which was capable of containing about four gallons. Having asked through the People's Journal for information regarding the famous tar pot, a local archaeologist writes as follows: 'I first heard of this unique though gruesome tar pot quite 35 years ago, my informant being a shepherd who had thus used the gigantic skull in question a few years previously. Since I came to reside in the parish of Cavick a few years ago I have made inquiries, the result of which has fully verified this Shepherd's story. The skull, which had been found in the ruins of a broch or Pictish tower on the banks of Cassley, was undoubtedly used by the shepherds of Glencassley as a receptacle for tar for the purposes of marking sheep with the owner's mark. I cannot say what the capacity of the skull in gallons might have been, but it was large enough to go over the head down to the ears of the largest bearded man in that part of the country, and the bone was half an inch in thickness. Having been reminded of the unseemliness of using a human skull as a tar pot the shepherds replaced it in the broch. The broch was, however, shortly*

afterwards explored by a gentleman who is described as an English antiquary, and as the skull has not been seen since it is conjectured that he took possession of it. Another correspondent who signs "Anaeus," says; A shepherd who lived at Croich had the famous tar pot in use for a number of years. I quote the following from a magazine in my possession: 'In this area a human skull of extraordinary dimensions was discovered about 40 years ago. A shepherd named Paul Irvin used it for a long time as a tarpot. Mr. Irvin, who frequently handled the skull, solemnly declares that it was capable of holding half a gallon of the tar and that the bones were of an enormous thickness.' [49]

A GIGANTIC HUMAN SKULL.

A gentleman travelling recently through Glen-easley, in the parish of Croich, Ross-shire, had his attention drawn to a large cairn which was said to have belonged to the giants which inhabited this glen at the beginning of the Christian era. He was also informed that about twenty years ago a giant's skull was found in the cairn, which the inhabitants of Glenmasley used as a tarpot, and which was capable of containing about four gallons. Having asked through the *People's Journal* for information regarding the famous tarpot, "A Local Archaeologist" writes as follows :— I first heard of this unique though gruesome tarpot quite 55 years ago, my informant being a shepherd who had thus used the gigantic skull in question a few years previously. Since I came to reside in the parish of Croich a few years ago I have made inquiries, the result of which has FULLY VERIFIED THIS SHEPHERD'S STORY. The skull, which had been found in the ruins of a broch or Pictish tower on the banks of the Cassley, was undoubtedly used by the shepherds of Glenmasley as a receptacle for tar for the purpose of "buisting," or marking, sheep with the owner's mark. I cannot say what the capacity of the skull in gallons might have been, but it was large enough to go over the head down to the ears of the largest headed man in that part of the country; and the bone was half an inch in thickness. Having been remonstrated with on the unseemliness of using a human skull as a tarpot the shepherds...

ORKNEY AND THE ISLANDS

Beyond Britannia, where the endless ocean opens, lies Orkney.
- Orosius, 5th century AD

The Orkney Islands are an archipelago between the northern tip of Scotland and the Shetland Islands. It consists of some seventy islands, with about twenty of them currently inhabited. The earliest known permanent settlement is the *Knap of Howar*, a Neolithic settlement on the island of Papa Westray, which dates from 3700 BC. The village of *Skara Brae*, Europe's best-preserved Neolithic village, is believed to have been inhabited from around 3100 BC. Other sites include the *Standing Stones of Stenness*, the *Maeshowe* passage grave, the *Ness of Brodgar*, and the *Ring of Brodgar* which was built in 2500 BC.

The Ring of Brodgar, with the Ness of Brodgar on the thin strip of land, and behind that the Stones of Stenness and Maeshowe. Photo H. Newman.

THE GIANT CUBBIE ROO

Cubbie Roo is Orkney's best-known giant. Like his contemporaries throughout the British Isles, he is intimately connected to megaliths and ancient construction. His permanent residence was the island of Wyre but his exploits took him throughout the isles. This is shown at *Cubbie Roo's Burden* (a chambered cairn) and the *Cubbie Roo Stone*, both on Rousay.

Rousay tradition has it that Cubbie Roo once proposed to build a bridge between Rousay and Wyre, and for that purpose he was carrying a cubbie-load of stones when the fettle or shoulder strap broke at this spot, and the mound of stones represents his fallen burden. Actually, it is a pre-historic chambered burial cairn which had collapsed in upon itself. In Rousay, by the Sourin shore is the monolith known as *The Fingersteen*. This stone is said to be marked with Cubbie's fingermarks, when he threw it from Westray. According to tradition, he was aiming for another giant on Kierfea Hill, but missed. The *Cubbie Roo Stone*, lying in the Evie hills, also bears the marks of Cubbie's fingers. This is in addition to two other stones on Rousay, one at Marlayer, the other on the hill near Onziebust. Meanwhile, a boulder lying in Stronsay was also said to have been hurled by Cubbie, this time in an attempt to wake a brother giant who was too slow to rise in the morning.

Midhowe Broch, Rousay, 200 BC and 200 AD. Photo H. Newman.

MEN OF GIGANTIC SIZE

This short passage from 1875 gives an introduction to the folk tales of Orkney, as well as some possible living and deceased remains of giants.

> *Among the Orkneys, at the opposite extremity of the island, but equally within the range of Ossianic narrative, both as to time and place, discoveries of the same sort are recorded, -According to Brand, gigantic men inhabited these islands. Great bones, compared in size to those as a horse, backbone and thigh, have been found in Sanday, and one giant, in later days there, 'was of so tall a stature that he could have stood upon the ground and put the capstone upon the chapel which no man now living by far could do.' Another of these giants, as remarkable apparently for strength as for stature, 'died not long since (1701), whom for his height, they commonly called meikle man of Waes.'*[50]

10FT GIANTS AND MUMMIES AT MAESHOWE?

Maeshowe is a Neolithic chambered cairn located on the mainland of the Island of Orkney, Scotland. Believed to have been constructed in 2800 BC, the mound is 115 feet in diameter and rises to 24 feet in height and has a complex of passages and chambers built of carefully crafted flagstone slabs weighing up to 30 tons apiece. Like Newgrange in Ireland it has a winter solstice alignment and is covered in an earthen mound. According to the *Orkneyinga*, (a 14th century manuscript) Maeshowe was looted by the famous Vikings Earl Harald

246

Maddadarson and Ragnvald, Earl of More in about the 12th century. The modern opening of the tomb was by the antiquarian James Farrer in July 1861. He and his workmen discovered the famous runic inscriptions carved on the walls which proved that Norsemen had broken into the tomb at least six centuries earlier. *The Orcadian Newspaper*, July 20, 1861, reported some other gargantuan finds:

Maeshowe Tomb. Photo H. Newman.

> *Ever increasing interest continues to be felt here in the excavations at the tumulus of Maeshowe, and no wonder, for it is reported in the beginning of the week that two female mummies had been discovered, and also the skeleton of a gentleman ten feet long.*

The report goes on to state that *"no-one as far as I have heard, has been favoured with a sight of the lady mummies or the long gentleman."*[51] We were able to locate a rare personal excavation copy of Farrer's work at Maeshowe entitled *Notice of runic inscriptions discovered during recent excavations in the Orkneys,* published in 1862. The report gives no mention of these finds and leaves us concluding that the mummies and the giant ten-foot-tall skeleton may either have been

NOTICE

OF

RUNIC INSCRIPTIONS

DISCOVERED DURING RECENT

EXCAVATIONS IN THE ORKNEYS

MADE BY

JAMES FARRER, M.P.

PRINTED FOR PRIVATE CIRCULATION
1862

Left: Interior of Maeshowe when first excavated. Right: Front cover of James Farrer's personal excavation report. Overpage: Artist's impression of the discovery.

whisked away into a private collection, or it was simply exaggerated by the local newspaper.

The three side chambers were thought to have contained skeletons, although the account is unclear on this. But what is certain is that the bones were regularly taken out and presented to the community as a reverence to their ancestors, or as part of initiations.[52] The bones even got mixed up between individuals which was recorded by DNA analysis.

The mention of two mummies next to a giant skeleton seemed initially like a 'tall story' concocted by someone's overactive imagination. However, mummified remains have been found not too far from Orkney (and in other parts of the British Isles) dating from around 2700 BC.[53] In 2001, two mummified Bronze Age bodies were found buried under a prehistoric house at Cladh Hallan on the Hebridean Island of South Uist, not too far from Callanish Standing Stones. These dated to 1440 - 1130 BC. It was later revealed that the bones had not been immediately interred after death, so might have been display pieces, like those of Maeshowe. Astonishingly, the bodies were made up of six different individuals to look like one skeleton. The mummification process included placing the bodies in bogs for about a year before they were recovered.[54]

14-FOOT-TALL MAN FOUND IN HILL NEAR A LAKE

Another document of interest, this time written in 1529, Jo Ben's *Descripto Insuluarum Orchadiarum* (Descriptions of Orkney) reports that the bones of a 14-foot-tall man were found in a tomb *"on a little hill near the lake."*[55] Whether this is the same account from Maeshowe or another remarkable find is unclear, but both descriptions seem quite insistent that skeletons of spectacular dimensions were found.

THE STONE GIANTS

Getting turned to stone is a regular occurrence in British folklore that is also found on Orkney, often associated with giants:

> *A specific type of legend has become attached to a number of Orkney's solitary standing stones. This legend dictates that once a year, usually New Year, these stones - said to be transformed giants - move from their resting place to nearby bodies of water where they dip their heads down and 'drink' the water. The best known*

of these stones are the Yetnasteen and the Stane O' Quoybune, but the giant connection includes even the Ring of Brodgar. The motif of the petrified giant is clear when it comes to the Yetnasteen on the island of Rousay. This monolith takes its name from the Old Norse 'Jotunna-steinn' and simply means 'Giant's Stone.'[56]

The recurrent theme here is that giants are perpetually linked to megaliths, alignments and landscape features.

THE DWARFIE STANE

The Dwarfie Stane is a megalithic chambered tomb carved out of a massive block of Devonian Old Red Sandstone located on the island of Hoy. It is the only Neolithic rock cut tomb in Britain, being 28ft long, 13ft wide and 8ft tall and is thought to be at least 4000 years old. It was looted by making an opening in the roof of the chamber. In legend, it was related to both giants and dwarfs.

The Dwarfie Stane on Hoy, Orkney.

It was popularized by Walter Scott's novel *The Pirate* published in 1821 who revealed the local giant lore of the site:

> *The legend in the late sixteenth century was that one giant was imprisoned here by another and gnawed his way out through the roof, though when Martin Martin visited the site around 100 years later he heard the tradition that a giant couple had found shelter there... within this stone there is cut out a bed and pillow, capable of two persons to lie in. At the other opposite end, there is a void space cut out resembling a bed, and above both these there is a large hole, which is suppos'd was a vent for smoke.*[57]

GIANTS OF ZETLAND (SHETLAND)

The Shetland Islands (formerly called Zetland) are a subarctic archipelago in the Northern Isles of Scotland fifty miles northeast of Orkney. The first evidence of Mesolithic activity dates back over 6000 years and is abundant in ancient stone ruins, boasting over 5000 archaeological sites. It is also rich in giant lore:

> *In the popular mind [standing stones] are usually attributed to the giant race, of whom a faint tradition lingers. It was the giant of Roenis Hill who, in his combat with the giant of Papa Stour, threw a stone at his opponent in the distant island, which fell short and is now known as the Standing Stone of Busta. Similar tales are told of other standing stones.*[58]

There are other stone sites associated with giants as well on the Shetlands, like the *Giant's Stones of Hamnavoe*. The monument consists of two standing stones (and a former third upright, now removed) and a number of low or prostrate stones nearby. These represent a ritual monument of prehistoric date, that may be the remains of a small stone circle. Other sites of interest are the *Giant's Grave* standing stones and *Trowie Knowe* (Troll Mound) Cairn at Beorgs of Housetter. The monument comprises two standing stones and the remains of a cairn from the Neolithic period, built between 4000 and 2500 BC. The two standing stones are both rough undressed blocks of red granite. Like many of Shetland's prehistoric monuments this cairn has become the focus of local legends, including trolls and giants because it is supposed to mark the spot of a giant's burial. We also have nearby the *Giants Garden*, a massive boulder field.

THE GIANTS AND THE MERMAID

One of the myths that endures on the islands is that of the giants Saxi and Herman who once lived on the most northerly island of the Shetlands, Unst. According to legend they were brothers both deeply in love with a beautiful mermaid they observed swimming in the seas around their homeland, temptingly out of their reach. The giants tried desperately to gain her attention which eventually elicited a response from the lovely sea creature: she would give herself to whichever of them managed to swim to the North Pole with her. The mermaid set off swimming, and the love-struck giants without hesitating strode into the sea and followed her, never to be seen again. Saxi and Herman are commemorated on the island still, at a great hill called *Saxa Vord on Unst*.

CROFTER'S FIND A GIANT'S GRAVE

This interesting account from 1951 speaks of an unusual find on the Shetland Isles that appears to be the unearthing of some bones, and the uniform of a 7ft tall gentleman with coins dating from 1683.[59]

Crofters find a giant's grave

Two crofters cutting peat on Mainland, largest of the Shetland Isles, unearthed what appears to have been a uniform for a man nearly 7 ft. tall.

It consists of a short jacket or waistcoat over a three-quarter length coat with 25 cloth buttons on each side of the front and seven buttons at the waist. Wide breeches were attached to long stockings and hide shoes.

A double-brimmed cap, resembling a forage cap was found, and also human bones. There was also a small horn, a quill pen, a wooden bowl, and a horn spoon.

In a small knitted bag were three coins, one of them dated 1683.

THE STONE GIANTS OF LEWIS

Callanish is located on the Isle of Lewis in the Outer Hebrides and is one of the most impressive stone circles in the world. The thirteen stone blades form what Alexander Thom called a 'Type A Flattened Circle,' overlaid with a great crucifix of stones, with avenues heading in four directions. A 15ft menhir stands in the middle. It has become famous for the 18.6 year moon cycle observation as the lunar orb rolls along the *Sleeping Beauty* mountains to the south. The extreme position was last seen in June 2006. A giant legend about the circle is given here:

> *Until recently it was believed that at sunrise on midsummer's day the 'Shining One' walked along the avenue, his coming heralded by the call of the cuckoo, the bird of Tir-nan-Og, the Celtic land of youth. Another tradition is that of a priest-king who came to the island bringing with him not only the stones but black men to raise them. He was attended by other priests and the whole company wore robes of bird skins and feathers. These tales smack a little of Druidism but genuine folk tradition seems to lie behind the*

genuine name for the stones, Fir Chreig, or the 'False Men'. They are said to be the old giant inhabitants, who would not build a church, nor let themselves by baptized by St. Kieran when he came to preach to them, so he turned them to stone.[60]

Sunset shadows at the stones of Callanish.

Does the 'Shining One' mentioned have a connection to the 'Shining Ones' from the *Book of Enoch*, those archaic gods who taught the arts of civilisation? Tir-nan-Og was considered the home of the gods, an island paradise and supernatural realm of everlasting youth, that may have been an actual lost island somewhere in the Atlantic. The *"black men"* who arrived to raise the standing stones may be the same Black Giants discussed in the Wales chapter, who were also revered sorcerers and priests who had command of supernatural forces. The idea of giants being turned to stone is found in many megalithic sites in Britain, and is a Christian strategy to erase the old Pagan ways, although the lingering association between giants and megaliths could not be fully erased.

Another giant legend is brought to us by Donald Alexander Mackenzie in his *Giant Lore of Scotland*, p.10:

The Kewach was a giant and a real hero, a man not only of great size but great dignity...In the island of Eigg the Kewach lives in a cave. He is still remembered in Barra. On the north side of Loch Lomond, a similar hero was associated with a circular fortress on a promontory which...was called Giants Castle.

THE INNER HEBRIDES
A LIVING GIANT

The rather tall Angus MacAskill with a 6ft 5in friend.

Angus MacAskill was the tallest non-pathological giant in recorded history, born in 1825 on the Island of Berneray in the Inner Hebrides. He stood 7ft 10in tall and ended up weighing an astonishing 580 pounds (41.5 stone). His shoulders were 44 inches wide, and the palm of his hand 8 inches wide and 12 inches long; his wrists were 13.5 inches in circumference and his ankles measured 18 inches in circumference. His almost supernatural strength was well documented and he went to work for P. T. Barnum's circus in 1849.

MacAskill lifted a ship's anchor that weighed 2800 pounds (1.27 tonnes) to chest height, and had the ability to carry barrels weighing over 350 pounds (160kg) apiece under each arm. He was also able to lift a hundredweight with two fingers and hold it at arm's length for ten minutes. He appeared before Queen Victoria at Windsor Castle after which she proclaimed him to be *"the tallest, stoutest and strongest man to ever enter the palace."* MacAskill's feats of strength cannot be overlooked, because this kind of strength is exactly what has been described as a trait of the giants of old, so could MacAskill have had a genetic or ancestral link to them?

STEPPING STONES OF THE FOMORIANS

Fingal's Cave is a sea cave on the uninhabited island of Staffa in the Inner Hebrides. It is formed entirely from hexagonally jointed basalt columns within a Paleocene lava flow, and is an identical structure to the Giant's Causeway in Northern Ireland. It is roughly 75 feet high and 270 feet deep. In Irish myth the cave and the Giant's Causeway are linked in legend to the renowned Irish warrior *Fionn mac Cumhail* (Finn McCool) and was called 'the stepping stones of the Fomorians' in Irish lore. Fionn is credited with defeating Scottish giant Benandonner and building this causeway between Scotland and Ireland. The Celts had previously called the cave *Uamh-Binn* meaning the 'Cave of Melody' because of the magnificent acoustics found there.

The stunning Fingal's Cave of Staffa from the exterior.

ARRAN: ISLE OF THE GIANTS

The Isle of Arran lies off the coast of Western Scotland and is the largest in the Firth of Clyde and the seventh largest Scottish island, at 167 sq. mi. Arran has a concentration of early Neolithic Clyde Cairns, a form of Gallery Grave. The typical style of these is a rectangular or trapezoidal stone and earth mound that encloses a chamber lined with larger stone slabs. Several Bronze Age sites have been excavated here, including *Ossian's Mound* near Clachaig and a cairn near Blackwaterfoot that produced a bronze dagger and a gold fillet. In 2019, a Lidar survey revealed 1,000 newly discovered ancient sites in Arran including a cursus.[61]

One of the Machrie Moor Stone Circles on Arran.

Fionn mac Cumhail and his son Ossian are two legendary characters revered around Scotland, and Arran is no different. It is quite common to find that ancient circles of stone have become associated with mythical figures and giants. At Machrie moor, the double circle, *Suide Choir Fhionn* (Fingal's Cauldron Seat), is named after the legendary warrior giant Fionn, who was said to have used a holed stone to tie his dogs to while he ate his dinner inside the inner ring.

More connections between giants and ancient stone sites are to be found at two chambered cairns on Whiting Bay that are known as the *Giants' Graves*. The well-known traveller and author Martin Martin in the late 17th century described the traditions of the site:

> *The name of this isle is by some derived from Arran, which in the Irish language signifies bread. Others think it comes more probably from Arjn, or Arfyn, which in their language is as much, as the place of the giant Fin-Ma-Couls slaughter (or execution). The received tradition of the great giant Finn McCool's military valor, which he exercised upon the ancient natives here, seems to favour this conjecture; this they say is evident from the many stones set up in divers places of the isle, as monuments upon the graves of persons of note that were killed in battle.* [62]

Old Photograph of the Giant's Grave of Whiting Bay, Isle Of Arran.

These Giants' Graves are the remains of two Neolithic chambered tombs situated within 100 feet of each other. They stand on a ridge above the sea in a forest clearing, overlooking Whiting Bay to the south.

MEN OF GIGANTIC SIZE

Rev. James Headrick in 1807 described gigantic human remains and most notably a huge jaw bone that easily fitted around the face and beard of the discoverer. Headrick wrote:

> *I have seen many graves, in various parts of the Highlands, of extraordinary dimensions, and reputed to be graves of giants of the Fingalian era.* [63]

We also find a report from The *Glasgow Herald*, December 19, 1872, stating:

> *On a portion of the Tonderghie estate, in which the last-mentioned farm is detached parts of a human skeleton, consisting of a massive skull and leg-bones of a man, apparently of gigantic proportions, were, about a year ago, exhumed by some ditchers while draining a piece of wet bog land.*

A lengthy report from the *Tourist's Guide to Arran*, 1872, describes the traditions of locals regarding the supernatural forces surrounding the graves of the ancient heroes. This account is given in full in the *Sowers of Thunder* chapter but here is a summary:

> *Along the south end of Arran, close by the shore, is a series of monumental mounds, among the largest and most perfect of their kind in the island, from time immemorial reputed to be the graves of giants or Fingalian heroes, and held sacred by the peasantry on the pain of supernatural mischief or death. One of these on Margreeach Farm, by the Sliddery Burn, as we learn from Headrick, was bored long ago by some daring islander, who discovered a bone of huge dimensions in it, but which, from apprehensions of the consequences he was induced quietly to restore.* [64]

THE GIANTS OF SKYE AND RAASAY

The Isle of Skye is the largest and northernmost of the major islands of the Inner Hebrides of Scotland. Its breathtaking scenery has attracted visitors for aeons and unsurprisingly is the home of another giant legend. Situated atop Trotternish Ridge is the *Old Man of Storr* which is a 160-foot pinnacle rock formation named after its likeness to a man. The Storr is a title derived from the Norse words for 'Great Man.' Legend has it that the Old Man of Storr was a giant who resided on the Trotternish Ridge. When he was laid to rest upon his death, his thumb remained partially above ground. The Old Man of Storr towers over the Sound of Raasay at an elevation of more than 2300 feet.

Otta Swire shares a giant report from her book, *Skye: The Islands and It's Legends*:

> *The earliest parish church of Kilmuir was that of Kilmoluag, dedicated to St. Moluac and giving its name to the Kilmoluag river on whose banks it stood. It was succeeded by the church of Kilmhairi, which remained in his until long after the reformation. Indeed, the skull and thigh bones of Uisdean MacGillespie Chlerich, famous for his great size, were said to be on view on the window ledges of that church until 1827.* [65]

The Old Man of Storr.

We conclude this trip around the Scottish Islands with Otta Swire sharing more giant lore from her book *Inner Hebrides and their Legends*:

> *Raasay, though so small an island, has had a long and chequered history. In the very ancient days, it was called the Isle of the Big Men because it was believed to have been inhabited by giants - perhaps the Feinn. It seems to have been more generally thought however, that Raasay had been the last stronghold of a giant race, which once owned the isles. These people were seven to nine feet tall and their skeletons have been found in more than one ancient grave in the islands and on the western shore of Scotland, notably in Glenelg. But nothing, save the size of their skeletons and a general tradition of their size and power now remains.*

GIANT SCOTTISH SWORDS

We complete this chapter with a look at the giant swords of Scotland. The Highlanders were known for wielding what are called *Claymores* (from the Gaelic for 'Great Sword'), that were two handed. One of these swords was said to have been used by William Wallace (Braveheart). They were very heavy and much larger than the standard size.

From the 12th century onwards the Scottish Monarchy ruled the country by granting authority and land to members of the nobility. Gigantic ceremonial swords became one of the symbols of this new-found power. The finest example of Scotland's gigantic sword tradition was created for the Sempills of Elliestoun, near Glasgow. Sheriffdoms were prevalent at this time to maintain order and justice with the swords as the primary symbol of the sheriffs. The swords were also used in ceremonies for knighthoods. The Sempills were stewards of the Barony of Renfrew from the reign of Alexander II (1214-1249 AD), as well as being the Hereditary Sheriffs of Renfrewshire. It is believed that the Sempill Sword was carried before Mary Queen of Scots at the Battle of Langside.

Hugh next to the nine foot long ceremonial sword in the Edinburgh Museum owned by the Sempills of Elliestoun.

GAZETTEER OF GIANTS - SCOTLAND

DATE	LOCATION	SIZE	No.	FEATURES	PAGE
	DUMFRIES AND GALLOWAY				
1848	Cairnderry, Kirkcudbrightshire	Giant	1	Neolithic burial chamber	221
1778	Stone Circle (unknown)	Giant	1	Found in tumulus	222
1809	Cairnholy Burial Chamber	Giant	1	Tomb of ancient king	222
1778	Glenquicken Moor Stone Circle	Giant	3	Three found in total	223
1853	Chapel of Riddell, Glenquicken	Giant	1	Stone coffin dated 936 AD	223
1846	Kirkmabreck, Glenquicken	Giant	1	King with stone sphere	224
	EDINBURGH, GLASGOW AND THE LOTHIANS				
1838	St. Fort, Edinburgh	7ft 5in	2	Stone burial chamber	225
1826	Unknown, Edinburgh	10ft 6in	1	Giant king	225
1526	Unknown, Edinburgh	14ft	1	Called 'Little John'	225
1823	Unknown, Edinburgh	11ft 6in	1	Called 'Funman'	226
1855	Dalmeny, Lothian	7ft 6.5in	2	Found in stone lined graves	226
1855	Springfield House, Lothian	7ft 6in	1	Very wide chest	227
1783	Port Seaton, East Lothian	7ft	1	Crumbled to dust	227
1819	Paisley, Glasgow	8ft	1	27 ½ inch femur	228
1792	St Bride's Chapel, Glasgow	8ft 6in	1	Neolithic burial chamber	228
1941	Hunterian Museum, Glasgow	8ft 6in	1	Nazi bomb destroyed giant	228
	PERTH, ARGYLL, FIFE AND KINROSS				
1946	Dumbarton, Argyll	Giant	5	Many skulls and bones	229
1834	Ossian's Grave, Perth	7ft	1	Legendary king	230
1859	Carnlee, Perth	7ft	2	Stone coffin & big cairn	231
1820s	Westfield, Perth	7ft	1	Huge teeth	232
1800s	Lundin Links, Fife	Giant	1	Stone coffins near megaliths	232
1860	Muir of Ochil, Perth	7ft	2	Found in cairns	233
	THE MIDDLE COUNTRY - ANGUS				
1194	Red Castle	10ft	1	Found with 3ft dwarf	234
1838	Logie Pert, Montrose	Giant	3	Circular stone cairns	235
1620	Panmure Tumulus	Giant	1	Perfect condition	235
1835	Craigo	Giant	1	Huge ebony ring in burial	235
1620	Camus Cross, Carnoustie	Giant	1	Found in tumulus	236

DATE	LOCATION	SIZE	No.	FEATURES	PAGE
	The North - Inverness-shire, Ross and Cromarty				
1893	Dun Telve Broch, Glenelg	8ft-11ft	2	Iron Age Broch	240
1848	Cromarty Parish, Inverness	Giant	1	Massive skull	241
1851	Portmahomack, Inverness	Giant	4	Found in stone coffins	241
1871	Caithness, Parish of Reay	7ft	2	Sent to London	242
1390s	William Sutherland, Caithness	9ft 5in	1	Recollection of living giant	243
1850s	Dail Langwell Broch, Caithness	Giant	1	Huge skull & thick bones	243
	Orkney and the Islands				
1700s	Sanday, Orkney	Giant	2	Numerous bones reported	246
1701	Orkney Mainland	Giant	1	Recollection of living giant	246
1861	Maeshowe, Orkney	10ft	3	Giant & two mummies	247
1529	Orkney Mainland	14ft	1	Found on hill near lake	249
1683	Shetland Mainland	7ft	1	Wearing uniform	252
1800s	Berneray, Inner Hebrides	7ft 10in	1	Angus MacAskill	254
1871	Tonderghie Estate, Isle of Arran	Giant	1	Huge skull	256
1872	Sliddery Burn, Isle of Arran	Giant	1	Storms during excavation	257
1800s	Kilmuir, Isle of Skye	Giant	1	On display at church	258
1800s	Raasay, Isle of Skye	7ft - 9ft	?	'Isle of the Big Men'	258

Two views of the Towie stone ball discovered in Aberdeenshire. It is one of 425 similar spheres dating to the late Neolithic period and contemporary with the recumbent stone circles found in the area.

Illustration by Nicholas Cope

261

GRIANAN
OF AILEACH

GIANT'S
CAUSEWAY

Castle Caldwell,
p.278

Duchuil,
12ft, p.278

Antrim Giant,
12ft 2, p.298

Market Cross, Sligo, p.283

Giant's Grave,
Fermanagh, p.278

NEWGRANGE

Dysart,
10ft, p.295

HEAPSTOWN
CAIRN

Four Knocks,
7ft, p.293

Roscommon, 8ft,
p.285

Hill of Tara
p.290

Castlebar,
12ft 6, p.284

Knocksedan,
9ft, p.296

Louisburgh,
p.284

Leixlip,
10ft, p.295

STONE OF DIVISIONS,
HILL OF UISNECH

Cloneybrien, p.287

St Canice Round Tower,
p.297

Nenagh,
p.287

Scattery Island,
8ft, p.289

Ballyneety,
p.289

Kilkenny,
7ft, p.296

LOUGH GUR
GIANT'S GRAVE

Ballinahalla, 8ft 6, p.285

Ballynemony Castle, p.286

10

GODS AND GIANTS OF ANCIENT IRELAND

It is this air of mystery and sanctity, which has always attached itself to the British Isles, that the mainspring of Atlantean influence is made manifest. The 'Gods', 'Demons', 'Fairies' and 'Giants' that haunted the minds of our ancestors, sprang from the residual unconscious memories of the mighty races who went before them... Admittedly it is a history of incredible, even bizarre proportions; nevertheless, the core remains constant. The 'Days of the Gods' and their fantastic civilizations and conflicts were once an actual historical reality. - Anthony Roberts.[1]

In order to better understand the origins of the giants outlined in this book, we must look at the oral and written traditions that talk about a great cataclysm and flood that destroyed a former homeland. Survivors of this lost world are recorded as introducing to Ireland the high arts of civilisation, such as advanced geometry, mathematics, agriculture, geomancy, and megalithic construction techniques:

> *...it appears that the population of Ireland came from the West, and not from Asia - that it was one of the many waves of population flowing out from the Island of Atlantis...There are many points confirmatory of this belief. In the first place, the civilization of the Irish dates back to a vast antiquity.*[2]

THE LOST LAND OF HY-BRASIL

The mysterious island of *Hy-Brasil* appeared on maps from between 1325 to the 1800s. In Irish myth, it was covered in mist, becoming clear only one day every seven years. It was rumoured to be the home of the first 'Irish gods' of antiquity,

263

as well as the 'land of eternal youth' or even the location of Atlantis. On old maps it appears to be around 200 miles west of Ireland in the North Atlantic Ocean. According to Barry Cunliffe, Professor of European Archaeology at Oxford University, Hy-Brasil is the most intriguing of all the legendary islands of the Atlantic and could be connected to the origins of Irish civilisation:

> *The legend goes back much further, probably into pre-Christian times, appearing first in the seventh century in the Irish text known as 'The Adventure of Bran Son of Febal', which tells of Bran's visit to this Other World island supported on pillars of gold where games are played, people are always happy, there is no sorrow or sickness, and music is always to be heard - truly a land of the blessed.*[3]

Hy-Brasil on a map from 1325.

Numerous voyages searched for it, but most failed, except when in 1872 Robert O'Flaherty and T.J. Westropp claimed to have visited the island four times and even witnessed it vanish before their very eyes. The submerged remains may well be what is now known as Porcupine Bank, discovered in 1862 located 120 miles west of Ireland. At extreme low tide parts of it become visible and in 1830 a sea-chart had 'Brazil Rock' located in this area. It was not removed from British Admiralty charts until the 1870s.

IRELAND'S FIRST INHABITANTS

The Emerald Isle is home to a rich written and oral history describing ancient visitors arriving on its shores, possessing advanced technology, supernatural abilities and giants in their ranks. Successive waves of invasions are chronicled before and after the great flood in Irish history. Partholians, Nemedians, Fir Bolgs, Fomorian's and the Tuatha De Danann preceded the Milesians (who were thought to have arrived around 1000 BC). The two groups most connected with a lost Atlantic continent were the Fomorian's and the Tuatha de Danann.

The Fomorians were considered to be the earliest inhabitants of Ireland, living in the sea or on offshore Islands. They were described as monstrous and malevolent giants, masters of black magic who demanded child sacrifices on Samhain (Halloween). Many in their ranks were strange hybrids of different grotesque forms, a zoological nightmare that plagued the shores of Ireland. The Fomorians battled many invaders, including the Partholonians and the Nemedians, who both eventually died of plagues or fled the country.

The Fomorians, as depicted by John Duncan, 1912.

The next wave of settlers were called the Fir Bolg, who, according to the annals, were defeated Nemedians returning from Greece or Scythia (after being banished from Ireland by the Fomorians). They were sometimes described as giants and their name translates as 'men of the bag', because, as slaves in Greece, they were forced to carry clay and soil. They eventually intermarried with the Fomorians and together held the country until the arrival of the Tuatha de Danann.

Tuatha De Danann translates to the 'Children of Goddess Danu', and are recorded as being very tall, fair-haired, and possessed a high degree of knowledge of all the arts and sciences. They appeared on Iron Mountain within a 'strange mist' in much the same way the Watchers appeared on Mount Hermon in biblical tradition.

ISLANDS OF THE GODS

The Tuatha De Danann were said to be from four drowned island cities in the western sea called Murias, Gorias, Falias and Finias. They brought 'four treasures' (or jewels) with them to Ireland. The stone of Fal (Lia Fal) was from Falias and it would cry out beneath the king who took the sovereignty of Ireland. It was supposedly placed near the Hill of Tara in County Meath. The Spear of Lugh was from Gorias. It was reported that no battle was ever sustained against it, or against any man who held it. The Sword of Nuada was brought from Finias and no one ever escaped from it once it was drawn from its sheath. It was also described in the *Táin Bó Cúailnge* (8th century) as a glowing bright torch. From Murias was brought the Cauldron of Dagda from which no company ever walked away from unsatisfied.

The Tuatha Dé Danann as depicted in John Duncan's *Riders of the Sidhe* (1911).

The Tuatha De Danann vanquished the malevolent Fomorians at the legendary second battle of Moytura, which was the inspiration for the opening battle scene of *The Lord of the Rings,* based on Tolkien's understanding of mythological traditions. We are told that a great warrior helped the Tuatha De Danann win the battle, his name was *Lugh of the Long Arm.* This mysterious hero was reared in the west in a land of magic and upon appearing in Ireland brought with him the 'Wave Sweeper', the boat of the sea god Manannan Mac Lir (the Irish representation of Poseidon). This magical craft, which knew the mind of the owner, flew over land as readily as it traversed the sea. Manannan Mac Lir was said to travel from his home in the Isle of Man to *Tir nan Og*, the island realm of the gods in the Atlantic.

The Tuatha de Danann, painted by François Pascal Simon Gérard.

The mystic Edgar Cayce named Og, Posedia and Aryan as the three Atlantean islands that sank in a cataclysm roughly 13,000 years ago. Og shows up many times in the British Isles. Ogham was an ancient Irish script, Ogma was a great Irish hero, Gogmagog was a Cornish giant and in the Holy Lands, King Og of Bashan was a giant warrior king.

Oisín and Niamh travelling to Tír na nóg, by Stephen Reid in T. W. Rolleston's *The High Deeds of Finn* (1910).

ANCIENT TECHNOLOGY OF THE GODS

Lugh of the Long Arm was equipped with a magic sword called 'The Answerer' which was able to burn through metal of any thickness. Lugh was described as head and shoulders taller than any man, a giant who radiated a great light that dazzled all who beheld him. The descriptions of advanced weaponry and technology seem quite clear, albeit couched in the language and descriptions of the confused later scribes. The great enemy of the Tuatha De Danann were the Fomorians whose leader at the second battle of Moytura was the giant Balor.

THE EYE OF BALOR

Balor was claimed to be a wizard of the dark arts who had a tower on his stronghold of Tory Island (the inspiration for Tolkien's *Twin Towers*). He is claimed to have one eye like the giant Cyclopes of Greek mythology. However, it appears that his great eye was actually a terrible weapon of supernatural aspect. The Eye of Balor was described as being a gigantic, burnished, circular disc mounted on pulleys and derricks that could be rotated in any direction. When its lid was rolled up, a blast of devastating fire swept out and destroyed everything in its path.

Dún Bhaloir (Balor's Fort), The Anvil, Tory Island, Co. Donegal.

In the final battle against the Fomorians, Lugh kills the Fomorian leader with his spear, running it through the eye, which then wreaked havoc on the Fomorian ranks. The defeat for the Fomorians is final and they never return to power. They are pushed out to the islands off the coast of the mainland including Rathlin, some of the Scottish Isles and, according to Nennius, the Isle of Man.

The Isle of Man was the domain of Manannan Mac Lir, and it was considered the legendary last stronghold of the giants in the region.

THE MAGICAL WEAPONS OF CÚCHULAINN

A look at the greatest hero of Irish mythology, Cúchulainn, solidifies the point that an ancient, advanced technology is attributed to these refugees of from a lost world. He is described by Anthony Roberts in *Atlantean Traditions in Ancient Britain*, 1975:

> *Dating to the most remote antiquity, the memories of this giant among men can be seen to carry some disturbingly modern connotations, particularly in the detailing of his equipment and weapons. Whole epic cycles revolve around him and he is quite probably a distorted memory of a late Atlantean Lord whose technology was certainly never forgotten by the less advanced men who followed him.*[4]

Cúchulainn possessed numerous weapons described as magical and highly advanced. The 'Thunder Feat' was a metallic circular armament that could be changed to different settings to increase its destructive power. Manannan Mac Lir, the God of the Western sea, gifted him with other technology such as his silver ocean chariot, which turned the sea to foam and was heavily armoured and covered in silver. The sea god also gave him a metallic helmet which conferred invisibility on the wearer.

The traditions of Ireland are unambiguous regarding the home of the gods being lost islands in the Atlantic. They were not just a marginally sophisticated invading force, but sorcerers and mystics with supernatural abilities and technology.

CÚCHULAINN CARRIES FERDIAD ACROSS THE RIVER

Cúchulainn Carries Ferdiad Across the River, illustration by Ernest Wallcousins from Charles Squire, *Celtic Myths and Legends*, 1905.

IRELAND'S SACRED CENTRE

Soon after they arrived the Fir Bolg divided Ireland into five parts and according to the Lebor Gabala, *"is the division of the provinces of Ireland which shall endure forever, as the Fir Bolg divided them."* There are now four provinces in Ireland, with this fifth being the sacred centre or omphalos, where the Fir Bolg's geomancers decided to place it. Traditionally this is the location of the *Hill of Uisnech* in Killare, close to the geographical centre of the country. The *Stone of Divisions* is a nearby natural outcrop of rock that marks the sacred centre.

The Hill of Uisnech is also said to hide the burial place of Giant King Lugh and the spot where a Druid lit the first fire of Ireland. It has been discussed as being not only the precise centre of Ireland where the earliest rulers were crowned, but also may be directly connected to giants from Africa and the builders of Stonehenge. Killarus, the site of the original stones of Stonehenge in Ireland could well have been the Hill of Uisnech, as the village at its base is called Killare. Some commentators have speculated that it might be referring to Kildare. This area is east of the centre in Leinster Province and contains many henges, earthworks and megalithic sites.

Left: Punchestown Stone.   Right: Stone of Divisions marking Ireland's sacred centre.

The second tallest stone in Britain, the *Longstone of Punchestown*, standing at 23 feet is located here near a town called Naas.[5] Gerald of Wales in his *Topographia Hibernica* referred to the Punchestown Stone as part of an ancient pile of monoliths called *The Giant's Dance*, and reported how it and the other standing stones of the area had been brought by giants from Africa. Local legend claims that Ireland's tallest menhir was thrown, spear like, from Kildare's *Hill of Allen* by the mythical giant Fionn mac Cumhaill, either at his wife or in a throwing contest with another giant:

> *Fionn mac Cumhaill once used this huge stone as a toothpick. One morning after his breakfast Fionn was picking his teeth and while doing so he hurt one of his gums. He got so vexed that he flung the toothpick from him and it went all the way from the Hill of Allen and never 'lit' until it landed at Punchestown. There it stuck in the ground and there it remained all these hundreds of years until the present day.*[6]

After the stone fell in 1931, it was discovered that the base had been worked with a hammer stone, and erected in a setting of 13 smaller stones. The tool marks are reminiscent of what we see on two of the monoliths at Stonehenge (see p.54). In 1571, more accounts referred to this area as the origin point of Stonehenge:

> *There is no other record of anywhere called Mount Killaraus but, curiously, John O'Donovan, an exceptionally bright nineteenth century Irish cartographer and scholar, recorded 'certain stones on the plain of Kildare' that were 'in every respect similar to the others and erected in a similar manner' However, as O'Donovan lamented, "There is no doubt…that such stones existed … either near the Castle of Kildare or that of Naas, but I fear they have been long since destroyed.*[7]

THE NORTH: ULSTER

Ulster is one of the four traditional provinces of Ireland, located in the far north. Of Ulster's nine counties three are part of the Republic of Ireland and six make up Northern Ireland. Many ancient stone monuments exist here and a good deal are closely associated with giants. A passage from George Pepper in 1829 illuminates the Christian appropriation of ancient Irish sites from the earlier Pagan cultures:

> *There is scarcely a parish in Ireland without its giant's grave, its cairn, its cromlech and sepulchral tumuli. The Pagan Irish supposed that the spirits of their departed heroes, and Druidical sages, resided in these tombs; so that they were uniformly regarded with reverential respect by the living. On the introduction of Christianity, the*

missionaries observing the superstitious attachment of the people to these monuments, preached the sublime truths of the gospel in the fanes of the Druids, and raised their churches over the graves of royal heroes, gallant knights, and celebrated Druids, which had the effect of propitiating the prejudices of the people, and enlisting their passions under the banner of the cross.[8]

GIANT PLACE NAMES

In Belfast, Northern Ireland, a henge monument at Ballnahatty called *The Giant's Ring* was built in 2700 BC. The massive site is approximately 600 feet in diameter and covers seven acres. At the centre of the ring is a megalithic passage tomb.

Antiquarian illustration of The Giant's Ring, by William McComb, 1861.

In southern Ulster is the *Giant's Leap Wedge Tomb*, alternately called the Giant's Grave, residing in County Cavan's Burren Forest Park. The structure is 25 feet long and has 5 massive cap stones, 3 of which cover the main chamber. There is a giant legend connected to this site as well. According to Harold Johnson who lives close by in Blacklion, a local titan was trying to impress a lady but failed

Burren giants leap wedge tomb. Photo by Anthony Weir.

in his final attempt to jump the nearby chasm. That chasm is called the Giant's Leap.

THE GIANT'S CAUSEWAY

The most famous Irish giant legend comes from Ulster County. *The Giant's Causeway* is located in County Antrim in Northern Ireland. The area has roughly 40,000 interlocking basalt columns, a result of natural geologic processes. Legend has it that long ago when giants roamed the land, an Irish giant Fionn mac Cumhaill from the Fenian cycle of Gaelic mythology was challenged to a fight by the Scottish giant Benandonner. Fionn accepted the challenge and built a land bridge, the Giant's Causeway, across the North Channel from Ireland to Scotland, where there are identical basalt columns at Fingal's Cave on the Scottish Isle of Staffa.

The Giants Causeway.

In one story Benandonner is defeated, and in the other Fionn realises that his opponent is much larger and stronger than him. His wife then dressed him as a baby and put him in a cradle, and when the Scottish giant saw how huge the infant was he assumed his father Fionn must be of enormous stature. He fled in haste back to Scotland and in the process destroyed the Giant's Causeway so he could not be followed.

GRIANAN OF AILEACH

A spectacular ancient monument connected to the Second Battle of Moytura is the stone ringfort of Grianan of Aileach located in Donegal. It is said to have been built by the Dagda, king of the Tuatha De Danann, erected around the grave of his son Aedh who was killed by Corrgenn, a Connacht chieftain.

Aerial view of the Grianan of Aileach. Photo H. Newman.

The building of the Aileach is described in the 15th century manuscript, the *Book of Lecan*:

> *And he ordered these a rath to build, Around the gentle youth:*
> *That it should be a rath of splendid sections. The finest in Erin.*
> *Neid, son of Indai, said to them, He of the severe mind, That the*
> *best hosts in the world could not erect a building like Aileach.*[9]

There is disagreement about the approximate dating of the Aileach but it is known that the location was used for a long period of time. The site also once had a tumulus with 10 horizontal stones surrounding it of a much earlier age (described in 1835).[10] It was believed that a Stone Age burial was contained within the tumulus.

THE GIANTS OF ULSTER

The local legends of giants have become vindicated by a recent scientific study which offers an interesting twist on the giant phenomena of Ireland. In a *BBC News* article from October 12, 2016, we learn this:

> *Mid Ulster identified as 'giant hotspot' by scientists. The study*
> *found that 1 in 150 people in Mid-Ulster County were found to*
> *carry the AIP or 'giant gene', compared to one in 1000 in Belfast*

and 1 in 2000 in the rest of the UK.

The AIP gene causes excessive quantities of growth hormone to be produced by the pituitary gland causing pituitary gigantism or acromegaly. Marta Korbonits, Professor of Endocrinology at Barts and the London School of Medicine Queen Mary, who led the team that discovered the link between the AIP gene defect in Irish populations and gigantism in 2011 stated, *"This is probably the highest proportion of giants in the whole world in that little part of Northern Ireland."* Korbonits and her team were able to calculate that the AIP gene defect traced back at least 2500 years. The team found the variant in Charles Byrne (1761–1783), the famous Irish giant who grew to be 7ft 7in tall, as well as in 18 other families in Ireland. Byrne hailed from a remote part of northeast County Tyrone, Northern Ireland, called Littlebridge, not far from the shores of Lough Neagh. Until recently his skeleton could still be seen at the Hunterian Museum in the London headquarters of the Royal College of Surgeons. Some accounts refer to him as being 8ft 2in to 8ft 4in tall.

Left: The Giant Charles Byrne along with the Giant Knipe Twins.
Right: The skeleton of Byrne, the height of 7ft 7in.

MORE HISTORICAL GIANTS

In the *Transactions of the Royal Irish Academy,* 1892, we are told of Cornelius McGrath and other modern giants:

Upon his death his body was carried to the Dissecting House in the College; where his skeleton, on account of its extraordinary size, will amuse the curious and fill posterity with wonder. He was not the only person of a gigantick size born of late years in Ireland, for in the third volume of Lowthrop's Abridgement of the Philosophical Transactions, the late Dr. Molyneux has given an account of one Edward Malone, whom he measured in this city, with his shoes off, and who was seven feet seven inches high.

Cornelius Magrath, the Irish Giant.

Cornelius McGrath was equal in stature to Daniel Cajanus, the Swedish Giant, who was seven foot and eight inches high.[11]

James Paris du Plessis remarked that while in London in 1696 he saw, "*A handsome, well-proportioned giantess who was seven feet high without her shoes, who was born in the Isle of Portrush, not far from the wonderful Causeway in the most northern part of Ireland.*"[12] Other giants of the time in Ireland were mentioned as well, in *The Annual Register for 1760* it was reported that James MacDonald, who attained a height of 7ft 6in, died at his home near Cork at the great age of 117 years.

AN ARMY OF GIANTS

James Kirkland, known as the 'Soldier King' was one of the tallest soldiers recruited by Friedrich Wilhelm I, King of Prussia for his regiment of tall soldiers often called the 'Potsdam Giants'. Kirkland was believed to have been around seven feet tall and he was accompanied in the regiment by other oversized soldiers from Ireland. He came from Ballygar, Co. Longford.

Left: James Kirkland was 7ft tall.
Right: Patrick Cotter stood at 8ft 1in.

Patrick Murphy (1834-1862) was born in Killowen, Co. Down and lived on a farm in the Mourne Mountains and stood at 7ft 3in. James Hugh Murphy Jr. (1842-1875) was from Waterford, Ireland. He was known as the *Irish Giant* and the *Baltimore Giant*. He toured with P.T. Barnum and was over 8ft tall. Patrick Cotter, who died in 1806, stood at 8ft 1in (other reports put him at 8ft 7in).

Charles Byrne had a pituitary disease condition but it appears that others like Kirkland and MacDonald (who lived to be 117 years old), did not suffer from the debilitating disorder. The author's mother, Margaret Toomey-Vieira came from County Cork and she often spoke of her over seven-foot cousin, who possessed enormous strength and lived an otherwise normal life. Angus MacKaskill, the Scottish Giant was also a titanic 7ft 10in tall and equipped with herculean strength with no health-related issues. The question is, why does this small section of Northern Ireland, a place filled with giant-lore, possess the highest percentage of gigantism in the world?

The traditions state that the giants of old were robust, long lived and renowned for their legendary feats. Could the AIP gene be a trait based on a genetic rejection, that turned into a disease, as interbreeding of people of great stature with normal sized humans occurred?

THE GIANT'S GRAVE AND A GIANT JAWBONE

More historical discoveries in the area are found in the book, *Lights and Shades of Ireland*,[13] in 1850:

...it was once a land of giants - that her men like the renowned

giants of old, had bedsteads of iron, measuring many cubits, or that their spears were like 'weavers beams' and that six fingers and six toes were on each hand and foot. But the armoury and bones that have been dug up in that island testify that a race of greater men than now inhabit it were there, and that the supposed superstition of the incredulous peasants respecting giants, is nothing less than a false coin made from a true one.

The story continues with this description of the finder placing a gigantic jawbone over his face, to emphasise the colossal size of the jaw:

In the county of Donegal, near, Fermanagh, there is a curiosity called the 'Giant's Grave.' This grave is entered by a low aperture, formed by a projecting block of stone, resting upon two others... the people must have had a knowledge of mechanics, or else have had giants in those days to have elevated and fitted these stones. A peasant living near, being asked if he would open the grave refused, because some had tried to do so and found their feet sticking fast to the spade. There is a tomb of the giant's armour-bearer on an eminence near, which has been opened and an earthen urn containing some ashes, supposed to be the heart, was found, and bones of great size. The lower jaw-bone was quite perfect and so large that it went with ease over the jaws of the biggest headed labourer present.

HUMAN SKELETON OF IMMENSE SIZE

This discovery from Castle Caldwell included huge teeth and a skull that was 'on display' in 1832: [14]

> A human skeleton of an immense size was last week found by some labourers while working in the demesne of the Rev. J. B. TUTHILL, at Castle-Caldwell. The skull was the 18th of an inch thick, and the teeth, which were quite perfect, an inch long above the prong. Mr. TUTHILL has the skull preserved in his parlour.—(*Ballyshannon Herald.*)

SAINT PATRICK AND THE GIANT

Glyn Daniel, Professor of Archaeology at Cambridge University, suggested that Saint Patrick was Ireland's inaugural archaeologist. In 1972 Prof. Daniel gave a lecture on *Megaliths in History* where he discussed early Christian legends in medieval Irish manuscripts. He cited the following passage from the 9th Century *Vita Tripartita Sancti Patricii*, that seems to indicate that St Patrick was invoking spirits during the excavation of a giant skeleton from a long barrow:

> *And Patrick came to Duchuil to a great grave, of astounding breadth and prodigious length which his familia had found. And with great amazement they marvelled that it extended 12 feet, and they said, 'we do not believe this affair, that there was a man of this length'. And Patrick answered and said, 'if you wish you shall see him', and they said 'we do.' And he struck with his crozier a stone near its head, and signed the grave with the sign of the cross, and said 'Open, O Lord, the grave'. And the holy man opened the earth and the giant arose whole and said: 'Blessed be you, O holy man, for you have raised me even for one hour from many pains.' Speaking so, he wept most bitterly, and said, 'I will walk with you.' They said, 'We cannot allow you to walk with us, for men cannot look upon your face from fear of you. But believe in the God of Heaven and accept the baptism of the Lord, and you shall return to the place in which you were. And tell us of whom you are.' [And the man said] 'I am the son of the son of Cas, son of Glas, and was swineherd to King Lugar, king of Hirota. The warrior-band of the son of Mac Con slew me in the reign of Coirpe Nioth Fer, 100 years ago today.' And he was baptised and confessed to God, and he fell silent, and was placed once more in his grave.*

Above: Stained glass depiction of St Patrick, one of history's earliest giantologists.
Overleaf: Dan Lish's illustration of St Patrick discovering the skeleton.

Anthony Roberts discussed more details in *Sowers of Thunder*:

> *This giant was more than twelve feet in height and was wearing rusted armor, while alongside the body there lay an axe and a long sword in the last stages of decomposition. It is doubtful whether the saint recognized the full import of the find, for such wonders abounded during the early Christian era, but it is recorded that Patrick was impressed by the size and obvious age of the mound's occupant, to the extent of ordering the quick disruption of the bones and permanent dispersal of the evidence of ancient weaponry.*[15]

CONNACHT
BATTLEGROUND OF THE IRISH GODS

King Bres, from a 1910 illustration, at the First Battle of Moytura.

Connacht is the western Irish province of the original four provinces of the country. It is an area rich in archaeological sites and has been the place of many battles. In the *Mythological Cycle of Irish Mythology*, the *Cath Maige Tuired* (The Battle of Magh Tuireadh) is the name of two saga texts that describe separate battles fought in ancient times. In the first conflict, often called the First Battle of Moytura, the Tuatha De Danann defeat the Fir Bolg army to seize control of Ireland. It is thought that this battle took place on the border of County Galway and Mayo close to Lough Corrib.

In the Second Battle of Moytura thirty years later, the De Dananns defeated the dark sorcerer Balor of the Evil Eye and the Fomorians. This battle, according to legend, was fought in Sligo at a place today called Moyritta, which is abundant in ancient stone sites associated with the encounter.

BURIAL PLACE OF THE GODS?

Heapstown Cairn, also known as the *Giant's Grave*, is one of the most stunning monuments in the area that was once described as being *"higher than a house"* with a monolith at its peak.[16] It was constructed sometime around 3000 BC as a passage grave and, even though much has been removed for later construction projects, it still stands over 20 feet high and 190 feet in diameter. The legend is that the stone mound covered a healing well sacred to the Tuatha De Danann who used it to treat their warriors during the Second Battle of Moytura. The Fomorians were said to have taken boulders from the River Drowse and thrown them into the well to prevent the Tuatha De Danann from using it, thus creating the stone mountain. The site has never been excavated.

Heapstown Cairn. illustration by W.F. Wakeman, 1878.

Labby Rock (from 'leaba' which means bed) is another monument given a name from the legendary battle. The capstone weighs a staggering 70 tons and it was thought to have been built around the same time as Heapstown Cairn. It is believed to be the grave of King Nuada and his wife Macha of the Tuatha De Danann. A monolith called the Eglone Stone, has an old legend of a former giant who was turned to stone by a magician due to a dispute. It resides in Moytirra, near Highwood, overlooking Lough Arrow in County Sligo.

Wakeman's watercolour of the Eglone, a massive standing stone at Highwood village.

Wakeman's 1879 watercolour of the Labby Rock.

SLIGO - A GIANT'S BONES UNEARTHED

An account from Sligo may confirm an ancient battle with giant warriors in the area. The report is from 1898 and is included here in full: [17]

> **SLIGO.—A Giant's Bones Unearthed.—**After all Finn McCool and the Galway giant seem not to have been the only mighty men of Ireland. Sligo, like Louth, can now claim to have produced an Irish giant, although his history is forgotten. His bones have been discovered at the Market Cross, Sligo. Several men were engaged in digging a drain across the street when they succeeded, much to their astonishment, in unearthing a skull, jawbone (with teeth in good condition), and various other parts of the human anatomy. The bones were exceptionally large, and show that the man who in the long ago was laid to rest there must have been of giant statue. The discovery created quite a sensation in the town, and much speculation was indulged in as to the supposed age of the remains. Colonel Wood Martin, the eminent Sligo historian, has given it as his opinion that the body must have been buried over 250 years ago, and that the "giant" may probably have met his death in battle in those stirring times. The remains were taken in charge by the police, and interred in Sligo cemetery.

Sligo is home to many ancient stone monuments, often associated with giant heroes such as Fionn Mac Cumhaill and his wife Sadbh. The Easkey River Valley, west of Sligo, has over a dozen known sites. *The Giant's Griddle* is a Neolithic dolmen that is said to have been the cooking spot of Fionn and the Fianna Warriors.

In the same area is the split rock known as *Fionn Mac Cumhaill's Stone.* Legend has it he threw it from the Ox Mountains during a giant's stone-throwing competition. Fionn was reportedly so furious that he did not reach the ocean with his throw that he split the stone in half with his sword. Local tradition states that if you walk through the crack three times it will snap shut on you. Another ancient site in Sligo is the *Magherashanrush Court Tomb,* also called the *Giant's Grave,* built in 3000 BC. The 100-foot-long, three chambered construction, is reported as the legendary resting place of a giant.

Left: Fionn Mac Cumhal's stone, the Split Rock Right: Magheraghanrush Court Tomb.

William Wakeman's illustration of the Giant's Griddle, from the 1870s.

This following insightful legend is interpreted by the authors of the passage speculating about the purpose of megalithic construction, suggesting it was to bring peace and prosperity to the land:

> *According to a tradition from County Mayo Ireland, tells of a stone-built giant's grave on a wild mountainside. If anyone were to dig into the grave, the mountainside would immediately be changed into a fertile plain, and a key buried in the tomb would open the gate of a beautiful city at the bottom of a nearby lake. Also, the discoverer would have at his disposal 'a great golden treasure'. This allegory tells us that this current is the key to the well-being of the people, the fertility of the land, and spiritual riches.*[18]

Giant Skull in a Bog

In *Folklore, A Quarterly Review, Volume 28,* from 1917, it describes a giant king of the area and a skull as big as a 'poteen'. This is basically a small barrel:

SKELETON OF ANOTHER IRISH GIANT FOUND.

> *Fiachra, King of Irros in the late fourth century, was evidently a reputed giant, striding his horse over the great chasm at Dun Fiachrach. He was a special patron of the O'Hara's, even floating casks of wine to them, and protected forts and ancient hawthorns. The followers of Fionn are all reputed to be giants, one person, in 1838, used to tell as a fact that a human skull 'as big as a poteen' had been found in a bog near Louisburg.*[19]

OSCAR WILDE'S FATHER AND THE 12½ FOOT GIANT

On September 11, 2010 this article appeared in the *Connaught Telegraph*:

An inscribed headstone, found near Breaghwy House, Castlebar, recorded the amazing fact that close on 300 years previously Mayo men grew to the gigantic height of over 12 feet. The headstone, which weighed over 2 tons, created considerable interest at the time and recorded for posterity the size of a skeleton. The inscription on the headstone read: 'The above stone was found at Brefy A.D. 1732. The coffin of Genan contains a skeleton 12-foot-long, Glenmask, Ireland, A.D. 1681. This monument is erected to show the antiquity of the Irish character and the size of mankind in those dark ages, 1756.' [20]

Sketch of William Wilde by J.H. Maguire, 1847.

The original account comes from Sir William R. Wilde in 1866, the father of Oscar Wilde. He wrote on a variety of subjects including Irish Mythology. He was the first to notice the stone was built into the wall and it had an inscription on it about a 12½ft skeleton (rather than '12ft' in the above-mentioned report). We have been in touch with the Irish Land Commission and still have not been able to locate the stone as it was moved in recent years.

SKELETON OF ANOTHER IRISH GIANT FOUND

To add some more life to the legends we continue our journey through Connacht with an eight foot tall skeleton reported in the *Liverpool Daily Post*, August 22, 1918:

Turf cutters, when at work on Murroe Bog near Elphin, Roscommon, found the skeleton of a man measuring eight feet in length. The remains were some distance below the surface and had lain in there apparently for many years. Other skeletons have been found in the neighborhood.

MUNSTER PROVINCE

Munster is one of the four original provinces of Ireland located in the southwest of the country. St Patrick spent several years in Munster where he founded Christian churches and ordained priests. It was Patrick who supposedly drove the 'serpents' out of Ireland, a reference to banishing old Pagan religions and the ancient Druidical orders.

The Rock of Cashel, Co. Tipperary, historical seat of the Kings of Munster.

A MEDIEVAL GIANT

A giant from the 1200s is recorded from Galway in the *Bacchus Express*, 2 July, 1910, revealing *"a complete skeleton, measuring 8ft 5½ in."*

> An interesting discovery has been made at Ballinahalla, near Moycullen, Galway. Some workmen in the employment of Mr. James Lardner came upon a complete skeleton, measuring 8ft. 5½in., and subsequently unearthed an old sword bearing the following inscription in Gaelic: "Donach O'Keefe, A.D. 1231."

GIANTS ON THE RIVER AWBEG

Here we have a report of giant remains being found on the River Awbeg in Cork in 1873:

> *About a mile from Castle Saffron, on the Awbeg, is the ruined castle of Ballynemony... as they were digging the foundation of a barn, several large gigantic human bones, and, in particular, a great skull, were discovered; but by the negligence and incuriosity of the workmen, they were thrown into the rubbish and not preserved.* [21]

The Hermit Giant

Remaining in Cork we have a local legend of a giant and his grave. In Crosshaven there is a site known as *Curaghbinnie's Cairn* or the Giant's Grave, a Bronze Age monument dated to 2000 BC. The legend of the giant named Binne is that he made the Currabinny Woods his home. The grumpy giant kept within his cover of trees to shelter himself from outside interference. One night a loud party in Crosshaven, only separated from Currabinny by the narrow Owenabue River, disturbed Binnie from his slumber. In a fit of anger, he hurled a boulder across the Owenabue River and it landed where the water's edge meets the village. The boulder is now on display in the village square with the inscription, "*This boulder, which according to local tradition, was flung by a giant from the hills of Currabinny, to land on the green in Crosshaven.*"

A King Murdered and a Giant Found

In County Tipperary we have a report of giant remains at a slate quarry in 1825:

> *At the slate quarries of Cloneybrien, county Tipperary now working the mining company of Ireland, a human skeleton was found last week, under an immense heap of stones. There has been for years a story told by the country people, that the king of Leinster was murdered in that spot, and, as it is usual in such cases, a heap of stones was created each passenger throwing one as he went by: in the removal of them large flag was discovered, which induced the workmen to make a search, when a human skeleton, of extraordinary size, was discovered underneath. is added some, and believed many, that various costly emblems of royalty were also found under the tomb.*[22]

Skeleton of Gigantic Proportions

From 1902 we have this account from East Munster where the 'Ormond Fair' once took place:

> *An extraordinary discovery is reported from Nenagh. Some workmen were occupied in raising sand from a pit on the side of a hill at the back of the police barracks at a place called Beechwood, and when they came to the depth of about thirty-feet they discovered a human skeleton of almost gigantic proportions. A*

physician who took a photograph expressed the opinion that the bones belonged to a prehistoric age.

In the same county we find the following legend. On the summit of a hill in *Giantsgrave Townland*, stands a 10-foot stone pillar. The stone has two crosses carved into it and has several legends associated with it. The old Irish name for the hill is *An Cnoc Air a bhfuair Fionn Fios,* or 'The hill where Fionn Mac Cumhaill received his prophetic knowledge'. Fionn, of course was a legendary giant of Ireland. The other legend is that the stone was hurled from the summit of Slievenamon by a giant and landed in its present location.

Giantsgrave pillar stone.

AN ENDLESS AMOUNT OF GIANT'S GRAVES

In Limerick there is a site known as the *Giant's Grave Lough Gur*. It is a wedge-shaped gallery grave and dates to circa 2500 BC belonging to the final phase of megalithic tomb building (which includes passage tombs, portal dolmens and court cairns). It has two chambers, a main chamber and a porticoe. The main chamber is covered with four cap stones.

The Giant's Grave of Lough Gur.

More reports from the region continue to reveal prehistoric inhabitants of extraordinary size. *The Berkshire Chronicle* tells us from August 3rd, 1850:

> *An immense human skeleton was discovered some days back in Scattery Island, A surgical gentleman saw the bones and pronounced them to be the remains of a giant eight feet high.*

Next, we include the original account from the *Londonderry Standard* June 1st, 1854 that revealed a "*skeleton of extraordinary size.*"

> DISCOVERY OF A SKELETON.—On Thursday last, as some labourers were excavating at the foot of the ruined castle of Ballineety, in county Limerick, they found a human skeleton of extraordinary size, as if of a warrior taking his last repose, with his helmet on. It is supposed to be the remains of the English officer who commanded Villiers' dragoons when Patrick Sarsfield Earl of Lucan, then defending Limerick against the Dutchman with 500 chosen men of his yellow regiment, cut them to pieces.

LEINSTER PROVINCE
HOME OF THE GODS

Leinster province is the last of the four ancient provinces that we will visit in our tour of Ireland. It comprises the ancient kingdoms of Meath, Leinster and Osraige. The most famous and awe-inspiring ancient sites in Ireland are seated in Leinster province, specifically in the Boyne Valley. *The Hill of Tara, Newgrange, Knowth* and *Dowth* are marvels of complex construction that harken back to the time of the gods. We will briefly discuss these sites because of the important connection they have to the mythological races and our investigation into giants.

Hill of Tara aerial view. Photo H. Newman.

The Hill of Tara is an ancient ceremonial and burial site. The *Dumha na nGiall*, or Mound of the Hostages dates to as far back as 3200 BC. The *Lebor Gabala Ereen* or *Book of Invasions* is an 11th century manuscript which tells us that Tara was the seat of the high kings of Ireland. Intriguingly, in Irish mythology, Tara is said to have been the capital of the mystical Tuatha De Danann, who possessed giants in their ranks.

Mound of the Hostages, a Neolithic passage tomb on the Hill of Tara, Ireland

A GIANT VISITS THE HILL OF TARA

There is an interesting story that concerns the Hill of Tara, from the 14th century manuscript, *The Yellow Book of Lecan*. It is told that a giant stranger came to the court of the High King at Tara bearing a branch from which grew three fruits, an apple, an acorn, and a hazelnut. The stranger's name was *Trefuilngid Treochair*, meaning 'of the three sprouts'. He was described thusly:

> *As high as a wood was the top of his shoulders, the sky and the sun visible between his legs, by reason of his size and his comeliness. A shining crystal veil about him like unto raiment of precious linen. Sandals upon his feet, and it is not known of what material they were. Golden-yellow hair upon him falling in curls to the level of his thighs.*[24]

The 14th century manuscript, *The Yellow Book of Lecan.*

He requested of Conan Bec-eclach, a just High King, that all the men of Ireland be assembled, and from them he selected seven of the wisest men of knowledge from each 'quarter' of the land, and also seven from Tara. Trefuilngid Tre-eochair taught them all about their history and heritage, and shared with them his knowledge, but during that time, not a drop of wine or morsel of food passed his lips, for he was sustained purely by the fragrance of the fruits of his branch. When his work was done, he gave the fruits from his branch to Fintan, the *White-Haired Ancient One*, who extracted seeds and planted them in each quarter of the land, and one in the centre, at Uisneach. The trees which grew from these seeds became the five sacred trees of Ireland.

The Stone of Destiny (Lia Fáil) at the Hill of Tara, once used as a coronation stone for the High Kings of Ireland.

THE ABODE OF THE TUATHA DE DANANN

Newgrange is a breathtaking passage tomb dated to 3200 BC. It is known for its megalithic art and its winter solstice alignment, where sunlight shines through a roof box and illuminates the inner chamber. Newgrange, Dowth and Knowth

Newgrange aerial photo at sunset. Photo H. Newman.

291

were sites all connected to the Tuatha De Danann in mythology, and remembered as their dwelling places. They were said to have erected Newgrange as a burial place for their chief, Dagda Mór, and his three sons. It was also said to have been the place where the great mythical hero Cúchulainn was conceived by his mother Dechtine. Eventually the Tuatha De Danann faded into the mists of history becoming known as 'Aes sidh,' or 'people of the mound'. Tradition states that they retreated to an underworld realm through passage graves like Newgrange, eventually becoming the Irish fairy-folk.

The interior of Newgrange with triple spiral carving.

A recent genetic study was highlighted in a June 2020 article titled *DNA Study Reveals Ireland's Age of 'God-Kings.'*[25] An elite social class existed amongst the Stone Age inhabitants of Ireland. Forty four genetic samples were extracted with key evidence coming from a 5000 year old adult male buried at Newgrange. The DNA revealed that his parents were most likely brother and sister. These types of incestuous relations are found in far flung places such as Peru and Egypt where the elite family is considered to be divine and avoids breeding with the common people. Tutankhamun's parents, for example, are thought to have been full siblings. This phenomenon was highlighted in our last book with the Adena people of Eastern America who practiced the same type of interbreeding, which kept their royal class intact. Adena elite males were often over 7 feet tall and the females over 6 feet tall, they elongated their skulls and, just like the royals of Ireland, they were buried in ancient mounds and monuments.

Another interesting find from the Newgrange skeleton was that he had familial connections with burials at the 'mega-cemeteries' of Carrowmore and Carrowkeel in County Sligo. The oldest tomb at the site is dated to 3700

BC, 500 years earlier than the construction of Newgrange. As we have already discussed, Sligo is a place of numerous giant legends and giant skeletal finds. This strange behavior makes sense within the context of a supernatural, god-like race founding Ireland with their descendants compelled to continue the divine blood-lines.

GIANT MYTH MEETS REALITY

Let us now turn our attention to gigantic remains that were found in 1951 at an ancient burial chamber located 10 miles southeast of Newgrange. The site is called *Four Knocks* and features some intricate rock carvings:

Traces Of Ancient Giants Unearthed In Ireland

By DERRY MORAN
United Press Staff Correspondent

DUBLIN — (UP) — Irish archaeologists have unearthed traces of a bygone race of "supermen."

The findings may provide scientific substantiation for legends of a race of seven - foot giants who inhabited the Island of Hibernia (Ireland) in its golden age, long before the dawn of history.

and terrifying rumblings coming from within.

In July of this year, however, the mounds surrendered peacefully and silently their hidden treasures to a team from the National Museum.

The burial chambers, at Four Knocks, Clonalvey, County Meath, are not far from the historic Hill of Tara, scene of the court of the ancient kings of Hibernia.

Irish archaeologists have unearthed traces of a bygone race of 'supermen.' The findings may provide scientific substantiation for legends of a race of seven-foot giants who inhabited the island of Hibernia (Ireland) in its golden age, long before the dawn of history. In a prehistoric burial chamber dating back to 2000 BC, they found human skeletons which tower head and shoulders over modern man, stretched in slab topped graves with offerings of food and ornaments beside them. Most are around seven feet in height, of extraordinary width of shoulder and massive bone construction.

In the yellowed pages of Irish folklore and mythology, seven-foot giants stride gloriously through a 'land of milk and honey,' battling strange monsters and preforming fantastic deeds of physical strength and endurance. In the remote rural areas, old folks still point out where, according to legend, the great warrior Cuchalainn crushed the imprint of his mighty foot in the rocky

Southern recess lintel inside the main Four Knocks chamber showing zigzag patterns.

bank of a river after a colossal leap across a roaring torrent. Throughout the nation, story tellers relate similar feats by other Irish warriors and heroes of the island's golden past. Now, for the first time, what may be concrete evidence has come to hand to support the legends of a race of supermen. The burial chambers, hidden for nearly 4,000 years under mounds of overgrown earth, first came to notice 100 years ago, when local inhabitants began a search for gold but were deterred by mysterious and terrifying rumblings coming from within.

In July of this year, however, the mounds surrendered peacefully and silently their hidden treasures to a team from the National Museum. The burial chambers, at Four Knocks, Clonalvey, County Meath, are not far from the historic Hill of Tara, scene of the court of the ancient kings of Hibernia.[26]

The account also featured in the *Eugene Register Guard*, December 25th 1950 and several other newspapers.

TWO 10-FOOT SKELETONS

Our first report of a ten-foot-tall skeleton comes from the *Dublin Freeman Journal*, in 1812: [27]

> *This extraordinary monument of gigantic human stature was found by two labourers in Leixlip churchyard on Friday, the 10th when making a kind of sewer near the Salmon leap, for conveying water, by Mr. Haigh's orders. It appears to have belonged to a man of not less than ten feet in height. It is believed to be the same mentioned by Keating-Phelim O'Tool, buried in Leixlip churchyard, near the Salmon leap, 1252 years ago. In the same place was found to be a large finger ring of pure gold. There was no inscriptions or character of any kind upon it, a circumstance to be lamented, as it might throw a clear light upon this interesting subject. Our correspondent saw one of the teeth, which was as large as an ordinary forefinger.*

The other description of a ten-foot-tall skeleton comes from the *Feilding Star* in 1914:

> *While men were digging foundations for labourers' cottages at Dysart, Louth, they unearthed three human skeletons in separate graves encased with stones. One skull is entire and measures, 18 inches from the crown to the chin. The leg bones are abnormally large. The remains apparently are those of a person 10 feet high, who is presumed to have lived in a prehistoric age.*[28]

A Giant's Skeleton

AN INTERESTING discovery was recently made in Ireland. During the course of some excavating work two men unearthed three human skeletons.

One was entire and measured eighteen inches from the crown of the head to the chin. The leg bones were abnormally large and altogether the skeleton appeared to be that of a person ten feet high. The teeth also were large, and the remains are supposed to date from a prehistoric age.

Another version of this account featured in
The Milwaukee Sentinel, June 28, 1914.

HILL OF THE FAIRY MOUND

Our last stop in Leinster comes from the book *Irish Pedigrees,* 1880:

> *It is stated by Ware, that the sepulchral mound at Knocksedan, near Swords in the County Dublin, was opened in his time, and in it were found the remains of a man of gigantic size: the skeleton measuring, from the ankle bone to the top of the skull, eight feet four inches; the bones of the skull were very thick, and the teeth of enormous size; the limbs were all very large in proportion, and it appears that this giant, when living, must have been nearly nine feet high.*[29]

The original report came from the Antiquarian Sir James Ware who described workmen digging gravel in the mound in 1639 and finding the skeleton. With the feet attached it would have stood at roughly 8ft 8in. The demolished burial mound at Knocksedan translates to *Hill of the Fairy Mound* in old Irish.

Fairies dancing in a ring near a large mushroom and a mound
with a doorway in a 17th century woodcut.

GIANT IN A BOG-OAK COFFIN

Moving south to Kilkenny we find that a seven-foot skeleton was found in a bog-oak coffin. There is also an historical account of a seventeen-foot skeleton found by a labourer in the same area in 1885, but was later found to be a giant elk. In this case we don't think anyone was burying elks in coffins. The report from 1864 reads:

I was also informed that several bog-oak coffins, with perfect skeletons lying at full length, had been discovered on the line of the removed side walls, forming the west boundary of John's-lane, and that there is reason to suppose further explorations would discover others, Under the surface of the lane already mentioned. One of these skeletons had been accurately measured, and was found to be fully seven feet in length. The bones, as they now appear, are all quite blackish, and of various sizes.[30]

ROUND TOWERS OF IRELAND

We end our trip around Leinster with a brief look at the enigmatic Round Towers of Ireland. The conventional opinion is that they were constructed in the Christian era c.700 - 1200 AD, although archaeological evidence suggests some were built on much earlier burial sites.[31] About 120 round towers are thought to have existed. Most are in ruins, while roughly twenty are in almost pristine condition, some with the original conical roof.

Glendalough Round Tower, County Wiklow.

The *Annals of Ulster* describes the destruction of 57 round towers by an earthquake in 448 AD. Giraldus Cambrensis c. 1146 – c. 1223, also known as Gerald of Wales, in confirming the account of the inundation of Lough Neagh in 65 AD, reports in his *Description of Wales,* 1863, p.70:

The fishermen beheld the religious towers (turres ecclsiasticas), which according to the custom of the country, are narrow, lofty, and round, immersed under the waters; and they frequently show them to strangers passing over them, and wondering at their purpose.

The purpose of these towers is another mystery. Philip Callahan writing in his book, *Ancient Mysteries, Modern Visions,* believes they may have been designed as huge resonant structures for collecting and storing electromagnetic energy coming from the earth and the cosmos. This would bring fertility to the landscape and enhance the consciousness of meditating monks inside them. The doors on them are often unusually high up so impossible to gain easy access to. Callahan suggests the lower portion of the structures were filled with stone and rubble for tuning the resonant frequency of the towers. He also points out that their placements mark stars in the northern sky at the time of the winter solstice, basically laying out a country-wide astronomical map. Orthodox explanations have them as bell towers, watch towers or quiet spaces for monks to contemplate. They are rarely connected to giants, except for this one account below.

A GIANT SKULL IN THE TOWER

This story comes from *The Cork Examiner*, November 5th, 1847, entitled, *Round Tower of St Canice*. According to the report a member of the South Munster Antiquarian Society discovered a huge skull:

> *Detached bones of a skeleton with gigantic skull, found at the west side portions of other skeletons having been found in the fine mould–none could be particularly ascribed to the skull.*

Painting of the tower where the gigantic skull was found by Edward Fanshawe, 1856.

THE IRISH GIANT OF ANTRIM

We end this chapter by examining the strange case of the Irish petrified giant. This image and story have made the rounds on the internet as proof of the existence of ancient giants. The authors have taken a deep dive investigating this case to figure out if there is any truth to the claims. The photo of the fossilized Irish giant was taken at a London rail depot, and appeared in the December 1895 issue of *Strand Magazine*, a British publication. The giant was claimed to have been dug up by a Mr. Dyer while prospecting for iron ore in County Antrim. Here is the account from the *Strand Magazine*:

> *Pre-eminent among the most extraordinary articles ever held by a railway company is the fossilized Irish giant, which is at this moment lying at the London and Northwestern Railway Company's Broad-street goods depot, and a photograph of which is reproduced here. This monstrous figure is reputed to have been dug up by a Mr. Dyer whilst prospecting for iron ore in Co. Antrim. The principal measurements are: Entire length, 12ft 2in.; girth of chest, 6ft 6.5.in.; and length of arms, 4ft. 6 in. There are six toes on the right foot. The gross weight is 2 tons 15 cwt.; so that it took half a dozen men and a powerful crane to place this article of lost property in position for the Strand Magazine artist. Dyer, after showing the giant in Dublin, came to England with his queer find and exhibited it in Liverpool and Manchester at sixpence a head, attracting scientific men as well as gaping sightseers.*

The *Strand Magazine* ran from January 1891 to March 1950 and featured both fiction and general interest articles. The 12ft giant disappeared soon after this photo was taken. We were able to trace its origins to an Italian stone carver named Giuseppe F. Sala, who was also involved in the infamous Cardiff Giant hoax in New York in 1869. The full story is beyond the scope of this book as we will cover it in its entirety in a future publication. We were able to track the exploits of Sala over several decades and prove his hoaxing of more petrified giants in Colorado, Los Angeles, New Hampshire and several in Australia. Now, the next time you see the photograph of the Irish Giant, you will know the truth.

Photo of the Antrim Giant from the *Strand Magazine,* 1895.

GAZETTEER OF GIANTS - IRELAND

DATE	LOCATION	SIZE		FEATURES	PAGE
	MODERN IRISH GIANTS				
1700s	County Tyrone, N. Ireland	7ft 6in	1	Charles Byrne	275
1700s	Tipperary, Munster	7ft 7in	1	Cornelius McGrath	276
1700s	County Cork	7ft 6in	1	James MacDonald	276
1800s	Ballygar, Co. Longford	7ft	1	James Kirkland	277
1800s	Waterford, Munster	8ft +	1	James Hugh Murphy Jr	277
1800s	Killowen, Co. Down	7ft 3in	1	Patrick Murphy	277
1700s	Kinsale, Co. Cork	8ft 1in	1	Patrick Cotter	277
	THE NORTH - ULSTER				
1850	Fermanagh, Co. Donegal	Giant	1	Huge jaw	278
1832	Castle Caldwell, Enniskillen	Giant	1	Perfect teeth & thick skull	278
1827	Duchuil, Ulster	12ft	1	Found by St Patrick	278
1895	County Antrim, Ulster	12ft 2in	1	Elaborate hoax	298
	THE WEST - CONNACHT				
1898	Market Cross, Sligo	Giant	1	Buried in local churchyard	282
1838	Louisburg, Co. Mayo	Giant	1	Huge skull in bog	284
1800s	Castlebar, Co. Mayo	12ft 6in	1	Linked to Oscar Wilde's dad	284
1918	Elphin, Roscommon	8ft	1	More found nearby	285
	THE SOUTHWEST - MUNSTER PROVINCE				
1213	Ballinhala, Co. Galway	8ft 6in	1	Inscription on sword	285
1783	Ballynemony, Awbeg, Co. Cork	Giant	3	Numerous discoveries	286
1825	Cloneybrien, Co. Tipperary	Giant	1	Possible royal burial	287
1902	Nenagh, Co. Tipperary	Giant	1	30ft below surface	287
1850	Scattery Island, Co. Limerick	8ft	1	Size confirmed by surgeon	289
1854	Ballyneety, Co. Limerick	Giant	1	Helmet & armour	289
	THE EAST - LEINSTER PROVINCE				
1400s	Hill of Tara, Co. Meath	Giant	1	Witnessed and recorded	290
1950	Four Knocks Tomb, Co. Meath	7ft	6	Many warriors unearthed	293
1812	Leixlip Church, Co. Kildare	10ft	1	Large gold torque found	295
1914	Dysart, Co. Louth	10ft	3	Huge skull & bones	295
1639	Knocksedan, Swords, Co. Dublin	8ft 8in	1	Found in 'Fairy Mound'	296
1864	Kilkenny, Co. Kilkenny	7ft	1	Found in bog oak coffin	296
1847	St. Canice Round Tower, Co. Kilkenny	Giant	1	Huge skull	297

11
Sowers of Thunder

And Jesus arose and rebuked the wind and said to the sea: Peace,
be still. And the wind ceased and there was a great calm.
- MARK 4:39

Worldwide traditions of giants talk about mighty beings often with powers beyond the realm of mortal men. One of these is the ability to control the weather and to place a curse over their graves so that great tempests would manifest when their burials were disturbed. The following accounts from the British Isles parallel the great legends of weather-manipulating gods from all over the planet.

J.M.W. Turner's drawing of Stonehenge with what appears to be sheep
struck dead during an electrical storm, 1829.

North America, Finland and Polynesia, for example, have creation myths and epics of giant gods battling one another, using lightning, thunder, and control of the elements as their weapons.[1] The Greek God Zeus was provided with what can only be referred to as a 'storm' technology' by the Cyclopes, the one-eyed titans from an era before the Olympian gods. The Cyclopes were renowned as great stonemasons and metallurgists from a prehistoric pantheon, who had a flair for controlling the elements. They are said to have given Zeus the power over lightning as he was able to throw 'thunderbolts' at his enemies. It was this he used to defeat the remaining titans. All over the world the connection between giants and weather control is quite ubiquitous:

> *The Norse Gods are manipulators of earthquakes, clouds, storms, lightning and snow and are often specifically referred to in the Eddas as 'sowers of thunder.' The god Thor (who himself inherited strains of giant blood) was always fighting these beings; his own connections with weather are not inconsiderable.*[2]

Thor's Fight with the Giants
by Mårten Eskil Winge, 1872.

Sowers of Thunder was chosen as the title of Anthony Roberts' classic book on the mythology and reality of giants in the British Isles forty years ago. His research hinted at this elemental control of the weather and since then we have compiled dozens of written reports that support his thesis. These include great storms or meteorological changes occurring during the excavation or

304

obliteration of ancient sites, usually the tombs of giants. Were these spells or curses left by the builders, or were the sites originally designed to affect the weather in some way? If so, this suggests a deep understanding of the elemental forces, as described in many myths worldwide. The Irish pantheon of gods were remembered from different eras, including the Fir Bolg who worshipped a lightning god called *Bolga*.[3] 'Gáe Bolg' or 'Gáe Bolga', means *lightning spear*, like the spear of Cúchulainn in the Ulster Cycle of Irish mythology. The Fomorians were also masters of the elements:

> *The word 'Fomor' has been translated as meaning 'giant' and the Fomorians are also sometimes designated as gods, masters of the fertilising powers of Nature, wielders of thunder and lightning, sowers of mist and rain. They are in fact thinly disguised guardians of the earth's natural energy systems often interpreted in a metaphysical form.*[4]

Throughout the world we have accounts of curses affecting those who enter, or steal from a sacred tomb. Tutankhamun is perhaps the most famous example, and many tragedies have befallen those desecrating sites, including disease, bad luck and even death.[5] The *Aughrim Wedge Tomb* in County Cavan, Ireland, is also known as a Giant's Grave and is said to have caused the downfall of one of Ireland's richest men, Sean Quinn. In 1992 the tomb was excavated and moved to the grounds of the former Quinn Group owned Slieve Russell Hotel in Ballyconnell. Soon after he had a string of bad luck, declaring bankruptcy, losing billions in revenue and putting hundreds out of jobs.[6] According to locals the removal of a Giant's Grave disturbed the fairy folk and instigated the wrath of the ancient giants. The tomb itself is intact on the grounds of the hotel.

In our book *Giants On Record*, in the chapter called *Curse of the Giant Hunters*, we explore some of these mysteries. For example, in the massive megalithic city of Nan Madol in the South Pacific *"several human bones of gigantic size"* were discovered in an ancient tribal ruler's tomb and skeletons up to nine feet tall were witnessed. The next morning, after an unusually stormy night, the Governor, who was part of the excavation, died. The natives were certain it was a curse that proved supernatural powers guarded the tomb. Also in the book, we feature several stories of renowned giant shamans who could control the weather. One of these was the powerful leader of the Penacook tribe in New England in the 1600s. Passaconaway was able to summon storms at will, and he even resorted to this when white settlers tried to arrest him. The intense rain and wind he manifested gave him the time to flee to the forest. He was said to have been over seven feet tall and lived to about 120 years old.

Indigenous peoples in America associated lightning with regeneration and saw it as a connector between the earth and the Milky Way or a 'Spirit Road.'[7] These beliefs are mirrored in the scientific understanding of lightning as an energy-balancing transfer between the positively charged earth and negatively charged thunderclouds.[8] Traditions all over North America talk of 'weather shamanism' often relating to sacred stones and the giants of old. In northern California, the Yurok tribe used a 'rain stone' to promote weather increases in their area. After centuries of use it was buried, as it was deemed as too powerful to control. However, in 1959 road works uncovered it and five inches of rain fell overnight. It was reburied and accidentally surfaced in 1966. This time there were excessive floods.[9] Similar stories of revered stones exist in Britain. In the western isles of Scotland, 'bowing stones' were kept wrapped in cloths and were unveiled to secure favourable winds for sailing. A black stone pig effigy unearthed on the Isle of Barra, resulted in such gale force winds, that it was returned and reburied.[10]

The connection between giants, ancient sites and weather shamanism in North America has revealed that the early shamans of numerous tribes could also magically control the weather. This Sioux myth highlights this reality:

> *It is said the thunderbirds once came to earth in the form of giants.*
> *These giants did wonderful things, such as digging the ditches where*
> *the rivers run. At last they died of old age, and their spirits went*
> *again to the clouds and they resume their form as thunderbirds.*[11]

Andrew Collins, in *Denisovan Origins*, revealed a tradition that suggests Algonquian-speaking tribes of the Great Lakes area spoke of the 'Animiki', god-like figures called 'thunderbirds' who were revered and feared in ancient times:

> *Algonquian language-speaking groups of the Great Lakes and*
> *St Lawrence River region of North America preserve traditions*
> *regarding the Animiki', generally translated into English as*
> *"thunderbird," "thunderers" or "thunder people." Primarily these*
> *are sky manitous (spirits), who bring forth lightning, thunder and*
> *rain storms. However, separate traditions held by these peoples*
> *talk about the Animiki' being shape-shifting giant birds that can*
> *assume human form by removing their 'feather blankets,' and even*

have mortal families.[12]

Collins goes on to discuss the 'Animiki' having *"the power to conjure thunder and lightning, the latter emitted from their eyes."* [13] If these giants of North America were as powerful as these stories claim, then placing a curse of some kind over one's burial site, may have been the norm in these cultures. It certainly seems it was no different in the British Isles:

> *Sometimes this awful labour [the removal of cromlechs in Wales] is accompanied by fierce storms of hail and wind, or violent thunder and lightning; sometimes by mysterious noises, or swarms of bees which are supposed to be fairies in disguise...In the prominent part played by storm - torrents of rain, blinding lightning, deafening thunder - in legends of disturbed cromlechs, and other awful stones, is involved the ancient belief that these elements were themselves baleful spirits, which could be evoked by certain acts. They were in the service of fiends and fairies, and came at their bidding to avenge the intrusion of venturesome mortals, daring to meddle with sacred things.*[14]

A Victorian sketch called *The Unexpected Consequences of Opening a Barrow.*
Here Peter Hutchinson's diggers are driven off a Devon burial mound by the
infuriated inhabitants, by L. Grinsell.

THUNDER GIANTS OF THE WELSH LANDSCAPE

At a *Giant's Grave* in Radnorshire (that has now been covered by a lake) historian Rev. Jonathan Williams in 1859 recorded, "*at a place called Abernant-y-beddau ... a huge stone set erect in the ground and bearing upon it this inscription - 'There are three graves in this pleasant clover glade / Owen, Milfyd, and Madog'*". Thornsley Jones in 1951 wrote a further account of the power of this site:

> *Some lead miners lodging in the vicinity last century, learning that the graves were reputed to be those of giants, resolved to dig up the bones and see. They started work with the sun shining brightly, but before they had proceeded far the sky became overcast and a thunderstorm broke over them. They stopped work, thoroughly frightened, and never resumed their task.*[15]

He goes on to describe the story about the unfortunate occupants of these graves as "tall men" whose end came with one of them angrily killing his two brothers, then committing suicide upon the hill.

The Giant Rhitta of Snowdon also held a spell over his burial place:

> *His grave can be seen today, a cairn five yards wide...It was whole until around 1920 when a reckless character by the name of Jack Six came to dig it, but he didn't dig much for a terrible storm of lightning and thunder came and he ran frightened to the house.*[16]

D.J. Evans, in his *History of Capel Seion*, when discussing Carreg Samson in Cardiganshire, stated: "*It is a huge stone. Attempts were made to move or piece it out many times. The tradition says that silver or some treasure is hidden underneath it, and also one dares not move it, lest suddenly come thunder and lightning.*"[17] Bedd of Owyddyn near Llanwddyn in Powys hides probable treasure: "*There is a place on the hills called Gwely Wddyn, where, the country people say, great treasures are hidden, but that every attempt to discover them have been frustrated by tremendous storms of hail and rain.*" Even a German miner had an allowance to find the treasure, but again failed miserably.[18]

Collected by school children in the Blaenrhymni region of Glamorganshire in 1911, a small book of local folklore revealed the true nature of a local giant, and him being a 'cyclops,' echoing other traditions around the world: "*The giant was a very strange creature because he had an eye in the centre of his forehead. Despite that, with his eye he could see the fairies and beyond* the clouds. *The wind, the snow, the thunder, the lightning, and the rainbow were obedient to him*."[19] The story goes on to be particularly gruesome to the point of a child killing and skinning a dog, and then wearing it to disguise himself. Again, shamanic traits are found within this allegorical tale associated with control of the weather. The cyclops motif is also found in the tales of Ysbaddaden (hawthorn or white thorn tree). Like Balor in Ireland his one eye got pierced by a poisoned spear and he died from his wounds.[20]

Coetan Arthur once existed in Llanllawer in Pembrokeshire but was destroyed in 1844 by a local peasant who was encouraged by the Rev. T. G. Mortimer not to do so because bad luck occurred to those who harmed Druid's altars. The tale runs that the desecrator did indeed have very bad luck, although no storms or weather changes were recorded.[21] In this next account from the area, during the 1940s, the weather was most definitely affected. A local farmer told Lewis Edwards that when he approached *Carreg-y-Bucci* (The Hobgoblin Stone), that sits upon an ancient mound near Lampeter, Pembrokeshire, with the intention of breaking it up to use the stone as gateposts, it was reported: "*there was a violent thunderstorm, the worst I have ever known. I ran for my life, but it followed me all the way home.*" Three other men had also been killed by lightning when visiting the stone.[22]

Llyn Cwm Llwch is a small lake that sits below the highest peak in South Wales called Pen y Fan, located within the Brecon Beacons of Powys. Legend has it that there is an invisible island in the middle of the lake. On May Day each year, a rock path would appear at the edge of the lake and lead visitors through a rock door to the magical island filled with fairies. It is said that centuries ago, a visitor to the island stole a sacred flower from the fairies, and they closed off the doorway forever. The locals heard about this story and dug a trench so they could find the doorway:

> *They were on the edge of arriving at it when an awful storm arose. They had a great dread when a giant arose from the lake and threatened to drown the town Aberhonddu which is miles away if they kept on with the work.*[23]

This echoes a passage from the *Second Branch of the Mabinogion* in which Bendigeidfran was carrying a cauldron: "*And I beheld a big man with yellow-red hair coming from a lake with a cauldron on his back. Moreover he was a monstrous man, big and the evil look of brigand about him.*"[24]

Llanymynech Hill is the fourth largest Bronze and Iron Age defensive settlement in Britain, covering an area over 140 acres. Skeletons with jewellery have been found within the now quarried site, that was thought to be haunted. Three local brothers heard that a giant had buried his wife under a large stone with a golden torc around her neck. It did not go well for the grave robbers: "*The neighbours will tell you how this vile act did not escape the vengeance of Heaven, but ended in the destructions of the perpetrators.*"[25] What exactly happened to them is unclear, but later, more discoveries were made: "*Gold torques, curiously wrought, bronze lance heads, iron swords, glass beads, coins, have all been found in the rocks. Skeletons of all sizes, many with broken skulls, are also to be found. Giants are also said to have lived in these caves and the large oblong holes on the top of the rocks are pointed out as the giants' graves.*"[26]

Grave of the Long Man (Bedd-y-gwr-hir) on the border between Breconshire and Monmouthshire is thought to be the resting place of a ten-foot-tall giant. An earlier version of the story relates how, while carrying the corpse of the great warrior to the preferred resting place, the weather suddenly changed:

> *The story was that a giant was being carried to Llanwenarth to be buried ...and that the bearers of the corpse, discouraged from going farther on account of its weight and the tempestuous weather, buried him here. The length of the grave between the two stones was thirteen and a half feet.*[27]

Another version from 1809 stated it was snow, not rain, that caused them to bury him at that certain spot:

> *Near the boundary line, between this parish and Monmouthshire, is a small tumulus, like those over graves in country church-yards, with a stone at each end, without any inscription, called Bedd y gwr hir, the giant or tall man's grave, but who the hero here interred was, or at what period his death happened is not known; the legends and tradition of the country inform us that a person of very extraordinary stature, above ten feet high, a chieftain of Blaenau Gwent, having been slain in the valley, was brought thus far by his friends, who were desirous of burying him honourably on his own demesne, but that a sudden fall of snow in the night, prevented their further progress, and compelled them to deposit the corpse here.*[28]

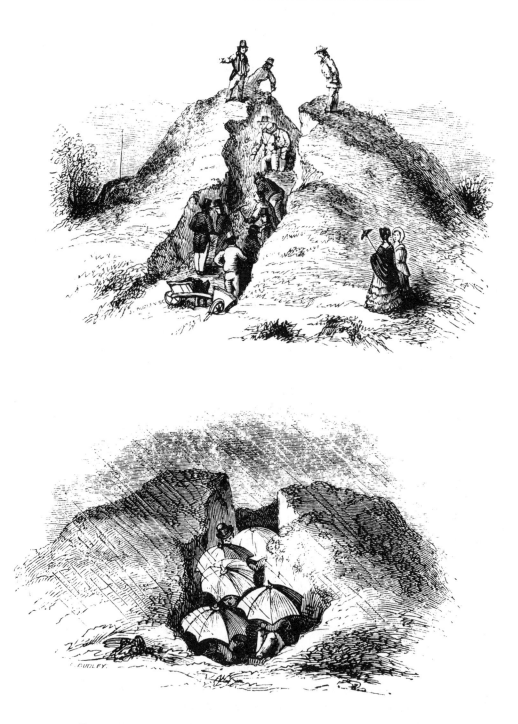

Excavation of a barrow in Aylesford, Kent in 1844 that was rudely
disturbed by the weather gods as seen in these illustrations.

The report concluded that:"...*we can fairly infer... that he was of very remote antiquity.*" According to a local researcher, the giant had six fingers on each hand. She also revealed that two small standing stones, at 13.5ft apart marked the length of the interred skeleton.[29]

We end this section of the Welsh 'sowers of thunder' with the giant of *Caer Drewyn Hillfort*. In early traditions, the Welsh titan built the fort as a stronghold for his wife who herded cows into it for milking. One of his more elaborate skills was that he could make the entire mountaintop invisible during thunderstorms.[30]

GIANT'S GRAVES ON THE ISLES OF SCILLY

Beyond the southwesterly tip of Cornwall lies a mysterious group of islands called the Isles of Scilly, the last remnant of the lost land of Lyonesse, and possibly one of the last strongholds of the giants:

> *There has always been a popular tradition that any form of desecration or interference with certain prehistoric stones or sites would result in earthquakes or thunderstorms. This lasted well into the eighteenth century when the great antiquarian William Borlase reported the incidence of severe thunderstorms while opening certain known Giants' Graves on the Scilly Isles.*[31]

William Borlase was an antiquarian and Rector of Ludgvan, Cornwall. He is remembered for his works entitled *The Antiquities of Cornwall* (1754) and *The Natural History of Cornwall* (1758). He excavated numerous cairns, cists and chambered tombs, known locally as *Giant's Graves*. He opened two of them on Buzza's Hill in St Mary's. He found little of interest, but the very next night a terrible storm erupted that caused much damage to the island:

> *This prompted the many islanders to question him as to whether he had offended the Giants who were buried in the graves he had disturbed. Even when he travelled to Bryher the next day, he found that he was already known and was once more quizzed, albeit politely. The islanders believed that, because the length of the chambers comfortably surpassed even the tallest of the people, they must have belonged to a separate race of giant people. Borlase realised that the chambers were disproportionate for single burials and took this to mean that they were designed for multi-occupancy though he failed to persuade the locals.*[32]

THE WEDDINGS OF STANTON DREW

The triple stone circles of *Stanton Drew* are in the domain of the 'Cangick Giants' that were discussed in detail in the *Stonehenge and Ancient Wessex* chapter. This account from Somerset certainly suggests that 'sowers of thunder' were guarding it. Legend states that the stones are the petrified members of a wedding party and its musicians, lured by the Devil to celebrate on the Sabbath, who were punished for their revels.

Featuring the third largest stone circle in Britain, the three circles at Stanton Drew were first noted by the 17th century antiquarian John Aubrey in a sketch from 1664. At about the same time, a Mr. John Wood visited it to make a plan of the site, but his survey was interrupted by a storm and the villagers accused him of having *"Disturbed the Guardian Spirit of the metamorphosed stones."* [33]

Left: William Stukeley's map of Stanton Drew, 1776. Right: The stones in 1845.

Further weather anomalies occurred during the Second World War. Major F.A.Menzies, a distinguished First World War army engineer and surveyor, claimed he witnessed unusual phenomena at the site in the 1940s. Having moved to France to study Feng Shui, he was an established geomancer, reportedly able to locate telluric currents and advise on ill health caused by geopathic stress. One day, he had an extraordinary experience which was later related to fellow surveyor, George Sandwich in 1952, a year before his death:

> *Although the weather was dull there was no sign of a storm. Just at a moment when I was re-checking a bearing on one of the stones in that group, it was as if a powerful flash of lightning hit the stone, so the whole group was flood-lit, making them glow like molten gold in a furnace. Rooted to the spot – unable to move – I became profoundly awestruck, as dazzling radiations from above caused*

the whole group of stones to pulsate with energy in a way that was terrifying. Before my eyes, it seemed the stones were enveloped in a moving pillar of fire – radiating light without heat – writhing upwards towards the heavens: on the other hand it was descending in a vivid spiral effect of various shades of colour – earthward. In fact the moving, flaring lights gyrating around the stones had joined the heavens with the earth.[34]

SILBURY HILL, WILTSHIRE

In 1849 an early dig at *Silbury Hill*, Europe's largest earthen mound, was shaken at its foundations in what was described as a remarkable tempest:

As a finale to the excavations, the night following work in unfavourable weather, a dramatic high Gothic thunderstorm set the seal on Merewether's [Dean Merewether of Hereford] Wiltshire sojourn. The event was 'much to the satisfaction of the rustics, whose notions respecting the examination of Silbury and the opening of the barrows were not divested of superstitious dread'. It must have been a spectacular affair. The Dean described it as 'one of the most grand and tremendous thunderstorms I ever recollect to have witnessed'. It 'made the hills re-echo to the crashing peals, and Silbury Itself, as the men asserted who were working in its centre, to tremble at its base.'[35]

Old illustration of Silbury Hill, part of the greater Avebury landscape.

LONG MEG STONE CIRCLE, CUMBRIA

Long Meg and Her Daughters is Britain's second largest stone circle and 'Long Meg' is thought to refer to a giantess, who is also linked with Westminster Abbey (see *The Northern Giants* Chapter). There is an interesting story of when Colonel Lacy, who expanded Lacy's Caves on the River Eden, attempted to blast the circle away, a huge storm broke out and the workmen fled in terror.

TUMULUS AT WEST WOODYATES, DORSETSHIRE

On the Wiltshire/Dorset border sits a charming mound group of all shapes and sizes looking like massive droplets of water (when viewed from the air). These Bronze Age tumuli held some remarkable secrets, and appeared to have been protected by some kind of prehistoric weather magic. *The Oakley Down Barrow Group* was excavated by William Colt-Hoare. In *The British Critic*, 1813, it reported: *"Such were the arms found with a skeleton of large dimensions, in a remarkable barrow, on the verge of Wiltshire, by the road leading from Salisbury to Blanford."* [36] We take this next passage from *The Early Barrow-Diggers* to summarise what happened on that eventful day:

> *To the excavation party it seemed that the gods disapproved of the unearthing of the skeleton, as a tremendous thunderstorm broke over their heads. Hoare remarked that the occasion 'will ever be remembered with horror and pleasure by those who were present.' Their only refuge was the trench cut into the mound, but*

Aerial view of Oakley Down Mound Group in Dorset. Photo H. Newman.

315

the lightning flashed on their spades, and the flints cascading down on them from the barrow summit forced them to leave their shelter and abide the pelting of the pitiless storm upon the bleak and unsheltered down.[37]

In another mound in the Oakley Down complex, the report tantalisingly describes another elongated skull, a rarity in England. It says *"The skull was pressed flat, and near it lay part of a deer's horn, perforated in the stem."*

THE ST BEE'S GIANT

The discovery of a thirteen-and-a-half foot tall skeleton and a six-foot-long sword in Cumberland is featured in full in *The Northern Giants* chapter, but the discovery not only caused severe weather, it also shook the earth:

After the giant was disinterred terrible storms affected the area, and the History and Antiquities of Allerdale records floods and a small earthquake.[38]

BEEDON BARROW, BERKSHIRE

Local traditions reveal that a burial mound called Beedon Barrow in Berkshire was the haunt of the fairy folk. Inside the mound was said to be a golden sarcophagus, containing the spirit of a supernatural entity. Parochial legends of the area warned people from interfering with the barrow, or these spirits might awaken. In the mid-nineteenth century some locals got together to dig into the mound, but were cautioned about the dangers. This did not stop them and they even got permission from the landowners. However, the excavation was interrupted by violent storms:

There is a large tumulus in Stanmore field... called Burrow Hill by the common people; who have a tradition that a man of that name was interred there in a gold or silver coffin. This barrow was opened during the month of April, 1815. The common people state that an attempt to open this barrow was made about fifty years ago, but the design was frustrated by a dreadful hailstorm, with lightning, which compelled the labourers to desist. Thunder being also heard during the second attempt in question, the excavators were universally considered as the disturbers of the atmosphere; those that remembered the previous event, remarking, that "the undertaking seemed not altogether pleasing to the Lord!

A terrific thunder-storm happening on the following day, the labourers were obliged to desist and take refuge in a neighbouring cottage; which had such an effect on the mind of one of the workmen employed, that he actually refused to come again. The recurrence of a thunder storm during this, the second attempt, was generally considered as remarkable; but such was its melancholy influence on this poor fellow that he became completely deranged, and never did a day's work afterwards; being confined in St. Luke's and other lunatic asylums for the remainder of his life.[39]

The author tells us of another folk story and an occurrence that took place at the mound:

Among other ridiculous stories and puerile superstitions respecting this tumulus, the peasantry relate that it is inhabited by fairies; and that a certain ploughman having broken his share, and gone home to procure some tools, found on his return that the plough had already been mended.[40]

Although his tone was a bit off, he revealed a story that is often associated with mounds and fairy lore. Anthony Roberts outlined the deeper aspect of these sites and their supernatural inhabitants:

It must be stated here that it was only from Elizabethan times that fairies were described as tiny, ethereal beings, flitting delicately from flower to flower and bathing coyly in buttercups. Before Shakespeare many were vested with a stature and strength as great, if not greater, than man, and they were looked to as powerful and magical beings, the possessors of occult secrets that were constantly sought after by the enquiring and ambitious humans who followed them.[41]

GIGANTIC BONES OF SLIEVEMORE, IRELAND

On Slievemore Mountain, Achill Island, County Mayo in Ireland, a series of ancient megalithic tombs dating to as far back as 3000 BC reveal an old tradition of giants. This report from 1909 is included in full in a classic 'sowers of thunder' experience:

Slieve Mor, within the island, at its northwestern extremity, justifies its name of "big mountain" by raising its head only a few feet above that of its neighbor and rival, Crochaun, which, almost, 2,000 feet high, with the tips of its toes in the ocean, has some of the finest cliff scenery in Europe....Farther inland there is a large space sacred to Irish archaeologists, containing in the shape of a Druid's altar various cromlechs and stone circles, some of the most interesting archaeological remains in Ireland. Keltic scholarship has not yet said the last word on these wonders of a dim antiquity, but there is a consensus of opinion to claim this, as a pagan cemetary 'Where lie the mighty bones of ancient men'.

Some years ago, while excavations were being made in the neighborhood, human bones of giant proportions were unearthed and were immediately taken by a gentleman of antiquarian tastes who happened to be about the place. On the evening of that day a storm, such as rarely visits even the storm swept island, was let loose. The frightened villagers assembled in council and sent a hurried deputation after the profaner to have the bones returned and laid to rest once more. [42]

One of the Neolithic tombs on Slievemore Mountain, Achill Island, County Mayo.

Regular excavations have taken place here over the last few decades so it would be fruitful to learn if these giant bones were ever measured or put on display.

THE GIANTS OF ARRAN ISLAND, SCOTLAND

A lengthy report from the James Beckett in 1872, describes the traditions of locals regarding the supernatural forces surrounding the graves of the ancient heroes. Ossian was the giant son of the legendary giant Fionn mac Cumhaill who is thought to be buried on Arran. These accounts may not give clear examples of great tempests arising, but the local beliefs seemed to have made their way down through hundreds of generations, with the clear message - *do not desecrate these graves or something very bad will happen to you:*

> *Along the south end of Arran, close by the shore, is a series of monumental mounds, among the largest and most perfect of their kind in the island, from time immemorial reputed to be the graves of giants or Fingalian heroes, and held sacred by the peasantry on the pain of supernatural mischief or death. One of these on Margreeach Farm, by the Sliddery Burn...was bored long ago by some daring islander, who discovered a bone of huge dimensions in it, but which, from apprehensions of the consequences he was induced quietly to restore.*
>
> *This cairn was partially removed some years ago by a modern Goth, who rifled the cells of their contents and strewed them over the fields,' etc.. The perpetrator of this unnecessary outrage, we further learn from the same authority, was struck with horror at the thought of his own sacrilege, and finally met with a violent accidental death in his distraction, which the people of the district interpreted, of course, as a judgement on his profanity.*[43]

Another mound in the area was thought to be the resting place of the two great heroes, but locals thought otherwise:

> *But they rather believe them to have been giants and necromancers, than men of ordinary stature, who acquired celebrity by the exertion of their natural powers...and that popular fear of their supernatural power protected such graves from violation, in case the remains of either should be disturbed.....But apart from all these, and close to Clachaig House, is a much larger double mound, distinct and lofty, like a small Danish camp, of oval form, held not only in the utmost traditional veneration by the natives, but protected from violation by traditional prophecies of death to whoever should disturb it. One adventurous person more than two generations ago, who dared to turn the turf of the upper mound, was repelled, it is said, by the glare of two terrible eyes from beneath; and Mr. Speirs himself, the present tenant's father, when proceeding to excavate on the same spot was actually deterred by the prayers and entreaties of the assembled people, imploring*

him to forebear, for that instant death would be the penalty of his sacrilege.[44]

Whatever is to be made of these ancient beliefs that disturbing the burial sites of the giant heroes unleashed supernatural retribution, it certainly was well accepted among the indigenous population.

THE GIANTS OF GLENELG, SCOTLAND

Glenelg is located on the west coast of Scotland, near the Isle of Skye. On the road into town is a sign announcing that it is twinned with Glenelg on Mars. The planet Mars that is. In reality there really is a Glenelg on Mars, an area near the landing site of the Mars Science Laboratory (also known as the *Curiosity Rover*). There were also giants in the area of Glenelg (on Earth) that when uncovered, had some startling meteorological results:

> *There is a place in Glenelg called the tall or big man's ridge. Tradition says that two of the Fingalians were drowned whilst crossing Caol-reathain and that they were interred there. The bones were found to be quite fresh and of an extraordinary size. No person ever saw anything to compare with them before and it is said no person could at all credit or even imagine the size of them but those who saw them.*
>
> *One gentleman who was present, the late excellent Rev. Mr. MacIver of Glenelg and father of the much respected present minister of Kilmuir, Skye, stood six feet two inches high and very stout in proportion and was altogether allowed to be one of the handsomest men of his day. Everyone was wonder-struck at the immensity of the bones; he took the lower jaw-bone and easily put his head through it. It is added that it was a beautiful day; but all of a sudden there came on thunder and lightning, wind and deluging rain, the like of which no man ever heard or saw. The people thought judgment had come upon them for desecrating the bones of the dead and interfering with what they had no right, so they closed the grave and desisted. Possibly some may think this bordering on the marvellous; but let no one gainsay the truth about it. There are many yet living who were present, all of whom declare that they 'shall never forget the day and the scene till the day of their death.'*[45]

In our previous book we had a section about giant jaws in North America. We have uncovered many accounts where the finder has slipped the jaw-bone over their face with ease in England, France, Canada, Mexico and all over the US, predominantly from burial mounds. The sudden tempest gives the story a mystical aura, but as has been made clear in this chapter, the giants punished those, if their peace was disrupted, by throwing the elements at them with a primeval force.

ANCIENT SITES AS THUNDERSTORM CAPACITORS

All these events seemed to have a profound effect on those that witnessed them. This ability to manipulate the megalithic burial so as to affect the weather if it was disturbed, may seem like a magical spell, so could there be some kind of sorcery involved? Quite possibly, but we propose that it may also be the remnants of a lost earth energy science utilised for scientific and spiritual purposes in the prehistoric world, that to the lay person, would certainly appear to be 'magic'. This tantalising quotes from author Tom Graves, in his classic *Needles of Stone* may contribute to solving the mystery of 'sowers of thunder':

> *The idea that the energy of thunderstorms might somehow be locked up or stored until some kind of reservoir reaches bursting point brings us back to the idea of barrows as energy-stores, and to an interesting piece of archaeological folklore. There's always been a folk-superstition that some kind of 'divine retribution' follows the 'desecration' of ancient sites, particularly barrows. If you look back through the records, you'll find that this superstition has a basis in fact, for in the case of some barrows a thunderstorm followed within hours or minutes of the opening of the barrow...but the effect on breaching them was exactly like short-circuiting some kind of 'thunderstorm capacitor.'* [46]

Graves also noted that even today people are setting off thunderstorms when investigating ancient sites:

> *Henges seem to be capacitors too, as a friend of mine found out: he and a group of fellow researchers were doing a ley-line survey of a stone circle and were soaked through by an 'instant rainstorm', which only lasted two minutes, when they accidentally triggered something in or around the circle. What is interesting, from the*

point of view of our hypothetical earth-acupuncture, is that the downpour started the moment they rammed a stake into the ground at the circle exactly on line with the ley they were plotting.

Standing stones and stone circles also produce unusual effects, and dowsers have found that underground water lines congregate at these sites and act as a lightning attractor, as outlined by Michael Poynder:

Energy builds up in the clouds as electricity and is released in lightning. The stone age standing stones and circles acted as conductors, attracting electromagnetism down into the underground water and activating earth's life-blood. When the water energy becomes too strong, in turn excess energy naturally flows back into the atmosphere from the earth to the sky. This 'balancing' of natural energy is fundamental to the fertility of the planet - true metaphysics. This was one way the Druid priests controlled the weather.[47]

W. Nesfield's 'Circle of Stones at Tormore, Isle of Arran', 1828. John Michell commented that "*The haunted atmosphere of the old stones, here enhanced by weird light from a thundery sky, is constantly referred to by artists at megalithic sites*" (From his book, *Megalithomania!*, 1982).

Standing stones may have acted as a lightning rod allowing the conductive path for current to reach earth, providing a Faraday cage, or zone of protection, because it is much more conductive than a human body. Many megalithic sites are made of granite, a blend of quartz and other silica crystals, making it a passive conductor. Because it is piezoelectric it actively creates charged pathways for the current to flow. Standing stones would glow with St Elmo's fire under high electrical stress and send active plasma streamers to draw current lightning and connect it to ground. Perhaps dolmens, megaliths and standing stones were also used as protection from these lightning and thunder deities in the sky.[48]

Further examples of how ancient sites attract lightning are at the mound sites of North America, that often had a stone or wooden pillar at their peak. According to a medicine man called Rolling Thunder, in Cincinnati lightning struck a wooden stake on top of a large mound and it appeared to charge up the water in the circular ditch surrounding it. The local Native Americans all rushed there to drink the freshly charged water for its healing properties.

Similar effects have been noted at Serpent Mound in Peebles, Ohio, where a standing stone once stood in the centre of the oval, the egg in the mouth of the serpent. Again, massive weather fluctuations are reported even today, and lightning often hits parts of the site. At the Buffalo Mounds of Minnesota, lightning strikes hit several of the mounds that were revered as sacred up until modern times. At the Junction Group earthworks, Ross County, Ohio, lightning strikes were recorded that induced magnetic anomalies in the geological layers. A few years ago at Cahokia, some archaeologists placed a pine pole on top of a mound, which was hit by lightning.

Stone struck by lightning in 1980 at the Ring of Brodgar in Orkney, Scotland.

Standing stones are not immune from lightning strikes with some accounts of them sometimes being split in two. At the Ring of Brodgar, a monolith was struck by lightning on 5th June 1980 and the sign that has been placed next to it says *"Such events may also have occurred in earlier times, and might account for the damaged state of several other stones in the ring."*

On the Isle of Lewis near the enigmatic site of Callanish, a prehistoric lightning strike has been recorded. Although only one standing stone still remains, Site XI or *Airigh na Beinne Bige* was once a full stone circle and overlooked the main Callanish monument. At its centre it has a large magnetic anomaly, caused by a 3000 year old lightning strike. Project leader Dr Richard Bates, of the School of Earth and Environmental Sciences at the University of St Andrews, said:

> *Such clear evidence for lightning strikes is extremely rare in the UK and the association with this stone circle is unlikely to be coincidental. Whether the lightning at Site XI focused on a tree or rock which is no longer there, or the monument itself attracted strikes, is uncertain.*[49]

It is highly likely that this stone circle was placed there to attract lightning. This does indicate the megalith builders had a deep understanding of meteorological phenomena, with the elite having the shamanic skills to be able to control it, which may have made up part of their religious beliefs at the time.

It seems the ancient giants of the British Isles developed this lost science over thousands of years, and no doubt passed the knowledge on from generation to generation. Whichever way you look at it, the legends record their stories and now modern research is starting to shed light on the lost science of the 'sowers of thunder.'

12

GEOMANCER GIANTS

Geomancy is integral to the modern interpretation of giants and fairy myths because of the intimate association such beings have always had with the oldest magical sites of prehistory.[1]

In every county of England, in the mountains of Wales, the provinces of Ireland and the Highlands of Scotland we find parallel giant myths and landscape legends that appear to be encoded information linked to the ancient art of *geomancy*. Recurring themes include dropping stones from an apron, throwing boulders across landscapes, the digging of huge ditches, the creation of hills and mounds, hammer throwing between giants, the legends of the worn shoe, gargantuan pebbles being dropped from shoes, and numerous references to astronomical alignments. In Cornwall, witches wove spells over the landscape, while their giant husbands built megalithic sites marking spots to honour their magic.

 Geomancy technically means 'divination of the earth'. The word comes from Ancient Greek *geomantela* that translates to *foresight by earth*. It is a translation of the Arabic term 'Ilm al-Raml, or the *science of the sand*. It is the discipline of reading the natural elements and frequencies of the earth. Geomancy can also be described like a 'knowing of the earth', an earth-gnosis, or perhaps we should say *geo-gnosis*. Anthony Roberts summed up the deeper meaning of geomancy in *Sowers of Thunder*:

> *Geomancy consisted of modifying certain features of the landscape to blend with the mystical energies emanating from the area's natural shaping and distribution of telluric forces. At the roots of geomancy lies geometry, and the geometrical relationships between all phenomena make up the determining patterns that*

assert geomantic reality in an intellectually definable form.[2]

Roberts' summarised the modern view of geomancy as 'Geomythics', a system that combined multiple disciplines. The general themes are outlined here:

> Geomancy: *Divination of the earth.*
> Geodesy: *Greek Geodaisia translates to 'Division of the earth'. The measure and representation of earth.*
> Geometry: *Earth measurement. Concerned with questions of shape, size, relative position of figures, and the properties of space.*
> Sacred Geometry: *Number in space. Laws of earth measures and nature.*
> Ley (Ley-Line): *Alignment of 4 or more ancient sites over any distance, rediscovered by Alfred Watkins in 1921.*
> Archaeoastronomy: *Alignment and orientation of sites to the Sun, Moon and Stars at certain times of year.*
> Earth Energy Lines: *Natural yin and yang energies emanating from the Earth, that move in random directions and curves, but often connect ancient sites.*
> Telluric Currents: *Alternating electric and magnetic current that form from the earth's magnetic field. They strengthen at sunrise and weaken at dusk. They often form into spirals.*

John Michell wrote about the modern use of geomancy/geomythics in his seminal 'earth mysteries' book, *The View Over Atlantis* (1969). In this, he discussed both eastern and western forms of the ancient artform that all had one underlying theme; plotting sites on an intuitive, yet scientific survey with the hope of maintaining a golden age of harmony overseen by the designers.

One of these geomantic secrets is the alignment of ancient sites across the land. These are what Alfred Watkins in 1921 termed 'leys'. He originally believed they were old paths and documented many of them in his book *The Old Straight Track*. The discovery occurred as a visionary experience and he "... *privately maintained that he had perceived the existence of the ley system in a single flash.*"[3] Prior to his death in 1935 he had stopped using ley in favour of 'alinements' (his spelling).

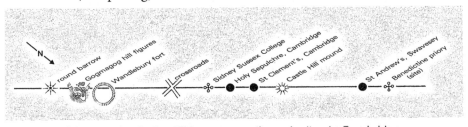

An example of a ley. This one goes through sites in Cambridge.

Hilltops also came into Watkins' equation and he saw Beacon Hills as 'initial points' where leys were surveyed from and used like a rifle barrel to set out the course of the alignment. This makes sense based on our findings, as many of the giants' strongholds are elevated areas, and when the myths are looked at in the context of Watkins' discoveries, they start to correlate to this theory that these were the genesis of the alignments. For example, we have the giant Bolster in Cornwall who would stride over great distances from Carn Brea to St Anne's Beacon, stepping over six miles between them. Carn Brae was a well known giant stronghold and a meeting place for these titans, a place where they discussed working on the landscape with other elemental beings:

> *Some tales say the giants regularly gathered within its ramparts...*
> *watched over quite naturally by the equally ubiquitous fairies or*
> *spriggans. These tutelary guardians were said to have had a vast*
> *hoard of treasure buried in a hidden cave under Carn Brea.*[4]

Watkins provided huge amounts of evidence to back up his theory that leys were the remnants of ancient trackways and his followers set up *The Straight Track Postal Portfolio Club* to coordinate fieldwork which documented more examples. It quickly unlocked a more sophisticated system then moved into the realms of astronomy and orientations to the specific sunrises and sunsets of the solstices and equinoxes. They also found grid patterns, parallel lines and giant landscape geometries on such a vast scale that brought into question how early Britons could have surveyed them.[5]

From as far back as 1911, it has been suggested that straight lines in the landscape may have an elemental force associated with them. In Ireland, traditions of fairies were often said to only move along in straight lines. This tradition was so strong, that people used to cut corners off their homes to allow a fairy path to have a clear way through. The fairies of old were not what we imagine them today. In Celtic tradition they were often very tall, fierce entities who were banished from paradise and inhabited the middle kingdom between the human realm and the otherworldly realm of the dead. They were said to live within the land, or at sacred sites and move around the landscape.

One of the most interesting parts of the straight line mystery is the tradition of *ghost roads* or *spirit paths*. With parallels to the fairy paths, these straight tracks are thought to have deceased spirits and ghosts trapped between this world and the next. Katherine Briggs noted that "*Ever since the first traceable beginnings of fairy beliefs the dead have been curiously entangled with fairies in popular tradition.*[7] Spirit paths were also straight but if buildings were placed along them, the front and back doors would need to have been kept open at night to allow the spirits (or fairies) to pass through without obstruction. Several giant burials featured in this book appear to have been placed upon such spirit paths.

These 'paths' can also be compared to Aboriginal lore. As we discussed in the *Devon and Cornwall* chapter, it has been noted that giant stories encode a supernatural route through the landscape of Britain. All of the geomantic myths form part of an oral tradition that guided our ancestors from one sacred site to another. It was as though it was a fundamental communication and memory tool that become part of the fabric of the early Briton's mindset. These traditions, given orally, were like the television, movies and novels of the day and the characters within them were renowned, just like we adore current-day celebrities. The alignment of sacred sites was part of the story and was not only encoding important geodetic information, it was also very useful to help navigate the wooded landscape of Albion. An example of this can be found in the story of three giantesses at Baglan, which lies between Neath and Port Talbot in Glamorgan, Wales:

> *In the lower parcel seems to be the ruins of two small castles, a third is said to be in Margam parish, reported to belong to three sister giantesses, the three standing upon equal or right line near east and west, and seemingly equal distance, but now scarce the names and ruins are preserved.*[8]

Another well known giant in Devonshire is called Maximajor, who lived on Dartmoor. Like Idris, he resided on an elevated stronghold and could even see across into Wales and down to the South Devon coast. Again, like his Welsh counterpart, this may be a metaphor for *line of sight* along a ley alignment, or surveying the greater landscape.

One strange local letter in a Welsh newspaper from 1851 describes what the writer calls *"the most magnificent Cromlech in Breconshire, sometimes called 'Kingstone', some days, which the giants, then living in the land, used to pitch from its present situation to a distant eminence, and back again."*[9] This strange fragment of folklore found in an obscure newspaper gives a tantalising insight into the lost science of prehistoric surveying.

One of the longest leys in England is the St Michael Line that stretches some 300 miles across southern England from Hopton on the Norfolk coast all the way down to the tip of Cornwall. It aligns to the Beltane (Mayday) sunrise and the Samhain (Halloween) sunset. It takes in many important ancient sites, including Avebury, Glastonbury and St Michael's Mount. Many of the sites upon this line reveal geomantic codes in their stories, and some are associated with giants and giantesses. There are a number of notable towns and villages with Og in their names along this line such as: Ogbourne St George and Ogbourne St Andrew in Wiltshire, Gog and Magog ancient oak trees in Glastonbury, the Gog Magog Hills in Cambridgeshire and tors in Cornwall said to be the home of Gogmagog and other Cornish titans. Therefore, could it have been the *Og Line* before it became the Christianised St Michael Line?

GIANTS AND DIVINING THE LAND

Some chroniclers state that the infamous Gogmagog was described as being so strong that he could uproot an oak tree and shake it like a *hazel wand*. This veiled reference is describing a dowsing instrument. Hazel has been used in divination for millennia and is still used today by farmers, electricians and geomancers to detect underground water, leys, earth currents, and in more recent times gas pipes, electricity lines and cables.

Where Gogmagog was thrown off the cliff by Corineus, "*Across the water, at Torpoint, is a small promontory called Deadman's Point, an interesting coincidence in that a giant of the ogreish, child-eating variety, who formerly enjoyed a bleak reputation in the Gorran Haven District, was bled to death by a wily doctor at Dodman's Point.*"[10] The term 'dodman' may be recognisable to ley hunters as this was the name given to the ancient giant surveyors of Britain. The Long Man of Wilmington in Sussex is thought to be a dodman.

The mention of the enigmatic hazel wand in the founding stories of Britain is also found in Wales relating to the giant Idris. He was one of the *Three Holy Astronomers of the Island of Britain*:

> Lhuyd notes two pairs of extra-ordinarily large skeletons found in the middle of the sixteenth century by peat diggers in a bog at the base of the southern ascent of the mountain at the farm named Llwyn-dol-Ithel, each buried in coffins with hazel rods.[11]

The use of this particular type of wood in esoteric training has been noted through history:

> The early Roman naturalist, Pliny, wrote of how to use hazel wands for divining underground springs. The rod of Moses was cut from a hazel tree by Adam in the Garden of Eden. Moses and

Aaron used hazel rods to bring plagues into Egypt. In the fourth century, St Patrick is said to have rid Ireland of snakes by drawing them together with a magic hazel rod and then casting them into the sea. [12]

Use of a hazel divining rod observed in Britain in the late 18th century
From, *A Tour in Wales* by Thomas Pennant (1726-1798)

It is known they were used in the seventeenth century in Somerset for more practical purposes:

In the lead-mining area of the Mendip Hills in Somerset in the 17th century the natural philosopher Robert Boyle... saw the hazel divining rod (virgula divinatoria) stoop in the hands of the diviner, who protested that he was not applying any force to the twig. [13]

If we return to the grave at Cadair Idris, it appears to be a *reburial* of some sort. Perhaps, even Idris himself, or one of the ancient chieftains who carried this title. Other discoveries in specific places in the landscape could be part of the same tradition of reburying bones to give a supernatural power to the area, often placed upon spirit paths. We have seen the effects of digging up graves, as though sorcery was involved when thunderstorms occurred, so the morbid ritual of the reburial of bones may have existed for thousands of years and been a tradition related to giants.

The bones of Christian saints were often considered sacred and were said to produce aromas and even spiritual energy called 'manna'. In ancient

Egypt, the integrity of the mummified body ensured ascension into the next world. The reburial of bones and hazel rods could also have been to *display* their culture hero, Idris, to the Druidic initiates. It could also have been opened for the bones to be touched during these initiations. A biblical tradition recorded the story of a man who died prematurely and was revived when he touched the bones of Elisha. Skeletons of saints were often buried oriented to the sunrise and even abbeys and churches were joined by straight lines, said to have been trodden originally by saints or angels.[14]

It has also been noted that many Neolithic and Bronze Age burials had particular bones (usually the femur or 'long bone') missing. This was used in rituals and would be the largest bone of the interred giant to show others to impress or remind incomers of their ancestors ownership of this territory. The mummified bog bodies found in Ireland date from between 2141 - 1910 BC and were specifically placed in bogs on the boundaries of territories that surrounded the inauguration hills of local kings. This tradition lasted for over 1000 years.[15]

Were the bones of Idris designed to remind future generations of the archaic landscape power the giant had over his domain? The placement of skeletons or bones may have been marking geomantic spots as well as being used for territorial purposes. If you could dig up a grave on your land, and prove they were your ancestor, you then maintain the rights to this particular area. If they were giant bones, even better. The locations where these burials were found may have a deeper meaning than previously thought, recognised by the ancient Feng Shui Masters of China:

> *The Chinese attached great importance to the influences which played over the bodies of their ancestors, believing them to control the future course of the family's fortune. Great dynasties were said to have arisen from the particularly favourable placing of an ancestor's tomb, and the first action of the central government, when faced with revolt, was to locate and destroy the family burial mounds of the rebel leaders.*[16]

GIANTS, DRAGONS AND SERPENTS

Llyn Cau, the beautiful glacial lake on Cadair Idris, is not only said to be bottomless, but the home of a *water dragon* which terrorised the locals. King Arthur is said to have captured it, tied it behind his horse and dragged it up

331

the mountain to release it in Llyn Cau.[17] Hundreds of other stories, including those of St George and St Michael, have the hero slaying the dragon, but what does it really mean and how does it connect to giants? These 'dragon lines' are recorded in tradition all over the world in different forms, such as serpents, worms and other types of snake-like monsters, even serpents with wings in American creation stories such as Quetzacoatl in Mexico.

Another interesting dragon reference is in this description of a giant called *Myfyr* who was one of three giant brothers in Shopshire near the Welsh border. He was often connected to Paganism and Druidry and in this story seems to be tracking earth currents up a mountain:

> *One wonders if Myfyr again is climbing this Maen Chwyf (shaking stone) and following the windings of the serpent? If so, remember that he is not following Druidism but some old paganism vanished from the earth long before Druidism, unfortunately, had its birth.*[18]

Piers Shonks of Hertfordshire was a giant dragon slayer and the symbolism found in his stories encoded multiple geomantic insights, not only working with the energies of the earth, but literally terraforming the landscape shifting millions of tons of earth and stone. Some of these we can see in the mega dykes, earthworks and massive stone circle complexes:

> *There is frequent mention of named giants being involved with the artificial structuring of the landscape (geomancy) and this is where the exoteric and esoteric aspects of giantism are seen to met and mesh. Tales are still told today delineating the giants' role in earth-moving, rock-piling, river-shaping and performing all the other numerous functions attributed to a geomantic civilization.*[19]

LANDSCAPE TRIANGLES OF THE GIANTS

A tantalising geomantic snippet regarding the domain of the giant Idris was recorded by Iolo Morganwg who stated that an area in Wales called the *Great Triangle* was in fact known as *Idris' Triangle* (Tryfal Idris). We also have this folk-story from 1860 detailing the laying out of a triangle in the landscape with large stones:

> *One day when Idris the giant of eternal memory was meddling with the stars in his seat [Cadair Idris], he threw three stones, one to the east, the point of the rising sun; another to the south,*

their highpoint in the heavens, and the final one to the west, 'The evening's lodging where lodges the sun'. This eastern stone is the standing stone on the high meadow. The other two include the one on the Hengae mountain near Aberllyfeni, and the other is Llech Idris.[20]

One of these standing stones still exists next to the A494 between Bala and Dolgellau. One of the others is built into a dry-stone wall on the hillside to the northeast, which lines up with this stone.[21] Also, a ten-foot-tall standing stone near the village of Bronaber has the name *Llech Idris*. It was said that the giant threw it here from the summit of Cadair Idris, suggesting again, a significant alignment. It is also precisely north of the burial place of Idris, where the giant skeletons were uncovered in 1685. This is 12.4 miles away which is 1/2000th of the earth's polar circumference, and one-tenth of the Lunation Triangle distance between Stonehenge and Lundy, suggesting precise measurement systems in the landscape. If we extend from the burial site of Idris further north past Llech Idris, we reach *The Giant's Staff* (Ffon-y-Cawr) standing stone in Conwy. This menhir stands alone near the Maen-y-Bardd burial chamber, a site it is linked with in legend.

In this image (below) it shows the triangle that incorporates three of the aforementioned sites. From bottom left: Cadair Idris' summit; bottom right the standing stone on Hengae Mountain, and Llech Idris at the top. Less than one mile west of this menhir is another *Barclodiad y Gawres*. This cairn, whose

Left: The Triangle of Idris and the meridian line to Llech Idris. Right: Llech Idris standing stone forms one corner of the triangle and also sits directly north of the giant skeletons unearthed below Cadair Idris.

name translates as *The Giantess' Apronful*, has the same name as one on Anglesey and is composed of a heap of boulders with a cist on the northern side.

The stories of Idris Cawr also continue in different parts of Wales, including the fable of the three stones that fell from his shoe in the *Lake of the Three Pebbles* at Graienyn. Notably, one of the pebbles that landed there was 24ft wide, 18ft across and 8ft tall. They must have been big shoes! These three pebbles almost certainly represent a triangular configuration of stones.

In his research connecting Stonehenge with Lundy Island and the bluestone site in Wales, Robin Heath discovered gigantic, geometric landscape triangles encoding astronomical and geodetic information, eerily echoing the secret traditions in the stories of Idris. Firstly, Lundy is 123.4 miles directly west from Stonehenge, located in the Bristol Channel and fits into a massive Pythagorean Triangle. On Lundy we find stone throwing giant legends and skeletons nearly nine feet tall. At Stonehenge we have multiple accounts from 9ft to 14ft skeletons and legends of giants in many forms. In Preseli, the source of the bluestones of Stonehenge we have numerous skeletons and legends and most specifically, we have *Carn Enoch* (Idris was also known as Enoch).

Robin Heath's landscape triangles of Britain. Both have Pythagorean geometry, two thousand years before Pythagoras was born. Both also have giants connected to all points on the two triangles.

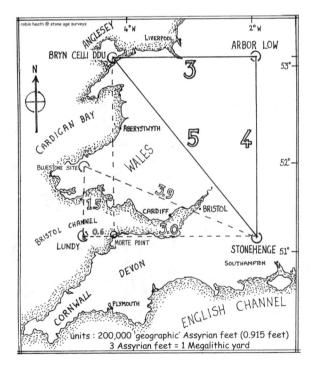

If we hover along the east-west alignment between Stonehenge and Lundy, we can see that both Glastonbury and Dunster Castle are on the horizontal line, both places where giants were discovered, and both linked with King Arthur, whose mythos was overlaid upon many of these earlier giant stories. Perhaps this is a secret tradition where the ancients utilised the power of their ancestors literally, by reburying their giant skeletons at important geodetic locations.

Furthermore, if we look at the other 3-4-5 Pythagorean Triangle, we have Bryn Celli Ddu in Anglesey, that we know a very long bone was unearthed from (see p.200), and Arbor Low stone circle, where nearby a giant skeleton was found (see p.139), and of course Stonehenge. Are these coincidences, or do these major stone sites, along with these burials, mean something more?

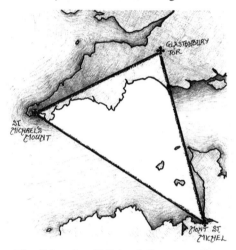

The Triangle of Michael discovered by Dion Fortune, and presented by Yuri Leitch at the Megalithomania Conference 2016.

Dion Fortune, in *The Goat-Foot God*, shares a map she named the 'Triangle of Michael' connecting Glastonbury Tor, St Michael's Mount, and Mont Saint-Michel in Normandy, France. Amazingly, she drew a segment of the Michael Line long before John Michell rediscovered it in its entirety in 1969. Yuri Leitch revealed this anomaly at the Megalithomania Conference in 2016. To the authors, this is significant as it not only includes traditional giant sites at each point of the triangle, it also encompasses the Devonshire coast where Brutus landed and defeated Gogmagog.

The alignment between Mont Saint-Michel and St Michael's Mount in Cornwall has legends of giants throwing stones between them (see p.83), and if you take the line between Mont Saint-Michel and Glastonbury Tor it reveals a stunning country-length alignment with many important sites. The authors were astonished how many of the places that fit perfectly on the line are associated with giant accounts and myths we feature in the book.

The alignment, projected from Mont Saint-Michel into England first enters the country at Portland on the Dorset coast. Then it passes through many powerful sites of sacred magical spiritual significance including the great hill-fort of Maiden Castle; Glastonbury Tor; Caerleon the legendary court of King Arthur; Lake Bala the sanctuary of the goddess Ceridwen and her magical cauldron; Great Orme's Head (the great dragon's head); to the Isle of Man; then up through the Western Isles of Scotland passing through the Isle of Skye to the standing stones of Callanish.[22]

The Ogmios Line, discovered by Yuri Leitch, based on the 'Triangle of Michael' from Dion Fortune. It shows a higher than expected number of sites related to giants. Caerleon was an Arthurian site that once had a 'Giant's Tower' at its centre.

QUOITS, PEBBLES & HAMMER-THROWING GIANTS

The *pebble in the shoe* is another recurring geomantic theme, although the pebble to mere mortals would be of gargantuan proportions. The giant Moll Walbee (known also as Maud de Breos), built the great castle of Paincastle in Wales and Hay Castle in Powys. She dropped a pebble nine feet long and threw it over three miles away.[23] Moll Walbee's stone is now placed inside Llowes Church in Powys where it has been since 1956. Arthur's Stone, a dolmen on

Cefn Bryn common, Gower Peninsula, West Glamorgan was merely a small pebble lodged in his shoe. At the time he was on his way to fight the Battle of Camlann so he dislodged it and threw it seven miles to where it now stands.[24]

Yr Ogof Cromlech in North Wales. One of many types of megalithic sites said to be constructed by giants.

The Long Man of Wilmington hill figure in Sussex appears to have surveying staffs in either hand. The two giants of Windover Hill and Firle Beacon used to throw boulders to each other in legend. It also marked a series of leys linking it with megaliths in the area (see p.106). Lundy Island also had a renowned boulder throwing giant who hurled stones from the Island to the Devonshire coast after he was exiled there. Two giants who were fighting on Benaughlin in Kinawley, Ireland, threw large stones five miles distant and their finger marks are still there to be seen on one of the stones.[25]

Benaughlin Mountain, or *Binn Eachlabrhra - Peak of the Speaking Horse.*

One strange local letter in a Welsh newspaper from 1851 describes what the writer calls:

> ...the most magnificent Cromlech in Breconshire, sometimes called "Kingstone", some days, which the giants, then living in the land, used to pitch from its present situation to a distant eminence, and back again.[26]

This geomantic mapping system features in other stories, like this one:

> The giantess Melangell sleeps on a 'Giant's Bed' at Pennant Melangell. The Rhuddgaer stones crossing Afon Braint are known as 'Giant's Stepping Stones'. The 'Giant's Stone' in Turton was thrown by a giant from Winter Hill.[27]

The traditions of 'quoits' or 'coetans' were particularly prevalent in the Western coastal regions of Wales where there are concentrations of megalithic sites. The Welsh titans were so strong they could throw boulders all the way to Ireland. The chief of the Aeron Vale Clan was said to have launched the largest quoit across the channel where it landed in Erin's Valley. The Trichrug giant of Cardiganshire played a godly game of quoits and again his attempt reached Ireland.

In London we have the legend of two giantesses hurling a hammer between Putney and Fulham who would respectively yell "Put it nigh!" and "Heave it full home!" where the names of Putney and Fulham were said to be derived from.[28]

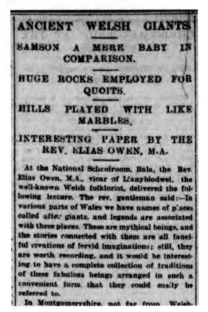

Article from the Western Mail, 27 December 1898 on the games employed by giants in Wales.

Another hammer myth is recounted in the parish of Gymyn in Wales. A ruined farmstead call Hammer's End is derived from the story of two giants throwing hammers and where the long hammer landed they built the local church, and where the short hammer landed they built the farmstead.[29] Once again we visit Trencrom Hill in Cornwall where the story of a builders hammer being thrown over great distances in straight lines back and forth has been recorded.

In one tale, it involved killing a giant's wife who got caught in the cross fire, but revealed that getting in the way of geomantic lines of force can be dangerous.

In the previously mentioned interview with Margaret A. Edwards from the Tanat Valley area of Montgomeryshire, it related " *...that a giant from one valley shouted at the other to borrow a hatchet. The other threw it and where it landed the river Llawenog started, and it was the hand of Enoch that threw the hatchet from Pen-yr-eryr."* It is also worth noting that the Tanat Valley is in the vicinity of Cadair Idris, the mythical home of Idris (who, as we'll soon see, is related to Enoch, the patriarch of the Bible).

The hammer and hatchet throwing mythos is found all over Britain and in Europe, including of course, Norse traditions with Thor. The earliest stories seem to relate primarily to giantesses, possibly a memory of the earlier earth goddess tradition that was in place going back to the Neolithic and Mesolithic eras. In the example from St Michael's Mount, where Cormoran's wife was ordered to carry stones in her apron, then got hit by a hammer by one of the male giants, this may represent a change in the archetypal role of the feminine transforming into the patriarchal mindset that appeared in the Bronze Age.

Another theme that turns up all over Britain is that of a giant firing an arrow to mark his own grave, rather than be imprisoned or killed randomly by his captors. This is found at Weston Church in Hertfordshire where a giant femur bone was unearthed (see p.107). At Brent Pelham, also in Hertfordshire, we have the story of the 23ft tall Piers Shonks (see p.111), and rather than an arrow, in East Anglia the giant Tom Hikathrift kicked a football rather than an arrow six miles away that eventually marked his burial place. He was also noted to throw hammers over vast distances. *It would seem obvious* that all these traditions are really talking about measuring the landscape in various ways encoded in myth.

Apronful of the Giantess

One of the most frequent myths we find is the giant or giantess striding across the landscape and dropping stones from an apron. The stories always have them carrying earth or stone in an apron and the string breaks, with the contents falling in a certain spot creating the specific ancient site:

> The 'broken apron strings' motif appears throughout British giant-lore as well as in Scandinavian, Indian and American legends, usually in conjunction with alignments of megalithic stones or, sometimes, linked to single free-standing boulders.[30]

On St Michael's Mount, Cormoran's wife was ordered to bring over white granite in her apron, but when caught bringing a different type of stone he then kicked her and broke her apron strings causing the boulders to fall. There are often particular types of stone mentioned in the stories, including white quartz, that is often found in graves of the ancient people, with later generations decorating their tombs with it, like Newgrange in Ireland.

We find similar stories of the apron throughout the British Isles. For instance, a giant called Rombold dropped great apronfuls of stones on Ilkley Moor in West Yorkshire. At Binsey Fell in Cumbria, a giant ghost is often seen wearing an apron. In Wales there are dozens of accounts linking sites across the country with giants, giantesses and Devils wearing aprons. Wade's Causeway, on the North York Moors has a giantess who carried stones for her husband Wade. In Cornwall at St Agnes Beacon, early traditions say Bolster's wife carried stones to the top of the hill.

Loughcrew in County Meath was once known *Sliabh na Cailleach*, or the 'Mountains of the Witch' (or 'hag'). Arguably the most important aspect of the site is the equinox alignment in Cairn T, where the sun illuminates symbols carved in stone at the back of the chamber on March 21st and September 21st. Several of the burial chambers were said to have been dropped in place by an archaic witch from her apron. However, if we look further back into the prehistoric lore, especially in Ireland and Scotland, we find the goddess Cailleach is represented originally as a giantess, a creator goddess who was associated with fertility and harmonising the landscape. Only later were the 'hag' and 'witch' labels applied to her in Christian times. Later still, and not only in Ireland, the giantess became 'the Devil' wearing the apron.

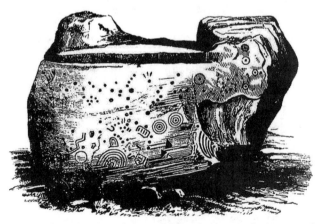

The 'Hag's Chair' at Loughcrew, with strange symbols carved on it.

At Loughcrew the giant goddess Cailleach was sometimes called *Garavoge*, who came from the northwest:

> *Determined now her tomb to build,*
> *Her ample skirt with stones she filled,*
> *And dropped a heap on Carnmore;*
> *Then stepped one thousand yards, to Loar,*
> *And dropped another goodly heap;*
> *And then with one prodigious leap*
> *Gained Carnbeg; and on its height*
> *Displayed the wonders of her might.*
> *And when approached death's awful doom,*
> *Her chair was placed within the womb*
> *Of hills whose tops with heather bloom.*

Jonathan Swift, c. 1720

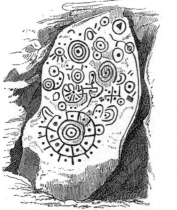

Carved stones from Loughcrew with symbols that may represent the divine feminine.

In some stories Cailleach had blue skin and one eye, not unlike the ancient stories of the cyclopes whom were also constructors or builders. She is an extremely ancient creator deity.

In some traditions she was known as the 'veiled one' and sometimes only had one eye and was blue-skinned. Whenever 'veiled' is mentioned in folklore, it is worth taking notice, as it may contain different levels of information. This does seem the case with these geomantic stories linked with giants. The one-eyed motif is also interesting as the Cyclopes of ancient Greece were known as builders of megalithic sites, associated also with metalwork and technical skills, sometimes using a hammer or other metal implements. Often, Cailleach is a hammer carrier, with which she would shape the landscape.

At *Barclodiad y Gawres* (Apronful of the Giantess) on Anglesey, it is said the giantess was Ceridwen, a Druidical goddess and her apron in Welsh is 'barclog' or 'arffedog', however the meaning of these relate to being taught by a *ward* or *keeper* of the mysteries. These mystical feminine themes repeat over and over again very specifically, suggesting an early mother goddess tradition may be connected with geomantic workings in the landscape, and may indicate that these stories are very old.

The carvings on the stones at both Loughcrew and Barclodiad y Gawres are similar, and anthropologists such as Marija Gimbutas have stated that these all represent various forms of the divine feminine, a goddess-inspired civilisation, who were eventually overthrown by a male dominant culture in the Bronze Age.

When tracing back the word apron to old latin, American researcher JJ Ainsworth found that it translates to 'mappa', meaning napkin and tablecloth as well as 'map', suggesting the apron had more than one meaning. Often old maps were made of leather or cloth, and could have been laid out over tables to show the map. Once it

Barclodiad y Gawres carved stone from within the chamber.

had been used, it could then be folded and bundled back up into an apron. If we apply this new information to these legends, we see how this illuminates a whole other meaning to these stories. The breaking of the strings may also represent exact measurements of where to place sites on the landscape. Knots are also sometimes mentioned in the old stories, and were known to be part of the technique of laying out the geometry of stone circles, the knot marking a particular length along a piece of string or rope. The use of knots on ropes was utilised to enable an accurate survey to take place, whether over a few feet or on a much larger scale.

In other prehistoric civilisations, the priestly elite often wore aprons for sacerdotal reasons, most notably in Egypt. The wearing of aprons in ancient cultures is widespread, and is still in use today by modern Freemasons.

In the images provided by JJ Ainsworth, we find the 'rosette' solar symbol, as well as the 'acacia' of 'wheat' symbol at Loughcrew. Right: Mason's Apron depicting the solar symbol and the wheat or acacia, symbols found at Loughcrew.

Aprons being worn by Freemasons may have a deep history and may have been worn by the early 'operatives', who were the actual stonemasons of the craft.[31] In certain initiations and meetings, the apron must be worn, even today:

> *A Masonic Apron is the most essential physical representation of a man's commitment to the Craft, and therefore should be made of the highest quality to represent the Brother who wears it.*[32]

Further aprons are found on Assyrian and Babylonian statues, with some on display at the British Museum. At the Minoan palace of Knossos in Crete, a famous snake goddess wears an apron, and other symbology relating to surveying is related on her head; a dove along with two serpents in her hands, possibly the surveying staffs of the ancient gods. In Greek tradition, releasing a dove signifies the genesis of a land survey, or a particular astronomical alignment. Similar themes are found throughout Europe:

> *In Scandinavia there is a triple goddess called the Mo Braido, the Mamma/Omma and the Kaelling/Karring. The Karring being the old crone figure who controls the winter storms, guards fresh water wells and has a black rod that spreads ice and frost. She also has a tinder box that produces the sparks of life. She creates the landscape by dropping boulders and stones from her apron and she lives in a cave.*[33]

New England has a giant called Maushop who was based on the island of Martha's Vineyard and recorded in other areas of Massachusetts. He was a stone-throwing, apron-wearing giant who shaped the landscape and dropped large stones at certain locations. In nearby Easton, the settlers remembered a story (that later became associated with the devil) of a giant who dropped stones from his apron, when the strings broke and created a local rock formation.[34] Geomantic principles, it seems, are hidden in folklore all over the world.

IDRIS AND THE FOUNDING OF GEOMANCY

Idris was the tribal leader of the giants of Wales and traditions of his astronomical and geomantic prowess are found all over the country. However, his name is also found in very ancient books of the Bible and the Qur'an. Muslims believe Idris was the second prophet after Adam and Islamic tradition has identified Idris with the biblical Enoch (and is also associated with Hermes) but is this the same Idris as the Welsh Holy Astronomer?

William Blake's Enoch Lithograph, 1807.

Exegesis narrates that Idris was among, *"the first men to use the pen as well as being one of the first men to observe the movement of the stars and set out scientific weights and measures."*[35] Further Muslim sources during the eighth century reveal that Idris had two names, Idris and Enoch, so can we conclude that Idris and Enoch are one and the same, and if so, how did the Welsh Idris end up being so closely associated with Enoch?

Idris/Enoch is also remembered in Islamic tradition as the first scribe and introduced the technical arts of civilisation, including the study of astronomy. He was also said to have introduced 'construction techniques' and at some point buried his knowledge beneath the earth. With Göbekli Tepe demonstrating the earliest megalithic temple building, as well as being deliberately covered over and its placement in the heart of the 'Garden of Eden', can we conclude there are connections here worthy of more investigation? The Welsh Idris was routinely involved in constructing megalithic sites and placing monoliths in the landscape, often being astronomically aligned.

Hugh Evans, author of *The Origin of the Zodiac*, found by analysing the old Welsh language that 'Id' means 'Lord', from Udd, and Ris means 'course' from Rhys. 'Lord of the course' is the result which could mean someone who records stars and their movements. It is also interesting that 'En' translates to a 'being' or 'spirit', and 'Uch' means 'above' or 'over', so the name Enoch could also mean 'spirit above.' In *Genesis 5:24*, Enoch was said to have been 'taken up to heaven.' Gwynedd has an equivalent meaning to 'heaven' in old Welsh.

Idris was understood by many early commentators to be both a prophet as well as a messenger. This correlates to the Welsh Idris, who sat upon his mountain and surveyed his kingdom and the heavens and was said to have

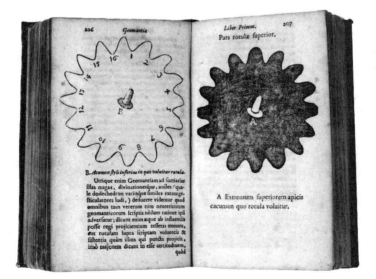

Relevant pages from Fludd's book on geomancy, 1687.

had the power to prophesise the past and future from his domain.

The genesis of geomancy goes back to the times of the biblical patriarchs and to our astonishment, Idris was its mythical founder. According to one Arabic Hermetic text Idris saw the angel Jibril (Gabriel) in a dream.[36] After asking for enlightenment, Jibril drew a geomantic figure and Idris requested to be taught the full science of the geomantic arts. He was also said to have written the first book on geomancy from what Jibril had taught him.

It appears that the origins of geomancy came from a very early, mystical source linked with Idris/Enoch, and we find evidence of his accomplishments in both the Bible lands and in Wales. In the next chapter we will outline further revelations about the primary 'Holy Astronomer of Britain', taking a step towards understanding the true origins of the giants of Albion.

13

GENESIS OF THE GIANTS

The mystery of where the giants of Albion originally came from may be found, in part, in many of the myths and legends we have encountered in this book so far. There are hundreds of accounts of giant bones, skulls and skeletons in the historical and academic record that also provide clues. We have unveiled hoaxes and exaggerations, yet we are still left with too many solid reports that correlate with the myths, and provide tantalising clues as to the real origins of the giants. The knowledge of the geomantic, astronomical and geometric transformation of the entire landscape was known about at the time of the Druids, yet their legacy goes back to a much earlier epoch.

BRITAIN'S EARLIEST GIANTS

Perhaps the most intriguing of all early hominids in Britain is that of Homo heidelbergensis who lived at Boxgrove in Sussex. The skeletal discoveries were described as *"tall and imposing"* and were dated to 500,000 years ago: [1]

> *These people were strong and muscular, with large brow-ridges and relatively large brains. They shaped tools with precision. Their handaxes and other stone tools have been found at several sites around Britain. Microscopic analysis shows that tool marks on animal bones lay beneath gnawing marks inflicted by wolves and hyenas. The people at Boxgrove probably had to butcher the animal carcasses while defending them from scavenging predators.* [1]

In Kent there are the gigantic Cuxton handaxes (see p.104) and the massive Furze Platt example from Maidenhead dating from this era (see p.44). We learn from esteemed anthropologist Lee Berger that a branch of Homo heidelbergensis were routinely over seven feet tall (in South Africa), with Prof.

Francis Thackeray of the University of Witwatersrand, claiming some were twice the size of normal human beings.[2] Were these the original inhabitants who ruled the British Isles at the very earliest of times?

> *According to one theory, Neanderthals, Denisovans, and modern humans are all descended from the ancient human Homo heidelbergensis. Between 300,000 to 400,000 years ago, an ancestral group of H. heidelbergensis left Africa and then split shortly after. One branch ventured northwestward into West Asia and Europe.[3]*

CRO-MAGNON MAN AND THE GIANTS OF EUROPE

Cro-Magnon man appeared suddenly around 44,000 years ago in Western Europe. Their remains revealed that they were very tall (up to seven feet in some cases, but often cited as being between 6ft and 6.5 ft tall). They had pronounced features, such as long broad skulls with a higher cranial capacity than modern man:

> *They are thought to have stood on average 176.2 cm (5 feet 9 ⅓ inches) tall, though large males may have stood as tall as 195 cm (6 feet 5 inches) and taller.[4]*

Cro-Magnon skulls and burials have been found all over Europe and Britain and seem to represent another wave of giant human beings in Britain. An upper jawbone of a Cro-Magnon was found in Kents Cavern in Devon in 1927 that has been dated to 44,200 - 41,500 years old, whilst the Red Lady of Paviland in Wales dates to 33,000 years ago (see p.213). Cro-Magnon traits did not die out quickly, and their characteristics were found in the Magdalenian populations in France until around 10,000 BC.[5] A question that is yet to be answered fully is where did they come from?

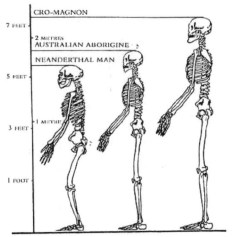

Cro-Magnon height chart.

Their beautiful cave paintings are found in Lascaux, France, and at Altamira in Spain, and other areas, but these are all on the western edges of

Europe. Their legacy is not found inland further east or south. It is found in the west, where seafarers would have arrived from across the Atlantic Ocean. The natural currents of the Gulf Stream are known to have carried migrants across the channel from at least 7000 BC,[6] but perhaps they came over much earlier. In *Giants On Record* we looked at the possibility that early humans may have migrated from the West, possibly America itself, because tools, skulls and other remains seem to match those of the Cro-Magnon in Europe, many of them contemporary with this date or older.

Wermer Muller in *America: The New World or the Old?* (1989) described four types of people existing in North America many thousands of years ago. These included the Salish, the Sioux, the Algonquin-speaking peoples, and a fourth group who were tall "white-skinned, bearded people." The main three Native American groups were terrorised by the bearded group and migrated to other areas at different times to get away from them. *"The white-skinned people moved eastward across the North Atlantic into what is now Scandinavia and western Europe."*[7] Muller dated this expansion to 42,000 BC, the exact time Cro-Magnon man turned up in Europe.

DENISOVANS IN EUROPE?

In *Denisovan Origins,* Andrew Collins and Greg Little suggested that the enigmatic Denisovans (who we discussed in our last book as probably being very tall and robust) may have had a presence in ancient Europe.[8] It is likely that their genes made their way to Britain, as the landmass was connected to Europe until around 8000 years ago. Their genes have also been found in Greenland and Iceland. The Chancelade skull found in France in 1888 may indicate what Collins and Little called a Neanderthal-Denisovan hybrid who

Cast of skull of Chancelade man.

appeared to have distinctive Cro-Magnon type features. He could even have been a Solutrean dating back between 20,000 and 17,000 years. The Finns (the Saami in particular) have a small amount of Denisovan DNA with seven percent of their ancestry from Northern Asia.[9]

In Oase, Romania, early human fossils may indicate this westward Denisovan trail. A skull which had archaic features, including a sloping forehead, large mandible (lower jaw) and molar teeth that *"exceed the dimensions of any human M3s (third molars) known from the last 500,000 years."*[10] The dating of this skull is thought to be around 34,000 years old, and although there are hints of Denisovan ancestry, the size of the teeth does stand out, as the two Denisovan teeth found in the Altai region of Siberia have twice the chewing surface of regular human molars.

Recently, an enormous human skull was found in China which has been named the 'Dragon Man', a description derived from the Long Jiang or Dragon River in the Heilongjiang province of China. The skull was actually found in 1933 by a farmer who hid it from Japanese soldiers in a well. He shared the location of the skull with his grandson in 2018, who brought it to Chinese scientists. While some believe it may be a new human ancestor, the consensus appears to be that it is a Denisovan skull and it has been scientifically established to be at least 146,000 years old. What is remarkable is that it is the largest Homo skull ever found. Chris Stringer, a highly respected anthropologist at the Natural History Museum in London stated on the BBC website, *"This is the biggest human skull I've seen – and I've seen a few."*

SOLUTREANS IN BRITAIN

Around 20,000 years ago, a group of hunter-wanderers called the Solutreans were present in Europe. At roughly 17,000 years ago we find evidence of them in Britain. It is thought they migrated westwards and northwards, possibly hugging the coastlines on their way. However, their distinctive stone tools have been unearthed in Ipswich, Southwald, Charsfield and Hoxne,[11] all nearer the east coast. Dr Bruce Bradley suggests they, in fact, went north via the Western coastline of Britain and reached Orkney. One thing is certain, these guys were known to be very tall and powerfully built, perhaps even carrying the Denisovan genes with them.

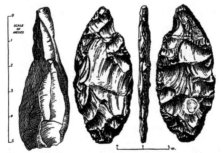

Solutrean artefacts found in Suffolk. Graving tool from Ipswich (left) and blade from Southwald (right).

The Solutreans carried the Haplogroup X gene that seems to come from nowhere, and has led many researchers to speculate that they came from a lost landmass in the Atlantic Ocean. Once again, Atlantis, Hy-Brasil and even Lyonesse may hold clues to this mystery. High concentrations have been found in Israel, with some researchers suggesting this may even be its origin point. Haplogroup X has been found occasionally in Western Britain, Europe and also in North America. This might have been the original Clovis culture of America who were wiped out by the Younger Dryas Impact Event circa 12,900 years ago:

> *It is interesting that a modern population with one of the highest percentages of the X2 clade, higher even than Native American populations, is found in the Orkney Islands off the coast of Scotland.*[12]

The ones that did make it to America, may have been remembered as invading giants, who became the elite in the Northeastern areas. They had knowledge of hunting, fishing and held secrets and the power over life and death,[13] perhaps founding, or at least influencing the Red Paint People and the early Algonquin speaking tribes. In *Giants on Record* we hypothesised that these Solutreans may have been the first recorded giants in the Americas, and may have bought the Haplogroup X gene to its shores. The Solutreans were described as a "powerful caste" and a "ruling elite" who dominated the pre-existing tribes of Europe, and quite possibly the Americas.

> *Some geneticists and archaeologists suggest that haplogroup X points to an ancient migration from Europe, in addition to ones from Asia.*[14]

New underwater research off the coasts of Orkney has pushed the dates of habitation there to much earlier than previously recorded. It is also here that the genesis of British stone circle construction was initiated around 3100 BC. Although the earliest dating of stone sites in the Orkney Islands is 3700 BC at the Knap of Howar on Papa Westray, there has been evidence found of much earlier Mesolithic activity:

> *The discovery on the Orkney Isles of various highly diagnostic stone tools of Mesolithic age could also help show that the Ness of Brodgar settlement has Mesolithic roots. One of the stone tools, an arrowhead found on the site of the Ring of Brodgar stone circle*

*northwest of the Ness of Brodgar site, displays a very distinctive
style of retouching to its tip. This appeared for the first time at the
end of the Upper Paleolithic age as part of the so-called Swiderian
tradition.*[15]

The Swiderians were descended from the Solutreans in Northern European
countries. Swiderian points have even been discovered in the Ness of Brodgar
excavations.[16] An early submerged henge was discovered off the coast of
Ornkey in the Bay of Firth, that should push the dating of large earthen or
stone structures much further back than ever realised. Archaeologists Richard
Bates and Caroline Wickham-Jones used sophisticated sonar technology to
detect the massive anomaly. It has been suggested that the early inhabitants of
Orkney came from a lost land now submerged under the cold coastal waters of
Britain that joined Britain to Northern Europe.

The Bay of Firth, Orkney, where the gigantic underwater henge was
discovered, with Cuween Chambered Tomb in the foreground.

SUNKEN ISLANDS AND LOST REALMS

Doggerland was a huge area of land, now submerged beneath the southern
North Sea. Rising sea levels flooded it sometime around 6500 - 6200 BC. It
covered a massive area from where Britain's east coast is now to the present-
day Netherlands, the west coast of Germany and the peninsula of Jutland in
Denmark. In 1931 a fishing trawler dragged up artefacts from this area that
revealed the archaeological potential that exists there. A recent published
report from the University of Bradford states that a massive tsunami devasted
Doggerland in 6150 BC:

Evidence of the catastrophic event has already been found in onshore sediments in Western Scandinavia, the Faroe Isles, northeast Britain, Denmark and Greenland but now for the first-time confirmation of the event has been found on the UK's southern coasts. The giant tsunami, known as the Storegga Slide, was caused when an area of seabed the size of Scotland (measuring some 80,000sq km and around 3,200 cubic km) shifted suddenly. This triggered huge waves that would have brought devastation to an inhabited ancient land bridge, which once existed between the UK and mainland Europe - an area known as Doggerland - that is now submerged beneath the North Sea.[17]

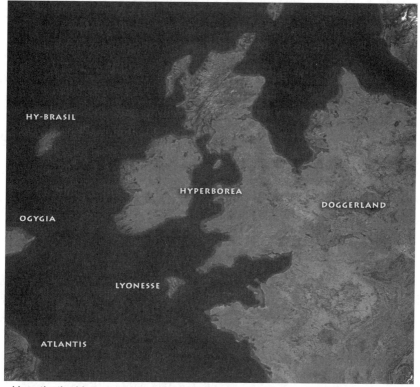

Hypothetical interpretation of the lost lands of Britain c.12,000 years ago.

If this destructive event flooded such an enormous landmass in ancient times, what other locations might have also been swallowed up by the sea? The legendary sites we will discuss here may be candidates for just such an occurrence. We have already discussed the notion that survivors from lost lands were memorialised as the bringers of culture to the British Isles. The traditions are crystal clear about this fact and they were often described as beings of giant stature.

The exceptionally tall Tuatha De Danann, were said to have arrived from four lost island cities in the Atlantic which were Murias, Falias, Gorias and Findias. The first recension of *Lebor Gabála Érenn* (The Book of Invasions) describes the Tuatha De Danann as having resided in "the northern islands of the world", where they were instructed in the magic arts, before finally arriving in Ireland. They brought with them advanced weapons, technology and megalithic construction expertise.

The sleeping prophet Edgar Cayce, in a well-known reading, stated that the island of Atlantis was situated in the Atlantic and that giants from 10 to 12 feet tall were part of its long-lived population. Like Plato, Cayce claimed that the island was destroyed in a cataclysm roughly 12,000 years ago. Science seems to have pinpointed this cataclysm as the Younger Dryas Impact Event that occurred in 10,800 BC. Evidence has recently turned up in Chile and South Africa expanding the field of this global destruction. It appears to have been a meteor strike to the North American ice sheet. This impact caused massive flooding and destruction on a worldwide scale. Plato in his *Laws, Book III*, argues that the destruction of Atlantis occurred ten thousand years before his time (10,600 BC). This is only 200 years off from the Younger Dryas Impact.

Tír na nóg was a mysterious island in the Atlantic described as a supernatural realm of beauty, abundance, joy and longevity. The giant sea god Manannan Mac Lir was said to travel there on a three-day journey from his stronghold on the Isle of Man aboard his mechanised silver sea vessel. Tír na nóg was also called the land beneath the waves, which may have been alluding to the fact that it is now submerged. Mac Lir was the god who ruled this region and was said to be the first ancestor of the human race and the god of the dead.

Aelian (175–235 AD), in his *Various Histories* shares the following about a mysterious lost land in the Atlantic:

> *Theopompus relates the particulars of an interview between Midas, King of Phrygia, and Silenus. Silenus told Midas: 'Europe, Asia, and Libya are islands washed on all their shores by the ocean, and there is but one continent, which is situated outside these limits. Its expanse is immense. It produced very large animals, and people twice as tall as those common to our climate, and they live twice as long.*

"Long lived" and "giant sized" are the consistent descriptions given to the inhabitants of these sunken lands by chroniclers and the great mystics, Rudolph Steiner, Madame Blavatsky, Manly P. Hall and Edgar Cayce.

Before the birth of Plato, Homer described a now lost enchanted island in the Atlantic which was called Ogygia in Book V of *The Odyssey*. He claimed it was the home of the nymph Calypso, the daughter of the Titan Atlas. Madame Blavatsky made the case that Ogygia may be Atlantis in her *Secret Doctrine*:

> However altered in its general aspect, Plato's narrative bears the impress of truth upon it. It was not he who invented it, at any rate, since Homer, who preceded him by many centuries, also speaks of the Atlantes (who are our Atlanteans) and of their island in his Odyssey. Therefore the tradition was older than the bard of Ulysses.[18]

The Alexandrian geographer Strabo (63 BC - 23 AD) proposed that Ogygia was located in the middle of the Atlantic Ocean. Plutarch (45 AD - 127 AD) informs us that, *"The Ogygian Isle lies far out at sea, distant five days' sail from Britain, going westwards, and three others equally distant from it."*

The mythical sunken land of Lyonesse was traditionally the abode of giants. It is thought to have once existed off the far western coast of Cornwall between St Michael's Mount and the Isles of Scilly, within a thirty mile stretch of ocean. The legendary Celtic tale of Tristan and Iseult (that may have been a precursor to Arthur and Guinevere), revealed that it was destroyed in a cataclysm. Likened to a Cornish Sodom and Gomorah, the dreadful crime Lyonesse's people committed is unclear but the vengeance was swift and decisive. In the dead of night there came a powerful storm followed by a huge wave which engulfed all before it. Much like the story of Plato's *Atlantis*, in one night the Land of Lyonesse disappeared below the waves never to be seen again. Elizabethan antiquaries collected reports recorded in and before the 16th century stating that Lethowstow, the Cornish name for this area, contained "fair-sized towns and 140 churches" but was suddenly engulfed by the sea:

> In the crystal depths the curious eye,
> On days of calm unruffled, could discern
> ...her streets and towers,
> Low-buried 'neath the waves.[19]

It is generally agreed that the Isles of Scilly were the high point of this lost island and a rocky outcrop called Seven Stones Reef, some eighteen miles west

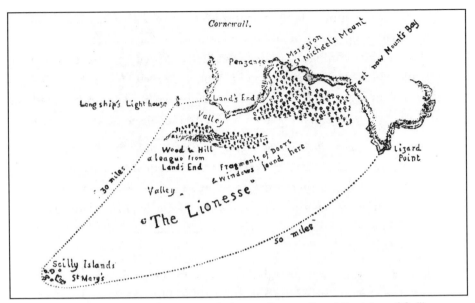

Map of the location of Lyonesse between Cornwall and the Isles of Scilly,
by Agnes Strickland, in *The Lost Story of England*, 1902.

of Land's End and eight miles northeast of the Isles of Scilly, may be evidence of
this lost land. The Isles of Scilly are rich in megalithic structures, some dating
back to the early Neolithic era. Anthony Roberts wrote about old folk tales of
the area:

> *Lyonesse (and the Scillies) is solidly associated with legends of
> the giants through both its stones and barrows. They are said
> to have ruled there at the 'dawn of time' and 'shaped' the stones
> 'into wondrous configurations'. Giants and megaliths lie at the
> foundation of the Lyonesse enigma, which must date back to at
> least 8000 BC, a time of especially heavy flooding around all the
> islands of Britain.*[20]

The dating of this lost island has become a heated debate over the centuries,
some saying it was in archaic times or even the Bronze Age when the land
between St Michael's Mount and Cornwall was flooded, whilst others say it was
in the 11th or 12th century AD. Whenever it sank, its legend lives on and giants
were at the heart of its creation, and possibly its destruction.

The final land we will look at may not have been lost at all. The ancient
Greeks spoke of a mythical island in the far north that was the abode of giants.

THE GIANTS OF HYPERBOREA

Corroboration of the British Isles being inhabited by giants is brought to us by the ancient Greeks, who believed it was the fabled land of *Hyperborea*, known for generations as the 'Land of the Giants'. We can see how aspects of their knowledge may have been inspired by ancient British megalithic builders and the later Druids. In Greek mythology the Hyperboreans were a race of giants who were said to have resided "beyond the north wind." This is spoken about by Diodorus Siculus:

> *In the regions beyond the land of the Celts there lies in the ocean an island no smaller than Sicily. This island, the account continues, is situated in the north and is inhabited by the Hyperboreans, who are called by that name because their home is beyond the point whence the north wind (Boreas) blows; and the island is both fertile and productive of every crop, and has an unusually temperate climate.*[21]

The island was said to have an aura of perfection, with the sun shining twenty-four hours a day. This unusual passage has raised questions about where this island really was, with most speculators saying it was Greenland or the North Pole. The god Apollo was said to have flown to this region every nineteen years to rejuvenate. This is an analogy of the 18.6 year cycle of the moon. If we look at the times of the extreme lunar maximums, every 18.6 years in the very

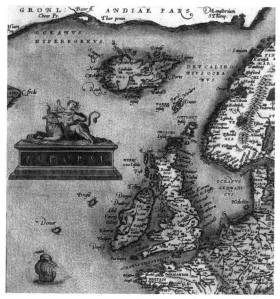

Map by Abraham Ortelius, 1572. At the top left *Oceanvs Hyperborevs* (The Sea of Hyperborea) separates Iceland from Greenland. To the left of Ireland *Brasil* is located, also known as 'Hy-Brasil'. Beyond that is another mysterious island called *Demar,* which could also be Ogygia.

northerly latitudes, such as the Outer Hebrides and Orkney, this could be what was being described, as at these times the nights never go fully dark.

It is also said that Medusa was banished to these lands and there are indications (within fragments of texts) that say that Pythagoras came from Hyperborea.[22] Hecateaus of Abdera also wrote that it contained:

> ...a magnificent sacred precinct of Apollo and a notable temple which is adorned with many votive offerings and is spherical in shape.[23]

This description has been identified as Stonehenge.[24] In his 1726 work on the Druids, John Toland specifically identified Diodorus' Hyperborea with the Isle of Lewis, and the spherical temple with the Callanish Stones.[25] Greek legend also talks about the giant Boreades, the descendents of the fabled Hypoboreans. This particular account is found in the writings of Aelian describing ten-foot-tall inhabitants:

> This god (Apollon) has as priests the sons of Boreas (North Wind) and Chione (Snow), three in number, brothers by birth, and six cubits in height (about 3 metres). The Boreades were thus believed to be giant kings, around 10 feet (3 m) tall, who ruled Hyperborea.[26]

ENOCHIAN TRADITIONS IN BRITAIN

One possible connection regarding giants arriving in the British Isles, is the theory that Enoch may have been transported here by the Watchers from the Garden of Eden. According to the *Book of Enoch* and the Old Testament, the Watchers were an offshoot of the original *Annanage* (Annunaki), referred to as 'serpents' and 'Shining Ones', who founded Kharsag (The Garden of Eden) remembered as semi divine beings. They worked the land and kept themselves away from the humans by staying in 'high places'. Successive waves of them arrived on Mount Hermon.

A few hundred years into the project, some 200 more arrived described as the Grigori, who were said to be much taller than the original Annanage. They appeared to be a different breed and were often dismissive of their Annanage leaders. They became known as 'The Watchers' and eventually rebelled against the laws of their rulers and bred with human women, creating the Nephilim (which translates to 'Fallen Ones' or 'Giants'). These Nephilim became out of control, even resorting to cannibalism. The Nephilim became revered builders and megalithic architects as they absorbed the knowledge of the Watchers.[27] This culture eventually merged into the Canaanites, who later became the Phoenicians.

A painting by Giovanni Lanfranco 1624, showing what could be a Nephilim giant.

The big question is, did they make their way to Britain, and were the stories of Albina's giant offspring, actually referring to these Nephilim giants? Hu Gadarn, a giant king of Wales also has similarities to the Nephilim. He was said to have arrived after a great flood from Asia Minor (modern day Turkey), with high technology, skills in surveying, agriculture, metallurgy, stone quarrying and was part of a 'serpent cult' (see *Giant Lore of Wales* chapter).

In our last book we looked at evidence of the Nephilim being in Europe, as the knowledge they were taught by the Watchers seemed to evolve into the megalith building culture of Brittany and the British Isles. This was know about in the 11th century:

> *The first people to dwell in this land were the British who came out*
> *of Armenia and they first took the southward part of Britain.*[28]

New DNA research has revealed that there were migrations from Turkey (that Armenia was a part of) all the way to Britain. We know that the so-called Beaker People were from Europe and began to arrive around 2600 BC, but this research suggests something quite unique, and older. An inundation of migrants from what is now Greece and Turkey arrived in Britain some 6000 years ago and virtually replaced the existing hunter-gatherer population that had been living in peace for thousands of years.

A report in the Scientific Journal *Nature*[29] suggested they overwhelmed the native population, and little DNA from the original Mesolithic Britons remains. However, one interesting fact is that they did not arrive on the east or south coast, but travelled up the western part of Britain, much like their earlier Solutrean cousins. One thousand years before this (c.5000 BC) these farming migrants were living in continental Europe, and were building the gigantic megaliths of Spain, Portugal and Brittany (and there are many). This may have delayed their advance into Britain. Rather than take the short route of the English channel, they braved the wild Atlantic. It seems Cornwall or the Isles of Scilly were the most obvious locations for their arrival, but because the report did not conclude exactly, perhaps they went further north to Lundy Island and Southern Wales. Certain traits of early incomers to Britain were noted by author Comyns Beaumont:

> *Another feature to be observed of the Giants is that nearly always they are described as red-haired, as were the Adamites and Edomites, names derived from the Hebrew adom or edom, red or ruddy...The same applies to the Phoenicians of like stock, whose name is but a derivation of the word meaning red or carroty, relating to a ruddy-faced people.*[30]

The megalithic sites of Canaan (the homeland of the Nephilim) that were adopted by the inhabitants of Israel are strangely similar to those found in Britain. These included sacred springs, standing stones, hillforts, dolmens and a type of stone circle called a *Gilgal*, a Hebrew name that stands for 'circle of standing stones.' Gilgal is mentioned 39 times in the Bible and these Canaanite constructions have many similarities to European and British megalithic sites. One example is Atlit Yam, a submerged stone circle off the coast of Israel dating to 6900 BC that has cup-marks and was built over a

Atlit Yam submerged stone circle.

360

natural water spring. Cup-marks are found at the 11,600 year old Göbekli Tepe site in Turkey, and at numerous sites in Britain.

A Gilgal located a few miles northeast of Jericho was said to be the first place where the Ark of the Covenant was installed.[31] The prophet Elijah was also recorded as repairing one of these circles:

> *And Elijah took the twelve stones...And with the stones he built*
> *an altar in the name of the Lord; and he made a trench about the*
> *altar...and he filled the trench also with water.*[32]

The above quote is particularly interesting because is clearly states that a henge was built around the stone circle, a style thought to be unique to Britain. The Canaanite sanctuaries also contained an altar, a standing stone called a messebhoth and a wooden pole, called an asherah, named after the goddess of sunrise and dusk. The asherah may have been a surveyors pole, as depicted in the Long Man of Wilmington Hill Figure in Sussex.

Also in Israel is the 'Stonehenge of the Levant', a gigantic series of concentric stone walls with a tumulus at its centre in the Golan Heights. It dates back to at least 3200 BC and was thought to have been built by giants with the name *Gilgal Rephaim* (Wheel of the Giants). It also contains dolmens, another megalithic structure that may have originated in this area and spread to Britain.

Gilgal Rephaim is an ancient megalithic monument in the Golan Heights.

THE BOOK OF ENOCH

In the *Book of Enoch*, we find the patriarch was trained as an astronomer by the Shining Ones or Watchers, residents of the Garden of Eden (Kharsag). It described him being flown to mountainous regions of high latitude where he was shown megalithic sites:

And I saw in those days how long cords were given to the Angels, and they took themselves wings and flew, and went towards the north. And I asked an Angel, saying unto him: 'Why have they taken cords and gone off?' and he replied: 'They have gone to measure.' And these measures shall reveal all the secrets of the depths of the Earth.[33]

There have been suggestions that Enoch may have travelled to Britain and was involved in either building stone circles, or recording their astronomical science.[34] The third-century BC translators who produced the *Septuagint* in Koine Greek rendered the phrase "God took him" with the Greek verb *metatithemi* (μετατίθημι) meaning moving from one place to another. From *The Book of Enoch*, it says:

They lofted me up and placed me on what seemed to be a cloud, and this cloud moved; and going upwards I could see the sky around and, still higher, I seemed to be in space. Eventually, we landed on the first heaven and, there they showed me a very great sea, much bigger than the inland sea where I lived.

The description seems like a valid account of being taken up into the sky on a small aircraft of some sort by the Shining Ones (Angels). He was also taught by these beings the arts of astronomy, geomancy ("peaceful order of the world") and other disciplines:

Some of the Angels study the movements of the stars, the sun and moon, and record the peaceful order of the world...There are Angels who record the seasons of the years.

In *Genesis 6* it states:

Enoch walked with the Shining Ones. Then he disappeared because the Shining Ones took him away.

The text continues:

And from thence I went towards the north to the ends of the earth, and there I saw a great and glorious device at the ends of the whole earth. And here I saw three portals of heaven open in the heaven: through each of them proceed north winds: when they blow there is cold, hail, frost, snow, dew, and rain. And out of one portal they blow for good: but when they blow through the other two portals, it is with violence and affliction on the earth, and they blow with violence.

This passage seems to suggest that Enoch was taken very far north possibly to the Orkney Islands or Callanish in the Outer Hebrides. A fascinating piece of folklore from Callanish reads as follows:

Until recently it was believed that at sunrise on midsummer's day the 'Shinning One' walked along the avenue, his coming heralded by the call of the cuckoo, the bird of Tir-nan-Og, the Celtic land of youth. Another tradition is that of a priest-king who came to the island bringing with him not only the stones but black men to raise them. He was attended by other priests and the whole company wore robes of bird skins and feathers.

What we have here is a direct reference to a 'Shining One', including a tantalising description of feathered robes. The Angel Uriel appears to have transported Enoch to Britain and it seems he, or Enoch (who may have worn such robes) were recorded in legend. In the early Sumerian texts, Uriel is one of the primary Watchers, who descended into the Garden of Eden. The Shining Ones were also regularly described as wearing feathered robes, and often giant in stature, and were even portrayed as having wings, hence the Angel association. When Enoch was summoned, this is what he wrote about the experience:

I awoke to find, in my room, two very tall men different from any that I had seen in the Lowlands. Their faces shone like the sun, and their eyes burnt like lamps; and the breath from their mouths was like smoke. Their clothes were remarkable - being purplish (with the appearance of feathers), and on their shoulders were things that I can only describe as 'like golden wings.'

The God Ashur from Nimrud, 865-850 BC in what looks like Enoch's flying machine.

Ninurta pursues Anzû, (883 - 859 BC). Both 'winged' beings.

Further 'feathers' are recorded on numerous statues and artefacts from the 3rd millennium BC in Sumeria, and mentions of wearing feathered robes are hinted at in various texts, such as the *Testament of Amran* where it states that the Watcher Belial was adorned in a cloak *"many coloured but very dark."*

In *Sun, Moon and Stonehenge*, Robin Heath revealed that the *Book of Enoch* encodes not only the latitude of Stonehenge, but it precisely describes astronomical observations from within the stone circle.[36] In Chapter LXXII, the *Book of the Heavenly Luminaries*, the Angel Uriel trains Enoch in the arts of observing the sun and moon through 'portals' or gaps between stones, and hints that it is a circular device based upon a 364 day calendar. There is a long description of what Enoch saw, but Heath summarises it here:

> *Astronomically, the text is specific, and the Sun and Moon are described rising within six portals in the east and setting within six portals in the west (There are also many windows to the right and left of these portals). For each month following the spring equinox, Enoch tells the reader in which portal the sun rises and sets and, more importantly, he tells us the specific ratio of day to night lengths during the first day of that month.*

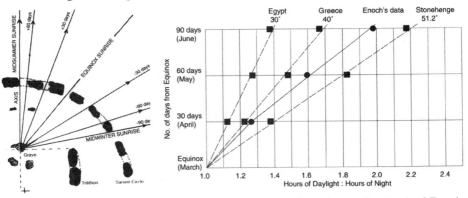

Heath's analysis of the alignments within Stonehenge based upon the *Book of Enoch*.

An unexpected reference comes from the *Book of Taliesin*, where the bard was asked to explain who he was in the presence of a king. He responded in verse with these particular lines standing out:

> *I know the names of the stars from north to south;*
> *I have been on the galaxy at the throne of the Distributor;*
> *I was in Canaan when Absalom was slain;*
> *I conveyed the Divine Spirit to the level of the vale of Hebron;*

> *I was in the court of Don before the birth of Gwdion.*
> *I was instructor to Eli and Enoch.*[37]

Taliesin is thought to be a sixth century bard and his comments on astronomy, his time in Canaan, and having taught Enoch about the heavenly sciences is noteworthy, even if it was referring to an earlier time. The authors wonder if the visit of Enoch and the Watchers was a big occasion that may have been greeted with pomp and ceremony. Could it be that the British megalith builders collaborated with the Watchers and Enoch then commemorated the visit? The memory of the 'shining one' at Callanish may also indicate this.

Enoch may have been more of a title than a name, related to the study of astronomy, surveying and various forms of geomancy as his name can translate to 'inaugurated', 'trained' or 'initiated', suggesting this to be the case, especially as there is another Enoch recorded as living before him in the same patriarchal line[38] who was the son of Cain:

> *God gave Prophet Enoch the knowledge of sciences, which he taught the people. God gave him the knowledge of Astronomy. ...Enoch was the first to set down the rules of city planning. It is said that Prophet Enoch built the two famous pyramids of Egypt.*[39]

At Göbekli Tepe in Southeast Turkey, nearly 12,000 years ago, circles of T-shaped pillars up to 18ft tall were being erected on the hills above modern day Şanlıurfa. Potentially 20 stone circles may exist, showing astronomical significance and carvings that represent astrological configurations, possibly the earliest known anywhere.[40] There are also cup-marks on the pillars, numerous holed stones and beautiful relief carvings reminiscent of Egypt and Peru. This was the world's first temple, and most certainly the first ever stone circles. Göbekli Tepe could have been linked with Enoch because nearby at the mystical town of Harran, the name of a Mesolithic mound carries the other name of Enoch - *Idris*:

> *Recent archaeological excavations at an occupational mound immediately to the south of Harran called Tell Idris (the 'mound of Idris'), the Arabic for the antediluvian patriarch Enoch, as well as the Greek Hermes, have revealed that the earliest inhabitants arrived in the area during the Pre-Pottery Neolithic B period, sometime around 10,000 years ago.*[41]

Harran is famous for its astronomical tower that stands close to the mound. It was here that the Sabeans taught the esoteric arts, including astronomy. The dating mentioned here is also of interest because at this time, huge wooden posts or totem poles were being raised at the location of Stonehenge, at exactly the same time as Göbekli Tepe was being closed down.

Left: Harran's Astronomical Tower was built near the Mound of Idris
Right: Reconstruction of Enclosures D and C at Göbekli Tepe.

In Pembrokeshire, near the Preseli bluestone site, one of the mountains has the name Carn Enoch, and on one of the rocks are what look like calendrical carvings. These are thought to be Neolithic tally marks. There are 28 to 32 marks of varying length coming out from a central groove in the rock. Tom Bennett in his book *3000 BC: Death and Rebirth Rituals in Neolithic Wales* (2019), suggests it is the genesis of a lunar calendar and shows the tidal height throughout a month, as can be viewed from that location. This early observation of the moon so close to the Preseli bluestone site is remarkable, as this gives an Enochian site, marked with calendrical information, a direct connection to Stonehenge via the nearby bluestone outcrop.

The calendrical tally marks at Carn Enoch.

Before we close this section of the chapter, we wondered if any other Enoch sites or legends existed in the British Isles that could give more credence to this theory. A tradition from the Tanat Valley area of Montgomeryshire was recorded in an interview with Margaret A. Edwards, where she described two giants, one with the name of Enoch, throwing a hatchet to one another.[42] This is only about 20 miles east of Cadair Idris, the mythical home of Enoch/Idris.

In Scotland we have three sites with the name Enoch. There are two chambered cairns in Dumfries and Galloway called *Mid Gleniron II* in the villages of *Galdenoch* and *Capenoch* dating back to the Neolithic era. In *Enochdhu* in Perth there is 7ft tall standing stone, and a Giant's Grave in the same area with the same name. There are more than expected, so does this indicate a memory of an important visit from a patriarch of the Bible?

THE SHINING ONES OF IRELAND

At some time in the third millennium BC, there arrived in Ireland a group of Sages who, in time, were to form the pantheon of the indigenous Old Irish, and the early Celtic invaders. In the later Gaelic tongue, they were described as - 'Dee in teas dana acus ands an teas' which may be translated as 'the men of science were gods and the laymen no-gods.' [43]

There are indications that Enoch may have visited Newgrange and was greeted by the Tuatha de Danann. According to The *Book of Enoch*, he visited during the winter months and described a unique construction:

And I went in till I drew nigh to a wall which is built of crystals and surrounded by tongues of fire: and it began to affright me. And I went into the tongues of fire and drew nigh to a large house which was built of crystals: and the walls of the house were like a tessellated floor of crystals, and its groundwork was of crystal. Its ceiling was like the path of the stars and the lightnings, and between them were fiery cherubim, and their heaven was (clear as) water. A flaming fire surrounded the walls, and its portals blazed with fire. And I entered into that house, and it was hot as fire and cold as ice: there were no delights of life therein: fear covered me, and trembling got hold upon me.

No other construction on the planet had 'walls of crystals' or a 'tessellated floor of crystal'. The evidence seems to suggest that not only did Enoch visit

Stonehenge, he also took a trip to Ireland's premier sacred site with his group of Shining Ones from the Bible lands. It was here they met up with the 'shining ones' of ancient Ireland, the Tuatha de Danann.

Left: Winter Solstice sunrise illuminating Newgrange. Right: The exterior crystal wall.

Enoch also described the climate with a clear reference to hail and snow, suggesting a possible winter solstice visit. The 'tongues of fire' appear to be describing the shaft of sunlight that illuminates the chamber on 21st December. It is worth noting that the winter solstice southeast alignment directly orientates to Stonehenge in Wiltshire, and the Summer Solstice sunset at Stonehenge orients to Newgrange over a 262-mile distance. This could be a fascinating coincidence, but the theory that Enoch was making observations at Stonehenge, and now possibly at Newgrange is compelling.

THE TRANSATLANTIC GIANT CONNECTION

Since we wrote our last book on American giants, we have been investigating further connections that may indicate proof of transatlantic contact between America and Britain going back long before Christopher Columbus made his claim. All over North America there are legends of giants who were described as fair-skinned with red or blond hair and often with beards. The oldest traditions may date back to 42,000 BC with a group of fierce, bearded giants who were remembered as crossing the Atlantic and heading to Britain and Western Europe, perhaps even being the original Cro-Magnons. This theory was first put forward in *America: The New World or Old?*, by Werner Muller, that has tantalising correlations with very old traditions in certain parts of the continent. Ross Hamilton shared a summary of these traditions with us.

The Yuroks, Iroquois, the Sioux and others tribes talk about a time before the 'Wagas' (benign red-and blond-haired, fair skinned people), and a rather disagreeable black-bearded race with "tawny" skin, who were excessively large in size. In esoteric circles they are speculated to be the 'Borean Race' and were purportedly sent out from Hyperborea between 40,000 - 50,000 years ago. They were good to the indigenous people at first and trained them in all manner of sciences, having them work on their behalf for thousands of years. The 'Boreans' lived in the higher elevations and were seen as semi-divine beings. They themselves were great miners, and located rich seams of gold throughout the Tetons and the Rocky Mountains depleting the massive veins of gold after many thousands of years of mining. This work they did themselves. The indigenous folk farmed for them to keep them well-nourished during their 'Great Labour'. After their work was completed, they became cruel landlords, so the Great Spirit sent the Red Beards to get rid of them. This is when many of them, now thought to be Cro-Magnons, fled the North American continent, arriving in Britain and Europe in 42,000 BC.

In more recent times it was recorded that two Native Americans arrived in Ireland on a wooden seafaring canoe in 1477. The Gulf Stream and easterly winds were known about by navigators since ancient times. Remarkably, 15 years before Columbus set sail to America, he actually met these two American Indians in Galway, Ireland.[44] This may have given him the final inspiration to sail west. However, the descriptions of these accidental migrants is of interest to our studies, because various translations of Columbus' notes (originally written in Latin in an earlier book called *Historia Renum* by Aeneas Sylvius Piccolomini), state the man and woman were "of wonderous aspect," "of admirable form," and "of great stature." This is identical to many other giant reports we have encountered. "Great stature" clearly means much taller than average and a strong, athletic build. For Christopher Columbus to witness live Native American 'giants' in Ireland is perhaps one of the most fascinating stories we have covered in this book.

Jean Merrien, a French authority on navigation who understood the nature of the winds and currents stated that "*Americans reached Europe long before any Europeans reached America.*"[45] He went on to say that you could reach Land's End in Cornwall in a straight line from Boston "*with almost the certainty of fair winds.*" However, going the other way, towards America, was much more of a challenge. We now have to ask, have we got the idea of migrations all mixed up? Did giants from America influence Britain going back to prehistoric times? Also, was there a genetic influence, perhaps from repeated arrivals in Ireland that triggered the 'giant genes' found in Ulster? On the Scottish Arran Islands, it is known that many people have Native American facial traits and some of the coastal myths echo those of parts of the east coasts of America.

Tall, red-haired, fair-skinned mummies were discovered in and around Lovelock Cave and Winnemucca, Nevada in the early 1900s. Giant petroglyphs were found next to Winnemucca Lake where some 14,800 - 10,500-year-old carvings can be seen. Ancient symbolism researcher JJ Ainsworth realised many of the designs were identical to those at Loughcrew in Ireland.

Both these cultures have legends of giants and giantesses, both have evidence in the anthropological record of having red hair and ruddy skin. The 'language of the giants' is found neatly carved, as though by the same hand, in both areas with exactly the same designs. Could the Irish have got to America at this early date? Or did the American giants cross the Atlantic, remembered as either the incoming Fomorians or the Tuatha de Danann in the Irish annals?

Left: The Equinox stone at Loughcrew in Ireland. Right: Hugh at the gigantic petroglyphs of Winnemucca Lake. Both have solar/flower carvings, 'wheat' icons, zigzags and other identical symbols.

The *Maritime Archaic Culture* (also known as the Red Paint People) of coastal Northeast America thrived from 6000 BC to 1500 BC. In *Giants On Record* we have numerous accounts of giant skeletons from Maine, one of them from a midden mound most likely constructed by this culture. They were famous for their burials utilising red ochre, a tradition that is also found in parts of Britain and Europe. The Red Lady of Paviland on the west coast of Wales, for instance (dating to c.33,000 years old) had red ochre used in its ritual burial. It has also been noted that the Gulf Stream went further north during the 'Red Paint People' era (before 3500 BC), with Orkney being one of the natural arrival points from America. Red ochre was popular in Scotland, especially amongst the Picts. The Algonquin-speaking peoples of Northeast America also have a memory of the arrival of white-skinned strangers:

The Sagamores, our chiefs, speak of these strangers who came to us by boat from the sea about two thousand years ago. They established their colonies on our territories trying to take us by force. But after they had destroyed their vessels, our Algonquin forefathers convinced them to live with us. They called themselves Kelts, I believe, and our legends speak often of them because of the many wars they fought with us. Our oral tradition also speaks of their respect for the laws of nature and how they learned to live according to our ways. They came on floating islands from lands across the sea and mingled with us. As time went on, they lost their pretensions to own the land and their ways became more respectful of their surroundings. But this does not mean that they did not have any influence on us either, for they nevertheless, although harsh, had a good language. This is how my forefathers learned to use some of their words to express things we found difficult in ours.[47]

Further west in the Mound Culture areas, more similarities are found with hillforts, earthworks and burial mounds that are too close to the old British style to accept there is no connection. Within these mounds hundreds of giant skeletons were unearthed. Marietta in Ohio has a main mound with a ditch around it, *identical* to a Bell Barrow in England.

Left: Great Mound at Marietta (1848) in Ohio, USA . Middle: Bell Barrow in Wiltshire. Right: Plan and cross-section of typical Bell Barrow in Britain.

Grave Creek Mound in West Virginia contained over eight feet tall skeletons and it was here that a remarkable tablet was unearthed with what look like Phoenician inscriptions. We detailed this discovery in our previous book, and although we wrote that this may provide evidence of a connection with the Middle East, it has since been found that the tablet's script may in fact be an ancient Colbrin, an archaic Welsh language.[48]

Further Welsh connections are found in North America. In Georgia, the Cherokee say white people built the forts and were called "Welsh." In North Dakota the Mandan Indians were also fair-skinned and had Welsh roots in their language.[49] In 1521, Spanish sailors witnessed the Duhare tribe of South Carolina who were described as gigantic white people with full beards, and the even-taller king of the tribe was called 'Datha' described as "*the largest man they had ever seen.*"[50] Duhare can be translated as "*du'hEir - place of the Irish.*"

> *The continent later known as America was referred to as 'Great Ireland,' by ancient geographers long before Columbus.*[51]

SCOTTISH LINKS TO THE TARIM BASIN MUMMIES

Another curious link to the giants of the British Isles may be found in China, where mummies have been unearthed reaching upwards of 6ft 7in tall. Since the late 1930s, numerous mummified remains have been found along the edges of the Tarim Basin in the region of Xinjiang. Some are over 4000 years old and are extremely well preserved, due to the hot, arid climate and the bitter dry winters. They had red and blond hair, thick ginger beards, fair skin and blue coloured eyes (as depicted on blue pebbles placed on a mummified baby, and in later cave paintings). Skin, hair, flesh and internal organs have remained intact and the burials often had two people in one grave, who are found with boots, trousers, stockings, hats and a particular style of fabric.

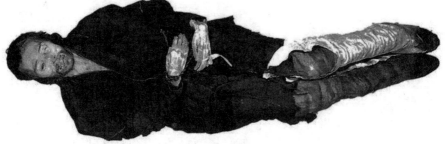

Cherchen Man.

Archaeologist Dr Elizabeth Wayland Barber was highly impressed by the near-perfect remains saying: "*The mummies appear to be neither Chinese nor Mongoloid in facial type, they look distinctly Caucasian.*"[52] One mummy was 6ft 7in tall and named Cherchen Man (due to the area he was found in), whilst his female companion was 6ft 2in. These extreme heights would have stood out anywhere in the world at this time, but to find them in a remote area of China is even more astonishing. Evidence points to them being western Europeans, but what were they doing in the region of the northern Silk Road, an ancient trading route?

> *Based on analyses of human remains and other archaeological materials from the ancient cemeteries (dated from approximately the Bronze Age to the Iron Age), there is now widespread acceptance that the first residents of the Tarim Basin came from the West.*[53]

The burial sites of Cherchen Man and his tribe were marked with stone structures not dissimilar to the dolmens of Britain, surrounded by standing stones and other artefacts. He had yellow ochre spiral and sunburst designs painted on his face. This is exactly the same design and materials used by the Picts of Scotland. The female mummy also had a yellow spiral and wore a long, conical 'wizard-like' hat like the Druids. The symbolism of the spiral is decorated in the fabrics, a design found carved in many ancient British megalithic sites.

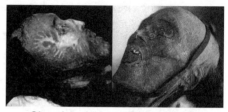

Cherchen man and woman with yellow ochre spirals on their faces.

When Dr Barber looked more closely at the textiles, she found that they also had remarkably well-preserved examples of lozenge designs on them, a feature found carved at megalithic sites all over Britain and Ireland. Perhaps the most amazing aspect of her analysis was that the designs were almost identical to certain Scottish tartans and plaids (and similar to Hallstatt culture textiles from the Bronze and Iron Age):

> *The dominant weave proved to be normal diagonal twill and the chief decoration was plaid, as in the woollen twill material of a Scottish kilt....Not only does this woollen plaid look like Scottish tartans but it also has the same weight and feel and initial thickness as kilt cloth.*[54]

Archaeologists at the Ness of Brodgar, Orkney, have discovered evidence of woven cloth imprinted on a piece of Neolithic pottery, that could be as much as 3100 years old.[55] Another documented example in Scotland was found at Flint Howe in Dumfries and Galloway, that showed a small impression of a plain-weave textile on a Late Neolithic/Early Bronze Age pot.[56] Together, these prove that textiles, and possibly the style of tartan found on the Chinese mummies, were being produced in Britain from a very early date.

Top Left: The Lozenge pattern woven into the Cherchen fabrics in China.
Top Right: Ness of Brodgar 'lozenge' motifs on Mainland Orkney, Scotland.
Bottom left: Clandon Barrow gold lozenge, Dorset. Bottom Middle: West Kennet
Long Barrow beaker pot. Bottom Right: Four Knocks, Ireland.

DNA testing confirms that the mummies found in Xinjiang's Tarim Basin are of European origin with Haplogroup H in their DNA:

> *We conclude that the west Eurasian component observed in the Xiaohe people originated from western Europe, and maternal ancestry of the Xiaohe people might have close relationships with western Europeans.*[57]

Brian Sykes, a former geneticist at Oxford University found Haplogroup H to be dominant in both Ireland and Wales so to find this connection with China is remarkable, suggesting indeed that a great migration took place over four thousand years ago carrying the giant genes of Britain with them. Further genetic research indicates that their arrival could have been much earlier with dates proposed by Dr. Barber anywhere up to 7000 BC.[58]

Further correlations between ancient China and Britain were unearthed in Scotland on the Isle of Uist in 2001. Two Bronze Age bodies were unearthed at Cladh Hallan that displayed clear signs of mummification dating to c1440 - 1260 BC. The two skeletons were made up of six individuals, some 195 years older than the main skeleton.[59] Further examination of thirty four Bronze Age bodies in different parts of Britain were found to have had some kind of mummification having taken place, suggesting this tradition was widespread across the whole country from the Neolithic era onwards.[60] Examples of mummification have been found at Neats Court in Kent, where the dead were 'dried' over a fire to start the process. Others have been found at Canada Farm in Cranborne Chase, Dorset. This time, the bones were drilled and joined to hold the skeleton together. Two out of the seven skulls had super-numerary teeth in their jaw, a strange anatomic anomaly that we will discuss in the next section. These stunning discoveries make it clear that the bones of important people in Britain (and the giants of China) were not only carefully prepared for burial, but honoured and displayed, long before they were finally interred.

Further Caucasian giants were found in China in 1995. Yingpan Man is a nearly perfectly preserved 2000-year-old fair-skinned mummy, 6ft 6in tall. This may suggest that the giant elite maintained their presence, and their genetics over many generations.

The layout of some of the burial sites brings this to mind as they were surrounded by what appear to be small dolmens and a circle of standing stones, whilst the Loulann site had similarities to the layout of Woodhenge. They erected concentric circles of wooden posts, with straight alignments spreading out from the circles, suggesting astronomical and geodetic knowledge.[61] It also seems that the placement of the burials, especially Cherchen Man with his wife and baby, were arranged in a specific way to mark a spot in this remote landscape, hoping one day they would be found, trusting that their legacy would be remembered. Perhaps with all the evidence presented here we can conclude that the giants of Britain really did exist, perfectly preserved, but 4000 miles away in a Chinese desert.

Cherchen Man's burial site, Tarim Basin.

A bizarre 9th-10th century fresco from the Bezeklik Grottoes near Turfan in the Tarim Basin region seems to depict a Western man with red hair, ginger beard, big nose, and large blue eyes. His pose is that of a teacher, perhaps even a Buddhist Monk:

The figure on the left is painted in a style that is not East Asian in character. The two circles on either side of his chin are remarkably unusual (piercings?). He features a strongly-set nose and clear blue eyes. His skin color is darker. His hair and beard color is reddish brown with brown shades. He stands looking at the younger monk, raising his hand in a teaching gesture as he speaks.[62]

Ninth century fresco from Bezeklilk, Tarim Basin depicting a Western European with red hair and a beard.

Old Chinese traditions refer to early spiritual leaders of that region as being tall, with fair skin, blue or green eyes, and also blonde and red hair. Pliny the Elder was a Roman scholar and historian, living between 23 - 79 AD and recorded evidence of Caucasians interacting with the East stating: *"they exceeded the ordinary human height, had flaxen hair, and blue eyes, and made an uncouth sort of noise by way of talking."* In one tradition, the king of these people is described as *"the tallest among the Barbarians in the western region."* [63] They had a tradition of flattening the skull, a form of cranial deformation. They tattooed their bodies and were recorded as having red hair and blue eyes. The king also had an unusual anatomic anomaly that is often related to giants, having six fingers on each hand and six toes on each foot. It was said he would not raise his children if they did not have the same anatomic anomaly.[64] Whoever these Tarim Basin giants were, they influenced the region for multiple generations.

DOUBLE ROWS OF TEETH AND ANATOMIC ANOMALIES

Are anatomic anomalies such as enormous jaws, extra teeth, extra fingers and toes telling us that there are genetic clues that can help us better understand the mystery of giants? The giant Irish hero of the *Ulster Cycle* Cuchulain is revealed to have had, *"seven toes on each foot and each hand seven fingers, the nails with the grip of a hawk's claw or a gryphon's clench."*[65] In Scotland, in the 15th Century manuscript *Eger and Grime*, we learn that the invincible knight

Greysteil had "*extra digits on his hands.*"[66] In Wales, the giant burial at *Long Man's Grave* belonged to a man of extraordinary proportions who lived in the area and had six fingers on each hand. From *The Giants of Cornwall*, B. C. Spooner tells us that, "*the giants of Tren the Mount had six fingers on each hand and six toes on each foot.*" Further afield in the Bible lands, the third son of Goliath is recorded to have had six fingers and toes, twenty-four digits in all.

A 2014 article from the *Ulster Medical Journal* exploring hereditary gigantism presents a link between the AIP gene that causes pituitary gigantism and polydactylism (extra digits). It is suggested that the AIP gene is connected with both conditions. The report states that the AIP gene lies on "*chromosome 11q13.3. The Bardet-Biedl gene, BBS1, is located close by on chromosome 11q13.2. Bardet-Biedl syndrome type I is characterized by rod-cone dystrophy, truncal obesity, cognitive impairment and postaxial polydactyly.*"[67] This technical terminology is marrying the two giant related conditions and revealing that they share almost the same genetic region. As previously noted, the highest occurrence in the world of the AIP giant gene is in Ireland's Ulster County. Could this be evidence to support the idea of ancient migrations of giant peoples from the west as recorded in so many ancient sources? It is clear that the giant gods and heroes of the past were described as having had this peculiar anatomical trait, but there is an even stranger anomaly recorded throughout history that is connected to these titans.

In our previous book we devoted an entire chapter to the strange cases of *double rows of teeth*, a condition widely reported across America in burial excavation accounts of enormous skeletons. Extra rows of teeth have also been historically associated with the giant Fomorians of Ireland and Scotland. A startling passage is to be found in the *Togail Bruidne Dá Derga* (The Destruction of Da Derga's Hostel), which is an Irish tale belonging to the Ulster cycle of Irish Mythology:

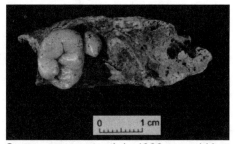

Supernumerary tooth in 4300 year old jaw from Boscombe Down near Stonehenge.

> Not one of these Formorians was found to fight him, so he brought
> away those three, and they are in Conaire's house as surities that,
> while Conaire is reigning, the Formorians destroy neither corn nor

milk in Erin beyond their fair tribute. Well may their aspect be
loathy! Three rows of teeth in their heads from one ear to another.[68]

Robert Temple in *The Sirius Mystery* noted this peculiarity and its association to giants: *"Growing a third set of teeth in ancient times was meant to be a sign of a supernatural hero... So perhaps there is a very rare genetic syndrome, where a man of abnormal size and strength with a third set of teeth occurs in the population."*
[69] This certainly might be the case as it is reported that those with extra teeth who are of giant stature, are also incredibly strong.

The Cranborne Chase Mummies of Dorset also had supernumerary teeth, and were honoured by the people who buried them. Only 0.1% - 3.7% of British population has this anatomic anomaly. To find it in two skulls out of seven suggests this was hereditary, and may have been a trait that was revered by the Bronze Age people. We found this directly connected to genetic giants in our previous book on the giants of North America. Near Stonehenge, a 4300-year-old skeleton found on Boscombe Down in Amesbury revealed extra teeth in the mouth of one of the burials.[70] At Drimsdale, South Uist (near where mummified bodies were found) a skeleton with extra teeth was discovered dating to the Iron Age.[71] These accounts of double rows of teeth or extra teeth have been made across vast geographic areas over thousands of years, which appears to be well beyond coincidence.

THE LAST GIANTS OF ALBION

The world is overflowing with myths and legends of giants and the British Isles is certainly no exception. As we have painstakingly documented in this book, giants are woven into the fabric of the landscape and our collective consciousness. Virtually every well-known megalithic site has giants in their myths, with a surprising number of skeletal remains unearthed at these sites bringing age-old stories to life.

Long before the megalithic age we have evidence going back half a million years at Boxgrove, huge artefacts between 400,000 and 300,000-years-old and Denisovan DNA reaching these shores, not only from Northern Europe, but also from America. The Solutreans left their genetic marker as they moved west via Orkney to America. The Watchers, Enoch and the Nephilim utilised the temperate weather to measure the sun, moon and stars at high latitudes, and African giants may have brought the trilithons of Stonehenge to Ireland in the distant past.

Whatever way you look at it, the powerful rulers of Albion made it known that they were once here, by terraforming the land, building massive megalithic monuments, surveying the country and defending their domain. Their memories lived on in legend, place names, creation myths, and in the graves that were unearthed, but the physical titans died out before records of them were written down. The time of the giants ran its course and the gene pool weakened as Britain became easily reached from every direction and the tall indigenous stock became degraded and bred out of existence, with the remaining titans being defeated by the Romans:

> *Caesar himself says that Gauls (including Celtae) looked with contempt on the short Romans. Strabo also says that Celtae and Belgae had the same Gaulish appearance, i.e. tall and fair...No classical writer describes the Celts as short and dark...Caesar's campaigns must have drained Gaul of many tall and fair Celts... though the tall, fair type is far from uncommon.*[72]

We hope this book has given some insights into the origins of Britain, its forgotten inhabitants and some clues as to who may have been behind the construction of sites such as Stonehenge. The memory of this fabled 'Island of the Giants' will live on in books such as this, the words of visionaries such as Anthony Roberts, and the archaic myths that penetrate the landscape. The giants were the protectors of Britain, and were said to be around when the world was young. Their spiritual presence remains within the land, and within the stone ruins. The stories and traditions of the 'Giants of Albion' have been with us for millennia, and will hopefully be with us for many more.

SKELETON OF GIANT IN LIMESTONE.

DERBYSHIRE QUARRY DISCOVERY.

GREAT AGE.

While some men were quarrying at

PRODIGIOUS SKULL OF A GIANT
Keswick Museum, England

Skeletons of Cornish Giants

The ruins of St. Piran's Oratory, near Perranporth, Cornwall, were the scene of a public ceremony on Saturday in connection with the protective work now being carried out. A body of trustees has been formed, and the ancient fabric, which has been brought to light by the removal of the surrounding sand, will be preserved by the erection of an enclosing building. During the work of excavation many skeletons have been found, some of them of human beings 7ft. in height. A feature of the skulls found is the perfect condition of the teeth. The remains have been discovered at a spot believed to have been a priest's residence.

A GIANT'S SKELETON.

THE complete skeleton of a human giant has been unearthed at Holbeach during excavations for the foundations of new houses. Every bone is in perfect condition, not a tooth is missing, and the skeleton measures no less than 8 feet 2 inches. Near the bones was found a curious key. Stukeley, the antiquary, who was born at Holbeach, records that at the spot where the discovery was made there formerly existed a chapel dedicated to St. Peter.

Giants of old

THERE is no fake about the giant skeleton of a man measuring 9ft. 2in. and probably at least 7,000 years old, which was found buried 10 feet deep near the village church at Swallow, near Market Rasen, and was reinterred on Saturday. It is the second giant skeleton which the

GIANT'S SKELETON FOUND

LONDON—(P)—A skeleton of a man seven feet tall and apparently hundreds of years old was found during quarrying operations near Ashby-de-la-Zouch, Leicestershire.

---

paper on giants gives a list of several, whereof the biggest is one found near Salisbury, 1719. Its length was 9 ft. 4 in. English, which is the largest human stature of which I ever heard. At Salisbury I remember in childhood a mound in a field, north of St. Edmund's Churchyard, called the "giant's grave." Is there any account of this skeleton, and where is it kept?—as a skeleton of that size was surely worth preservation. — E.L.G.

SEVEN-FOOT SKELETON.

The skeleton of a man, approximately seven feet in length, has been discovered at Orton, Waterville, Huntingdonshire, near Peterborough.

The landlord of the Windmill Inn, who has a bed of gravel in a yard adjoining was engaged with other men excavating the gravel when the remains were unearthed at a depth of 18 inches.

CLITHEROE.

The quarrymen employed by Messrs Carter and Co., Bellman Park Limeworks, Clitheroe, yesterday exhumed the skeleton of a man much above the normal stature, whilst "feeling" or baring the rock. In a declivity two feet below the sod lay the bones in a natural position, with head to the west and feet to the east. Unfortunately the workmen disturbed it with a pick, and some of the bones were smashed. A second skull, apparently buried with the body of the man, was found, but no trace of trunk bones could be found. The teeth and bones are in perfect preservation. Local antiquaries cannot name the period to which the man may have belonged, but placed in position, the bones show that his height must have been quite seven feet.

GIANT'S SKELETON FOUND

LONDON—(P)—A skeleton of a man seven feet tall and apparently hundreds of years old was found during quarrying operations near Ashby-de-la-Zouch, Leicestershire.

SKELETON OF ANOTHER IRISH GIANT FOUND.

Turf cutters when at work on Murroe Bog, near Elphin, Roscommon, found the skeleton of a man measuring eight feet. The remains were some distance below the surface, and had lain there apparently for many years. Other skeletons have been found in the neighbourhood.

Crofters find a giant's grave

Two crofters cutting peat on Mainland, largest of the Shetland Isles, unearthed what appears to have been a uniform for a man nearly 7 ft. tall.

It consists of a short jacket or waistcoat over a three-quarter length coat with 2 cloth buttons on each and of the front and seven buttons at the waist. Wide breeches were attached to long stockings and hide

A double-brimmed cap, resembling a sou'wester was found, and also human hair. There was also a small horn, a quill pen, and a horn spoon. A knitted bag were three coins dated 1685.

The man must have been a giant, as his lowest jaw, which was found intact, can be placed round the jaw of an average modern man.

NINE-FEET SKELETON.

Recalled by New Discovery.

EXCAVATED NEAR OLD CHURCH.

Some old remains which have just been discovered at Repton, Burton, deal with an extraordinary discovery which was made near the site of the old church in 1807.

It is recorded that when billards were being excavated and well out west, and on this being cleared it proved to be a square grotto of 15 feet. The big tomb designed and had fallen in, having been supported only by wooden pillars. In the centre of the chamber was the skeleton of a human being measuring 9 feet, and round him the skeletons of a hundred smaller persons.

ANCIENT WELSH GIANTS

SAMSON A MERE BABY IN COMPARISON.

HUGE ROCKS EMPLOYED FOR QUOITS

HILLS PLAYED WITH LIKE MARBLES.

INTERESTING PAPER BY THE REV. ELIAS OWEN, M.A.

At the National Schoolroom, Bala, the Rev. Elias Owen, M.A., vicar of Llanyblodwel, the well-known Welsh folklorist, delivered the following lecture. The rev. gentleman said:—In some parts of Wales we have names of giants called after giants, and legends are connected with these places. These are mythical beings, and with these giants the stories are all funny. These are mythical beings, and the creation of fervid imaginations; still, they are worth recording, and it would be interesting to have a complete collection of traditions of these folklore beings arranged in such a convenient form that they could easily be referred to.

In Montgomeryshire, not far from Welsh-

THE GIANT SKULL—From the fragments of the giant skull at the Museum, Keswick, Cumberland county, England, the size of the giant must have been in excess of 10 or about the height of the Phillistine Goliath. TUESDAY—"A U. S. PORT OF ENTRY ON THE SEA."

Relic of Giant Race THE skeleton of a man over in height, who probably belonged to giant race that lived from 3000 to ago, has been found by Lieut.-Colonel Cooke, a Southsea archaeologist, during excavations on Portsdown near Portsmouth.

The discovery of these remains interesting bearing upon the legend persisting, of giants who lived in

Ancient Supermen Dug In Irish Burial Chambers

DUBLIN—AP—Irish archaeologists have unearthed traces of a bygone race of "supermen."

The findings may provide scientific substantiation for legends of a race of mythical giants who inhabited the Island of Hibernia (Ireland) in its palmy age, long before the days of history.

In a pre-historic burial chamber dating back to 2000 B.C., they found human skeletons with lower legs and shoulders over modern man, stretched in state.

SAXON SKELETONS

REMAINS OF A GIANT AT GUILDFORD

Two skeletons, thought to be of Saxon times, were found by workmen while engaged in the construction of a tennis court in the garden of Upton, Guildown.

Guildford, the residence of Mr. Wood, the Town Clerk of Guildford. The remains were unearthed at a depth of about three feet, and are believed to be those of a man and a woman buried side by side.

SLIGO.—A Giant's Bones Unearthed.—After all Finn McCool and the Galway giant seem not to have been the only mighty men of Ireland. Sligo, like Louth, can now claim to have produced an Irish giant, although its history is forgotten. His bones have been discovered at the Market Cross, Sligo. Several men were engaged in digging a drain across the street when they succeeded, much to their astonishment, in unearthing a skull, jawbone (with teeth in good condition), and various other parts of the human anatomy. The bones were exceptionally large, and show that the man who in the long ago was laid to rest there must have been of giant stature. The discovery created quite a sensation in the town, and much speculation was indulged in as to the supposed age of the remains. Colonel Wood Martin, the eminent Sligo historian, has given it as his opinion that the body must have been buried over 200 years ago, and that the "giant" may probably have met his death in battle in those stirring times. The remains were taken charge by the police, and interred in Sligo cemetery.

Giants of old

THERE is no fake about the giant skeleton of a man measuring 9ft. 2in. and probably at least 7,000 years old, which was found buried 10 feet deep near the village church at Swallow, near Market Rasen, and was reinterred on Saturday. It is the second giant skeleton which the

GIANT'S SKELETON FOUND

LONDON—(P)—A skeleton of an seven feet tall and apparently hundreds of years old was found during quarrying operations near Ashby-de-la-Zouch, Leicestershire.

A GIGANTIC HUMAN SKULL

NOTES AND REFERENCES

CHAPTER 1 - ALBION

1. Thomas Bulfinch. *Bulfinch's Mythology*, 1913.
2. Geoffrey Ashe, *Mythology of the British Isles*, Guild, 1990, p.15
3. William Blake, *Jerusalem*, plate 27 'To the Jews'.
4. Anthony Roberts, *Sowers of Thunder*, Rider, 1978, p.115
5. Edward J. Wood, *Giants and Dwarfs*, Richard Bentley, 1868, p.28
6. Richard Barber, ed, 'The Giants of the Island of Albion' in *Myths & Legends of the British Isles*, Boydell Press, 2004, p.3
7. Ibid
8. https://lordmayorsshow.london/history/gog-and-magog
9. David Larkin. *Giants*. New York : H. N. Abrams, 1979, p.60
10. Barber, from Anglo-Norse Script, p.7.
11. Geoffrey of Monmouth, *History of the Kings of Britain*, c.1136
12. William Blake, *Jerusalem*.
13. Nennius, *Historia Brittonum* (History of the Britons), 828 AD
14. Geoffrey of Monmouth, *History of the Kings of Britain*, c.1136, p.53
15. Theo Brown, *The Trojans in Devon*. Report of the Devonshire Association. 1955, pp.68–69
16. It could also have been adapted from Bruiter's Stone, from where the medieval crier called out his bruit or news. It may also have been a brodestone, or boundary stone.
17. Anthony Roberts, *Sowers of Thunder*, Rider, 1978, p.115
18. Welsh Copy Attributed to Tysilio, Peter Roberts (trans.), *The Chronicle of the Kings of Britain*, 1811.
19. Ibid
20. Paul Newman. *The Lost Gods of Albion*. Sutton Publishing, 2000, p.102
21. www.plymouthherald.co.uk/plymouth-hoe-chalk-giants-plan-scrapped-lack-cash/story-27564974-detail/story.html
22. Paul Newman. *The Lost Gods of Albion*. Sutton Publishing, 2000, p.98
23. John Weever, *Ancient Funeral Monuments*, 1631. Thomas Harper. Cambridge University Press
24. Jennifer Westwood, *Albion*, Paladin Edition, 1985, p.29.
25. Edward J. Wood, *Giants and Dwarfs*, Richard Bentley, 1868, p.205
26. Paul Newman. *The Lost Gods of Albion*. Sutton Publishing, 2000, p.111
27. William Blake, *Jerusalem*., plate 27 'To the Jews'.
28. Eneas Mackenzie, *Historical, Topographical, and Descriptive view of the County of Northumberland*, 1825

CHAPTER 2 - STONEHENGE AND ANCIENT WESSEX

1. https://intarch.ac.uk/journal/issue55/4/ia.55.4.pdf
2. *Newhenge: Latest discoveries and interpretations from the Stonehenge Riverside Project team*. British Archaeology Report.
3. Christopher Chippindale, *Stonehenge Complete*, London: Thames and Hudson
4. *Camdens Britannia* 1772 in section on 'Name of Stonehenge'.
5. https://en.wikipedia.org/wiki/Stonehenge
6. www.liverpooluniversitypress.co.uk/books/isbn/9780859897341/
7. L V Grinsell, *Legendary History and Folklore of Stonehenge*, 1975
8. Samuel Danyel, *Stonehenge: A Poem*, 1624
9. Rev. Robert Gay, *A Fool's Bolt Soon Shot at Stonehenge*, 1725, in Rodney Legg, *Stonehenge Antiquities*, Dorset Publishing, 1986
10. Ibid p.24
11. archive.org/stream/tourinquestofgen00fent/tourinquestofgen00fent_djvu.txt
12. stone-circles.org.uk: Normanton Down Collyer, Chris. Retrieved 22 March 2009
13. *Wiltshire Archaeological Magazine*, Volume: Volume 70-1, p138.

14. M.C. Burkett, M.A., F.S.A., F.G. *Our Early Ancestors, an Introductory Study of Mesolithic, Neolithic and Copper Age Cultures in Europe and Adjacent Regions*, 1876
15. *Leland*, Collect, vol. iv. p. 141. b Leland, ubi supra
16. *Camden's Britannia* 1772
17. This was also reported in the following sources, *Oriental Literature, Applied to the Illustration of the Sacred, Volume 1* By Samuel Burder p.356, 1822. *The Family Magazine: Or Monthly Abstract of General Knowledge, Volume 1*, p.50, 1834. *Calmet's Dictionary of the Holy Bible* by Augustin Calmet, 1830, p.573. *Monumenta Antiqua: Or, The Stone Monuments of Antiquity* by Robert Weaver, p.131, 1840. *The Antijacobin Review* Issues 206-211 p.10, 1815, *A theological, biblical, and ecclesiastical dictionary* By John Robinson 1830
18. *St. Christopher, Martyr* by Father Francis Xavier Weninger, 1876
19. www.catholic-forum.com/SAINTS/golden234.htm
20. www.unexplained-mysteries.com/forum/topic/90029-st-christopher-a-giant
21. Jacqueline Simpson, Jennifer Westwood, *The Lore of the Land: A Guide to England's Legends, from Spring-heeled Jack to the Witches of Warboys*, Penguin, 2006, p.781
22. John Thurnam, *Examination Of A Chambered Long Barrow, At West Kennet, Wiltshire*, 1859
23. All these sites are documented at www.megalithic.co.uk
24. Edward J. Wood, *Giants and Dwarfs*, Richard Bentley, 1868, p.45
25. Peter Knight, *Ancient Stones of Dorset*, Power Publications, 1996, p.182
26. Geoffrey Ashe, *Mythology of the British Isles*, Guild, 1990, p.53
27. *Peter Roberts, the Chronicle of the Kings of Britain, trans. from welsh of Tysilio*.
28. *E. Waring, Ghosts and Legends of the Dorset Countryside, Compton Press*, 1977
29. Gary Biltcliffe, *The Spirit of Portland*, Roving Press, 2009, p.123
30. J.F. Pennie, *The Tale of a Modern Genius*, J. Andrews, London, 1827
31. Edward J. Wood, *Giants and Dwarfs*, Richard Bentley, 1868, p.42
32. Gary Biltcliffe, *The Spirit of Portland*, Roving Press, 2009, p.21
33. www.megalithic.co.uk/article.php?sid=11476
34. Jacqueline Simpson, Jennifer Westwood, *The Lore of the Land*, Penguin, 2006, p.308
35. *Bevis of Hampton*, unknown author, c.1324, https://d.lib.rochester.edu/teams/publication/salisbury-four-romances-of-england.
36. www.gatehouse-gazetteer.info/English%20sites/4673.html
37. Jennifer Westwood, *Albion*, Paladin Edition, 1985, p.77
38. www.sussexarch.org.uk/saaf/giants.html
39. E. W. Gray, *The History and Antiquities of Newbury and Its Environs*. p.219, 1839.
40. Ibid p.221.
41. www.mysteriousbritain.co.uk/featured-sites/the-aldworth-giants/.
42. www.ancient-origins.net/myths-legends-europe/life-and-legend-aldworth-giants-002756
43. E. W. Gray, *The History and Antiquities of Newbury and Its Environs*. p.232, 1839.
44. www.hungerfordvirtualmuseum.co.uk/index.php/19-people/family-history/888-hungerford-family.
45. Thomas Hearne, 18th century Oxford antiquarian.
46. Raphael Samuel, *Theatres of Memory: Past and Present in Contemporary Culture*, Verso, 1994, p144.
47. https://joyofmuseums.com/museums/united-kingdom-museums/london-museums/natural-history-museum-london/great-handaxe-from-furze-platt.
48. *Hampshire Antiquary and Naturalist*, v1, 1891, p136
49. John.L.Whitehead, *The Undercliff Of The Isle Of Wight Past And Present*, 1911. His quotes within the text come from: Westropp, /our. o,f Anihrop.Jnsi., vol. vii, 1878
50. See *Sowers of Thunder* p.124/125 and https://en.wikipedia.org/wiki/Grabbist_Hillfort (Giant's Chair)
51. www.bbc.co.uk/history/domesday/dblock/GB-296000-141000/page/6
52. Philip Rahtz, Lorna Watts, *Glastonbury Myth & Archaeology*, The History Press, 2009, p.55.
53. William of Malmesbury. *Antiquities of Glastonbury*, 1135 AD
54. Roy Norvill, *Giants: The Vanished Race of Mighty Men*, Aquarian Press, 1979, pp. 109, 111.
55. Rahtz, Watts, *Glastonbury Myth & Archaeology*, The History Press, 2009, p.55, from Gerard of Wales.
56. Giraldus, *Liber de Principis instructione, Distinctio I, folio 107b* (c. 1193) Translation by J. Colavito.
57. Virgil, Georgics 1.497. Virgil quote: H. R. Fairclough in the Loeb Virgil, 1916
58. T. Holmes in William Howship Dickinson, *King Arthur in Cornwall*, 1900.
59. *Encyclopedia Britannica*, 1967 ed., s.v. "Glastonbury"
60. Yuri Leitch, *Glastonbury and the Myths of Avalon*, Independently Published, 2019, p. 38
61. M. D. Cra'ster, *The Cambridge Region and British Archaeology*, 1984, p.5
62. www.somersetheritage.org.uk/downloads/publications/150years/HES_150_Years_Chapter_4.pdf.
63. Rev. Robert Gay, *A Fool's Bolt Soon Shot at Stonehenge* in *Stonehenge Antiquities*, p.8.
64. www.british-history.ac.uk/vch/som/vol1/pp206-218
65. Rev. Robert Gay, *A Fool's Bolt Soon Shot at Stonehenge* in *Stonehenge Antiquities*, p.34

66. Peter of Langtoft, *As Illustrated and Improv'd by Robert of Brunne*, 1810, p.501
67. ibid p.502, and Rev. Robert Gay, p.20-21
68. John Collinson, in *The History and Antiquities of the County of Somerset* and *Giants & Dwarfs*, p.34
69. www.geopolymer.org/archaeology/.
70. *Proceedings of the Somersetshire Archaeological and Natural History Society*: Volumes 10-12, January 1, 1861. p.14.
71. John Collinson, *The History and Antiquities of the County of Somerset: Vol. 3*, Edmund Rack
72. Edward J. Wood, *Giants and Dwarfs*, Richard Bentley, 1868, p.370.
73. Geoffrey Ashe, *The Landscape of King Arthur*, Henry Holt & Co, 1987. p.64.
74. Geoffrey Body, *A - Z of Curious Somerset*, The History Press, 2013, p.62.
75. Lornie Leete-Hodge, *Curiosities of Somerset*. Bodmin: Bossiney Books. p. 20.
76. https://hauntedwiltshire.blogspot.co.uk/2011/04/dunster-castle-sumerset.html
77. Anthony Roberts, *Sowers of Thunder*, Rider, 1978, p.125
78. Janet and Colin Bord, *Mysterious Britain*, Garnstone Press, 1972, p.101
79. Janet and Colin Bord, *Secret Country*, Book Club Associates, 1976. p.99

CHAPTER 3 - DEVON AND CORNWALL

1. John Michell, *The Old Stones of Land's End*, Garnstone Press, London, 1974
2. Jennifer Westwood, *Albion*, Paladin Edition, 1985, p.29
3. Edward J. Wood, *Giants and Dwarfs*, Richard Bentley, 1868, p.205
4. www.lundy.org.uk/index.php/about-lundy/archaeology
5. Robin Heath and John Michell, *The Measure of Albion*, Bluestone Press, 2004, p.46
6. A. F. Langham, quoting from Peter Levi, *The Flutes of Autumn*
7. Robin Heath and John Michell, *The Measure of Albion*, Bluestone Press, 2004, p.45
8. Ibid, p. 47
9. Geoffrey Ashe, *Mythology of the British Isles*, Guild, 1990
10. Robin Heath and John Michell, *The Measure of Albion*, Bluestone Press, 2004, p.46
11. *Sowers of Thunder*, p. 127 and *Holinshed's Elizabethan Chronicles of England, Scotland and Ireland*, 1578
12. *Fe Fi Fo Fum: The Giants in England*, p.64.
13. Roy Norvill, *Giants: The Vanished Race of Mighty Men*, Aquarian Press, 1979, p.119
14. Some of the multiple sources that reported this story are as follows, *The Living Age* - Volume 134 - Page 703, 1877. *The Wide World Magazine: An Illustrated Monthly of True Narrative*, Volume 17, page 396, 1906. *The Lost Science of Measuring the Earth* By Robin Heath, John Michell page 47, 2006. *Lundy Island: A Monograph, Descriptive and Historical* By John Roberts Chanter pgs 50-51 , 1877. *Chambers's Journal* - Volume 55 - Page 251, 1878. *Littell's Living Age* - Volume 134 - Page 703, 1877. *Devonshire Association for the Advancement of Science, Literature and Art* - pg 568, 1871. *The Age* - Apr 18, 1952. *Christian Science Monitor* - Apr 8, 1975. *Ellesmere Guardian* - Jun 9, 1906. *Christian Science Monitor* - Jun 14, 1966. *The Sun* - Feb 15, 1931. *Tuapeka Times* - Aug 25, 1870 and the *L.A. Times*, April 27, 1980.
15. www.gatehouse-gazetteer.info/English%20sites/871.html
16. John Lloyd Warden, *Coasts of Devon & Lundy Island*, 1895, p.207
17. *Taunton Courier, and Western Advertiser* - Wednesday 21 May 1856
18. www.legendarydartmoor.co.uk/maximajor_stone.htm
19. www.dartmoor.gov.uk/learning/dartmoor-legends/the-story-of-clotted-cream
20. Robert Hunt, *Popular Romances of the West of England: The drolls, traditions, and superstitions of old Cornwall*, 3d ed., rev. and enl. London, Chatto and Windus, 1903
21. Anthony Roberts, *Sowers of Thunder*, Rider, 1978
22. Robert Hunt, *Popular Romances of the West of England*, Chatto and Windus, 1903
23. www.stone-circles.org.uk/stone/lanyon.htm
24. John Michell, *The Old Stones of Land's End*, Garnstone Press, London, 1974, p.76
25. From a report of an excursion of the Penzance Natural History and Antiquarian Society, in the *Cornubian and Redruth Times*, 3rd September 1869
26. http://ancientwandlebury.blogspot.com/2009/03/circular-henges-ancient-megaliths-round.html
27. Fortescue Hitchins, *The History of Cornwall: From the Earliest Records and Traditions, to the Present Time*, Volume 2. Samuel Drew, 1824. p.524
28. www.intocornwall.com/engine/azabout.asp?guide=St+Goran
29. Lisa Bradley, Patrick J Morrison. *Ulster Med J 2011*;80(1):31-32, Commentary, *Giants of the British Isles*. November 2010
30. This version was also in the *Annual Register, Vol. IV*. p.88 and also featured in *Giants*, by Roy Norvill (1979). This version is from *Giants & Dwarfs* p.188.
31. https://web.archive.org/web/20140810012707/http://hypatia-trust.org.uk/2013/01/25/could-this-be-cornwalls-earliest-image-of-a-woman/

32. www.paranormaldatabase.com/reports/giants.php
33. www.britainexpress.com/attractions.htm?attraction=33
34. Anthony Roberts, *Sowers of Thunder*, Rider, 1978, p.98
35. Ibid, p.88
36. www.bolsterfestival.wordpress.com
37. www.paranormaldatabase.com/reports/giants.php
38. *The Giants of Trencrom, or Trecrobben*, in Robert Hunt, *Popular Romances of the West of England*, 1903
39. B. C. Spooner, *The Giants of Cornwall*, Folklore. 76 (1): 17, 1965.
40. Anthony Roberts, *Sowers of Thunder*, Rider, 1978, p.102
41. Hamish Miller and Paul Broadhurst, *Dance of the Dragon*, Mythos, 2000, p. 83
42. John Hobson Matthews, *Legendary Lore: A History of the Parishes of Saint Ives, Lelant, Towednack, and Zennor, in the County of Cornwall*. London: Elliot Stock, 1892.
43. BBC. *Fairy tale origins thousands of years old, researchers say*. BBC News. 20 January 2016.
44. William Bernard McCarthy, Cheryl Oxford, Joseph Daniel Sobol, eds. 1994. *Jack in Two Worlds: Contemporary North American Tales and Their Tellers*, illustrated ed. UNC Press Books. p. xv.
45. Tatar, *The Annotated Classic Fairy Tales*, p. 136).
46. *History of Jack the Giant Killer*. Glasgow: General Printed edition
47.www.paranormaldatabase.com/reports/giants.php?pageNum_paradata=3&totalRows_paradata=81
48. www.ancient-stones.co.uk/borders/001/007/details.htm
49. *The Giants of The Mount*, in Robert Hunt, *Popular Romances of the West of England*, 1903
50. https://en.wikipedia.org/wiki/St_Michael%27s_Mount
51. Peter Herring. *St Michael's Mount Archaeological Works, 1995-8*. Cornwall Archaeological Unit 2000.
52. ICTIS INSVLA at roman-britain.org
53. Joseph Jacobs, *Jack the Giant-Killer*. English Fairy Tales. London: David Nutt, 1890
54. Anthony Roberts, *Sowers of Thunder*, Rider, 1978, p.81
55. Ibid, p.84
56. Hamish Miller and Paul Broadhurst, *Dance of the Dragon*, Mythos, 2000, p.128
57. *Popular Rhymes* 1849.
58. B. C. Spooner, *The Giants of Cornwall*, Folklore. 76 (1): 19
59. *The Lord of Pengerswick and the Giant of St. Michael's Mount* in Robert Hunt, *Popular Romances of the West of England*, 1903
60. Fortescue Hitchins, Samuel Drew, *The History of Cornwall: From the Earliest Records and Traditions, to the Present Time, Vol 2*, 1824 p.327
61. Peter Herring, *St Michael's Mount: The National Trust. An Archaeological Evaluation of St Michael's Mount*. Cornwall Archaeological Unit. 1993. P. 95-96
62. *The Key of the Giant's Castle* in Robert Hunt, *Popular Romances of the West of England*, 1903
63. Charles Sandoe Gilbert, *An Historical Survey of the County of Cornwall, To which is Added, a Complete Heraldry of the Same*, 1817. p.189
64. John Thomas Blight, *A Week at the Land's End*, 1861, pp.19–21
65. Fortescue Hitchins, Samuel Drew, *The History of Cornwall: From the Earliest Records and Traditions, to the Present Time, Vol 1*. 1824. p.378-379
66. T. Bond, *Topographical and Historical Sketches of the Boroughs of East And West Looe*, 1823, in Wessex Archaeology Report 2009, p.29 (www.wessexarch.co.uk/sites/default/files/68734_Looe%20Cornwall.pdf)
67. M. Dunn, *The Looe Island Story*, Polperro, 2005
68. http://ads.ahds.ac.uk/catalogue/search/fr.cfm
69. Jacqueline Simpson, Jennifer Westwood, *The Lore of the Land*, Penguin, 2006, p.97

CHAPTER 4 - LONDON AND THE SOUTHEAST

1. Anthony Adolph, *Brutus of Troy and the Quest for the Ancestry of the British*, Pen & Sword Books, 2015
2. Patricia Monaghan, *The Encyclopedia of Celtic Mythology and Folklore*. Facts on File, Inc. 2004, p.55
3. www.britannia.com/celtic/gods/bran.html
4. www.celtsandmyths.mzzhost.com/bran.html
5.http://staging.waymarking.com/waymarks/WMCH3A_The_Giant_Suit_of_Armour_Tower_of_London_London_UK
6. Chris Grooms, *The Giants of Wales*, Edwin Mellen press, 1993, p.279, originally from *Peniarth 118, fol.827* by John Davies (Sion Dafydd Rhys, c.1590)
7. Ibid
8. Edward J. Wood, *Giants and Dwarfs*, Richard Bentley, 1868, p.33
9. Ibid p.217.
10. Ibid p.33
11. Ibid p.204, originally from *Survey of London Monograph 3*, Old Palace, Bromley-By-Bow. Formerly

published by Guild & School of Handicraft, London, 1901
12.www.british-history.ac.uk/survey-london/bk3/pp17-18
13. John Timbs, *A Picturesque Promenade Round Dorking, in Surrey*, 1823, p.158
14. Edward J. Wood, *Giants and Dwarfs*, Richard Bentley, 1868, p.189
15. Ibid, p.38
16. A. D. Shaw, M. J. White, *Another look at the Cuxton handaxe assemblage*. Proceedings of the Prehistoric Society, 69, pp. 305-314. Also at: https://dro.dur.ac.uk/5931/
17. http://news.bbc.co.uk/1/hi/sci/tech/5098748.stm
18. http://www.megalithic.co.uk/article.php?sid=2146412517
19. https://historicengland.org.uk/listing/the-list/list-entry/1014955
20. http://steyningmuseum.org.uk/prehistory.htm
21. http://news.bbc.co.uk/1/hi/sci/tech/7466735.stm
22. https://worldhistory.us/archaeology/boxgrove-man.php
23. https://archive.org/details/giantsinsouthafricabymichaeltellinger. See the *Origins of the Tall Ones* chapter in *Giants On Record* for the full analysis, and how they relate to the Denisovans of Siberia.
24. www.leyhunters.co.uk/tlh3.html
25. http://myths.e2bn.org/mythsandlegends/userstory2219-the-long-man-of-wilmington.html
26. www.paranormaldatabase.com/reports/giants.php?pageNum_paradata=3&totalRows_paradata=86
27. Jennifer Westwood, *Albion*, Paladin Edition, 1985, p.169
28. *Giants and Dwarfs*, p.132 and *Philosophical Transactions (1683-1775) Vol. 27 (1710 - 1712)*, p.436
29. *The New Monthly Magazine & Literary Journal*, Vol 2, p.272
30. Michael A. Cremo, *Forbidden Archeology: The Hidden History of the Human Race*, Bhaktivedanta Book Trust, 1993, p.410
31. Dio Cassius, 62.2.2-4=
32. Jennifer Westwood, *Albion*, Paladin Edition, 1985, p.131
33. *Journal of the British Archaeological Association for the year 1879*, in an article entitled 'Fen Tumuli', by Jonathon Peckover
34. T.C. Lethbridge, *Gogmagog: The Buried Gods*, Routledge, 1957, p.168
35. Ibid p.15
36. Ibid p.148
37. H. C. Lofts, *Notes and Queries*, December 26th, 1874.
38. *Cherry Hinton Chronicle* 1751-1899. P.59. Reported on May 27th 1854
39. Michelle Bullivant, *War Ditches, Cambridge, 1854-2008*. 2010, p.6
40. Anthony Roberts, *Sowers of Thunder*, Rider, 1978, p.123
41. Glyn Morgan, *Secret Essex*, Ian Henry Publications, 1982, p.85
42. Edward J. Wood, *Giants and Dwarfs*, Richard Bentley, 1868, p36
43. www.paranormaldatabase.com/reports/giants.php?pageNum_paradata=2&totalRows_paradata=86.
44. The Ontario Archaeological Society, Arch Notes. April-May 1977, p.37
45. www.bbc.co.uk/news/uk-england-norfolk-30175083
46. *Nature*, 1935. (Proc. Prehistoric Soc. East Anglia, 7, Pt. 3)
47. www.heritage.norfolk.gov.uk/record-details?MNF6312-Possible-Palaeolithic-handaxe-or-core

CHAPTER 5 - ISLE OF MAN AND MIDDLE ENGLAND
1. Train, Michell etc
2. John Michell, *At the Centre of the World*, Thames & Hudson, 1994, p.85
3. *Folk-lore of the Isle of Man*: Chapter I. *Myths Connected with the Legendary History of the Isle of Man*. Sacred-texts.com
4. W. W. Gill, *Third Manx Scrapbook*, 1938
5. Mave Calvert, *Earth Energies at Megalithic Sites in the Isle of Man*, 2017, p.11
6. www.gatehouse-gazetteer.info/Island%20sites/9009.html
7. Joseph Train, *An historical and statistical Account of the Isle of Man Volume 2*, 1845
8. Anthony Roberts, *Sowers of Thunder*, Rider, 1978, pp. 126-127
9. Joseph Train, *An historical and statistical Account of the Isle of Man Volume 2*, 1845
10. George Waldron, *Description of the Isle of Man*,1731, p.171
11. Ibid
12. The *Swarbreck Manuscript*, written in 1815
13. A. Henshall, F. Lynch, P. Davey, *The Chambered Tombs of the Isle of Man, A study by Audrey Henshall 1971-1978*, Pub. 2017, p.22
14. Anthony Roberts, *Sowers of Thunder*, Rider, 1978, p.120
15. Edward J. Wood, *Giants and Dwarfs*, Richard Bentley, 1868, p.45
16. Anthony Roberts, *Sowers of Thunder*, Rider, 1978, p.122

17. Janet and Colin Bord, *Secret Country*, Book Club Associates, 1976. p. 99
18 Anthony Roberts, *Sowers of Thunder*, Rider, 1978, p.123
19. Janet and Colin Bord, *Secret Country*, Book Club Associates, 1976. p.98
20. www.bbc.co.uk/shropshire/content/articles/2005/03/23/wrekin_giant_feature.shtml
21. Janet and Colin Bord, *Secret Country*, Book Club Associates, 1976. p.100
22. Anthony Roberts, *Sowers of Thunder*, Rider, 1978, p.92
23. *Yorkshire Archaeological Society*, J.W. Walker, Y.A.J., vol. 36, 1944, pp. 44-45
24. Anthony Roberts, *Sowers of Thunder*, Rider, 1978, p.92
25. http://midgleywebpages.com/littlejohn.html
26. J.W. Walker, *Y.A.J.*, vol. 36, (1944), pp. 44-45
27. www.paranormaldatabase.com/reports/giants.php
28. *Coventry Herald*, 24 April 1925
29. Robert Plot, *The Natural History of Staffordshire*, Oxford 1686 edition, p.331
30. Ibid
31. March 2016, BBC documentary, *The Vikings Uncovered*, presented by Dan Snow
32. Edward J. Wood, *Giants and Dwarfs*, p.124, and *Viking Repton*, by Barry M. Marsden, from 'The Vikings in Derbyshire', *Derbyshire Life & Countryside*, March/April 2007
33. Stephen Glover, *The History of the County of Derby*, 1829, p.457
34. James Pilkington, *A View of the Present State of Derbyshire, With an Account of Its Most Remarkable Antiquities*, 1789, p.424, and *Giants and Dwarfs*, p.39
35. Edward J. Wood, *Giants and Dwarfs*, Richard Bentley, 1868, p.44
36. Matt Sibson, www.thiswasleicestershire.co.uk/2012/09/bel-leicester-giant
37. Jacqueline Simpson, Jennifer Westwood, *The Lore of the Land*, Penguin, 2006, p.418
38. Ibid, p.437
39. Edward J. Wood, *Giants and Dwarfs*, Richard Bentley, 1868, p.127
40. https://stmaryskentchurch.org/legends-of-john-okent
41.Edward J. Wood, *Giants and Dwarfs*, Richard Bentley, 1868, p.39
42. *Giants and Dwarfs*, p.189 and *The Scots Magazine: Volume 25*, 1763, p.57

CHAPTER 6 - THE NORTH

1. www.blipfoto.com/entry/2853015
2. www.paranormaldatabase.com/reports/giants.php
3. Jennifer Westwood, *Albion*, Paladin Edition, 1985, p.382
4. Ibid p.421
5. Thomas Bulmer, *History, Topography and Directory of North Yorkshire*. S and N Publishing, 1890
6. Clive Kristen, *Ghost Trails of the Lake District and Cumbria*, 2014
7. Edward J. Wood, *Giants and Dwarfs*, Richard Bentley, 1868, p.45
8. Jacqueline Simpson, Jennifer Westwood, *The Lore of the Land*, Penguin, 2006, p.138
9. www.paranormaldatabase.com/reports/giants.php?pageNum_paradata=1&totalRows_paradata=87
10. Jacqueline Simpson, Jennifer Westwood, *The Lore of the Land*, Penguin, 2006, p.628
11. www.paranormaldatabase.com/reports/giants.php
12. Edward J. Wood, *Giants and Dwarfs*, Richard Bentley, 1868, p.234
13. Oliver Goldsmith, *The Vicar of Wakefield*, 1762
14. Edward J. Wood, *Giants and Dwarfs*, Richard Bentley, 1868, p.234
15. J. Nicolson, R. Burn, and W. Nicolson, *The history and antiquities of the counties of Westmorland and Cumberland: Volume 2*, 1777, p.410
16. www.stone-circles.org.uk/stone/giantsgrave.htm
17. *Giants and Dwarfs* p.31 and Samuel Jefferson, *The history and antiquities of Allerdale Ward, above Derwent, in the county of Cumberland: with biographical notices and memoirs*, 1842, p.331
18. James Fergusson, *Rude Stone Monuments*, 1872, p.156-157
19. *Archaeologia*, 1792
20.https://archaeologydataservice.ac.uk/archiveDS/archiveDownload?t=arch-2055-1/dissemination/pdf/Article_Level_Pdf/tcwaas/002/2000/vol100/tcwaas_002_2000_vol100_0007.pdf
21. C. Hulbert, *Legendary, Gothic, and Romantic Tales, in Verse, and Other Original Poems, and Translations*, 1825, pp32-33
22. Unknown newspaper source.
23. Charles Isaac Elton, *Origins of English History*, 1882, p.172
24. Robert Forster, *History of Corbridge and its Antiquities*, 1881, p.8
25. ibid
26.www.cwherald.com/a/archive/george-bott-recalls-keswick-museum-giant-s-shoes-among-the-odd-and-bizarre.237198.html

27. www.cwherald.com/a/archive/george-bott-recalls-keswick-museum-giant-s-shoes-among-the-odd-and-bizarre.237198.html
28. George Neasham, *The History And Biography Of West Durham*, 1881
29. Edward J. Wood, *Giants and Dwarfs*, Richard Bentley, 1868, p.126
30. T.F. Bulmer, Editor. *History, Topography and Directory of Westmorland*, 1885, p265
31. 'The Antiquities Column' in the Welsh newspaper *The Welsh Worker* dated Jan 4th 1886. (translation Chris Grooms in *The Giants of Wales*, pp.27-28)
32. *The Berwick Advertiser*, 24th October 1884. p2
33. Edward J. Wood, *Giants and Dwarfs*, Richard Bentley, 1868, p.39
34. Bozzi Granville described his visit in *The Spas of England and Principal Sea-bathing Places*, 1841
35. Joseph Jacobs, *More English Fairy Tales*, 1894
36. Edward J. Wood, *Giants and Dwarfs*, Richard Bentley, 1868, p.43
37. Also featured in *Giants and Dwarfs*, p.151-152
38. George Chalmers, *Caledonia; or a history & Topographical Account of North Britain, Vol 2*, 1888, p.45
39. taken from https://northeasthistorytour.blogspot.com/2011/05/fulwell-giant-cnz384599.html

CHAPTER 7 - GIANT LORE OF WALES

1. Also www.maryjones.us/ctexts/giants_wales.html - "Peniarth Ms. 118, fos. 829-837" ed. and trans. Hugh Owen. Y Cymmrodor. vol. 27. London: Honorable Society of Cymmrodorion, 1917. pp.115-152
2. Chris Grooms, *The Giants of Wales*, Edwin Mellen press, 1993, p.279, includes *Peniarth 118, fol.827* by John Davies (Sion Dafydd Rhys) c.1590
3. Anthony Roberts, *Sowers of Thunder*, Rider, 1978, p.55
4. William Forbes Skene trans., *The Four Ancient Books of Wales*, I, Edinburgh: Edmonston and Douglas, 1868
5. Anthony Roberts, *Atlantean Traditions in Ancient Britain*, Rider, 1977, p.57
6. Anthony Roberts, *Sowers of Thunder*, Rider, 1978, p.57
7. ibid.
8. ibid
9. Anthony Roberts, *Atlantean Traditions in Ancient Britain*, Rider, 1977, p.58
10. www.timelessmyths.com/celtic/welsh.html#Aranrhod
11. Anthony Roberts, *Atlantean Traditions in Ancient Britain*, Rider, 1977, p.59
12. The Mabinogion is based upon a 14th century manuscript known as 'Red book of Hergest'. The work is a collection of eleven tales of early Welsh literature and draws upon the mystical word of the Celtic people intertwining myths, folklore, tradition and history.
13. W. J. Thomas, 'Hu Gadarn', *The Welsh Fairy Book*, 1908, www.sacred-texts.com/neu/celt/wfb/wfb77.htm
14. www.maryjones.us/jce/defrobani.html
15. Chris Grooms, *The Giants of Wales*, Edwin Mellen press, 1993, p.187
16. Anthony Roberts, *Sowers of Thunder*, Rider, 1978, p.135
17. www.dunbrython.org/the-giants.html
18. www.bbc.co.uk/news/uk-wales-north-west-wales-11797766
19. www.paranormaldatabase.com/reports/giants.php
20. www.cadairidriswales.com/mountain-myths-legends-of-cadair-idris.html
21. William Davies, '*Casiglad o Len-Gwerin meirion*', Eisteddfod Transactons, 1898, p.220
22. R. Bromwich, Triad 28. From the triad created by Lolo Morganwg 'Myvyrian Third Series in 18th century, 1969, p.138
23. Glasynys's romantic poem '*Golygfa Oddiar Ben Moel Orthwm*' pp.37-39 (*The Giants of Wales*, p.189)
24. Hugh Evans, *The Origin of the Zodiac: Cadair Idris and the Star Maps of Gwynedd*, 2021
25. www.themodernantiquarian.com/post/144223/fieldnotes/dolddeuli.html
26. John Michell, *The Old Stones of Land's End*, Garnstone Press, London, 1974, p.240
27. https://the-history-notes.blogspot.com/2011/08/last-druids.html
28. www.genuki.org.uk/big/wal/AGY/Llanidan/
29. Chris Grooms, *The Giants of Wales*, Edwin Mellen press, 1993, p.16
30. Anthony Roberts, *Sowers of Thunder*, Rider, 1978, p.137
31. Chris Grooms, *The Giants of Wales*, Edwin Mellen press, 1993, p.14
32. Unknown newspaper source.
33. Anthony Roberts, *Sowers of Thunder*, Rider, 1978, p.137
34. https://en.wikipedia.org/wiki/Painscastle
35. www.dunbrython.org/the-giants.html
36. O 6-7, passim, pp.158-9
37. Chris Grooms, *The Giants of Wales*, Edwin Mellen press, 1993, p.131
38. Morgan Rhys, *Unpublished Transactions of Glamorganshire, Cambrian Journal, II*, 1855, p.p.68-72
39. Chris Grooms, *The Giants of Wales*, Edwin Mellen press, 1993, p.78

40. Chris Grooms, *The Giants of Wales*, Edwin Mellen press, 1993, p.65
41. www.britainexpress.com/attractions.htm?attraction=383
42. Jennifer Westwood, *Albion*, Paladin Edition, 1985, p.359
43. Anthony Roberts, *Sowers of Thunder*, Rider, 1978, p.55
44. www.sacred-texts.com/neu/celt/wfb/wfb39.htm - The Welsh Fairy Book by W. Jenkyn Thomas, 1908
45. John Castell Evans, *Yr Hen Amser Gyut*. NLW MS 10568C (*The Giants of Wales*, p.171)
46. Chris Grooms, *The Giants of Wales*, Edwin Mellen press, 1993, p.216
47. Anthony Roberts, *Sowers of Thunder*, Rider, 1978, p.138
48. John Leland, *Itinerary in Wales*, L. Toulmin Smith ed., 1906, p.119, col.76
49. Richard Fenton, *A Historical Tour Through Pembrokeshire*, 1811, p.160
50. J. Ceredig Davies, *Folk-lore of West and Mid-Wales*, Aberystwyth Welsh Gazette, 1911, p.270
51. www.earlybritishkingdoms.com/archaeology/brynm.html)
52. John Jones (Myrddin Fardd), *Llên gwerin Sir Gaernarfon, a collection of the folklore and traditions of northern Gwynedd, Wales*, Caernarfon, 1908
53. Chris Grooms, *The Giants of Wales*, Edwin Mellen press, 1993, p.85
54. Ibid p.90
55. Ibid p.87
56. Ibid p.228
57. Chris Grooms, *The Giants of Wales*, Edwin Mellen press, 1993, p.xiv and BL Harleian 4181:70a-71b
58. Sabine Baring-Gould, *The Lives of the British Saints: the Saints of Wales and Cornwall and such Irish Saints as have dedications in Britain*. Honourable Society of Cymmrodorion, 1907, pp. 292–293
59. Sion Dafydd Rhys, *The Giants of Wales and Their Dwellings*, Peniarth 118, fol.829-837, c.1590
60. https://en.wikipedia.org/wiki/Mitchell%27s_Fold
61. Chris Grooms, *The Giants of Wales*, Edwin Mellen press, 1993, p.62
62. Ibid p.72
63. Ibid p.155, trans. from YGELB 112.
64. Ibid p.75
65. J. Rhys, *Iberians, Canaanites and Romans*, 1886, p.143 (*The Giants of Wales*, p.77)
66. Thomas Gwynn Jones, *Welsh Folklore and Folk-Custom*, Metheun & Co, 1930, p.80
67. Chris Grooms, *The Giants of Wales*, Edwin Mellen press, 1993, trans. from OJC I:134
68. John Rhys, *Celtic Folklore, Welsh And Manx, Chapter VIII Welsh Cave Legends*, 1901
69. www.mirrortowonderland.blogspot.com/2013/11/brenin-llwyd-gray-king-welshscottish.html
70. Chris Grooms, *The Giants of Wales*, Edwin Mellen press, 1993, pp.93-94

CHAPTER 8 - THE GIANTS OF WALES

1. Owen Jones, *Cymru*, 1875, trans. from, Eds. John Rhys, G. Evans, *The Text of the Book of Llan Dav*, p.303
2. Edward Neil Baynes, *The Megalithic Remains of Anglesey*, Cymmrodorion Society's publications, 1911
3. Chris Grooms, *The Giants of Wales*, Edwin Mellen press, 1993, p.69
4. *Western Mail*, 27 December 1898, p.7
5. Rev. Elias Owen was also the author of *Welsh Folk-Lore: A Collection of the Folk-Tales and Legends of North Wales*, 1896
6. Angharad Llwyd, *A History of the Island of Mona, or Anglesey, Including an Account of Its Natural Productions, Druidical Antiquities, Lives of Eminent Men, the Customs of the Court of the Antient Welsh Princes, &c. Being the Prize Essay to which was Adjudged the First Premium at the Royal Beaumaris Eisteddfod, Held in the Month of August, 1832*, p.205
7. Robert Williams, *The History And Antiquities Of The Town Of Aberconwy*, 1835, p.51
8. Chris Grooms, *The Giants of Wales*, Edwin Mellen press, 1993, TW IV:332)
9. William howells, *Cambrian Superstition*, Thomas Danks Publishers, 1831, p.113
10. *Archaeoologica Cambrensis*, 1846, I,i:467 (*The Giants of Wales*, p.99)
11. www.bristol.ac.uk/news/2014/march/anglesey-dig.html
12. Chris Grooms, *The Giants of Wales*, Edwin Mellen press, 1993, p.131
13. Chris Grooms, *The Giants of Wales*, Edwin Mellen press, 1993, p.137
14. Edward J. Wood, *Giants and Dwarfs*, Richard Bentley, 1868, p.85
15. www.ancient-origins.net/ancient-places-europe/mold-gold-cape-finest-piece-prehistoric-gold-working-europe-001524
16. *Royal Coimission on Ancient and Historical Monuments in Wales and Monmouth*,1964. More info at www.gatehouse-gazetteer.info/Welshsites/1038.html
17. White Kennett, *Parochial Enquiries*, Oxford, 1695, pp.698-699 (*The Giants of Wales*, p.184)
18. Thomas Pennant, John Rhys ed., *Tours in Wales*, 1893, pp.380-381
19. Chris Grooms, *The Giants of Wales*, Edwin Mellen press, 1993, p.184
20. Glasnys, 1827, p.70, trans. from William Davies, *Eisteddfod Transactons*, 1898, p.139

21. *Cambrian Register* 1795, p.302 (*The Giants of Wales*, pp.187-188)
22. Chris Grooms, *The Giants of Wales*, Edwin Mellen press, 1993, p.80, trans. from ii:650
23. www.britainexpress.com/wales/mid/rhayader-st-clement.htm
24. Richard Fenton, *A Historical Tour Through Pembrokeshire*, 1811, p.22
25. Martin Phillips, *Folklore of the Afam and Margan Districts*, Guardian Press, 1933, pp.83-84
26. J. Lubbock, 'description of the Park cwn Tumulus', *Archaeologia Cambrensis*, Fourth series, II, 1871, p.171
27. Joshua Pollard, 'Wales' Hidden History, Hunter-Gatherer Communities in Wales: The Neolithic', 2001, In Prys Morgan & Stephen Aldhouse-Green. *History of Wales, 25,000 BC AD 2000*. Tempus Pub. pp.17–25
28. *Gower cave reindeer carving is Britain's oldest rock art*. BBC News Online website. 29 June 2012
29. Ewen Callaway, 'Archaeology: Date with history.' *Nature*, 485, pp.27–29

CHAPTER 9 - THE GIANTS OF CALEDONIA - SCOTLAND
1. S. McHardy, *A New History of the Picts*, Luath Press Ltd, 2011, p.31
2. ibid
3. www.worldhistory.org/article/776/tacitus-account-of-the-battle-of-mons-graupius/
4. www.ancient.eu/picts/
5. https://scotlandinmyheartsite.wordpress.com/2016/01/10/who-are-the-picts-scotlands-dna-at-last-finds-an-answer/
6. S. McHardy, *A New History of the Picts*, Luath Press Ltd, 2011, p175
7. Walter Bower, *Scotichronicon*, 15th century
8. Lorraine Evans, *Kingdom of the Ark*, Pocket Books, 2001
9. Donald Alexander Mackenzie, *Scottish Folk Lore and Folk Life*, Blackie, 1935, p.13
10. John Nicholson, *History of Galloway-From the Earliest Period to the Present Time*, 1841, pg. 17
11. https://dgartsfestival.wordpress.com/tag/big-man-walking/
12. *Kirkcudbrightshire Ordinance Survey name books* 1848-1851, Volume 42, p.10
13. Samuel Lewis, *A Topographical Dictionary of Scotland*, Volume 2, 1851, p.113
14. *The New Statistical Account of Scotland*, Volume 4, 1875, p.333
15. A. Burl, *A Guide to the Stone Circles of Britain, Ireland and Brittany*. Yale University Press, 2005, p.142
16. *Topographical, Statistical and Historical Gazetteer of Scotland*, Volume 2, 1853, p.266
17. Christian Maclagan, *Hill Forts, Stone Circles and other structural remains of Ancient Scotland*, 1875, p.123
18. *London Magazine or Monthly Gentleman's Intelligencer*, Volume 24, pg. 450, 1838
19. Alexander Low, *History of Scotland from the Earliest Period to the Middle of the Ninth Century*, 1826, p.72
20. Georges Louis Leclerc Comte de Buffon, Quoted in *A Natural History-General and Particular- Containing the History and Theory of the Earth*. 1817, p.169
21. *Encyclopedia Britannica*, Sixth Edition. 1823 p.700
22. *West Lothian Ordinance Survey named books, 1855-1859*, Volume 21, Parish of Dalemy p.10
23. *West Lothian Ordinance Survey named books, 1855-1859*, Volume 21, Parish of Dalemy p.8
24. Daniel Wilson, *Prehistoric Annals of Scotland, Volume 1*, 1863, p.72
25. John Sinclair, *The Statistical Account of Scotland*, 1795, pp.418-419
26. John Kobler, *The Reluctant Surgeon*, Doubleday, 1960, p.359
27. Comyns Beaumont, *The Riddle of Prehistoric Britain*, Rider & Co, London, 1946, p.31
28. Otta Swire, *The Highlands and their Legends*, Oliver & Boyd, 1963, p.275
29. James Taylor, *The Pictorial History of Scotland: from the Roman invasion to the Close of the Jacobite Rebellion, Volume 1*, 1859, p.22
30. James Browne, *A History of the Highlands and the Highland Clans*, A. Fullarton, 1857, p.11
31. *Fife and Kinross-shire Ordinance Survey name books, 1853-1855* Volume 26, p.14
32. *Fife and Kinross-shire Ordinance Survey name books, 1853-1855* Volume 96, p.57
33. *Perthshire Ordinance Survey Name Location, 1859-1862*, Volume 6, p.3
34. C. Maclagan, *The Hill Forts, Stone Circles and Other Structural Remains of Ancient Scotland*, 1875.
35. Ernesti, *Frustum*, 1838, p.369
36. *Forfarshire (Angus) Ordinance Survey name book 1857-1861*, Volume 71, p.69
37. *The New Statistical Account of Scotland*, Volume 5, 1835, p.265
38. William Marshall, *Historic Scenes in Forfarshire*, 1875, p.64
39. https://animalfolklore.wordpress.com/2014/11/25/the-grey-man-of-macdhui/
40. Donald Alexander MacKenzie, *Wonder Tales from Scottish Myth & Legend*, F. A. Stokes, 1917, p.22
41. Otta Swire, *Highlands and their Legends*, Oliver & Boyd, 1963, pp.42-43
42. Donald Alexander MacKenzie, *Wonder Tales from Scottish Myth & Legend*, Frederick A. Stokes, 1917
43. John Francis Campbell, *Popular Tales of the West Highlands*, 1893, pp.221-222
44. Otta Swire, *Highlands and their Legends*, Oliver & Boyd, 1963, pp.42-43
45. *Ross and Cromarty Ordinance survey name books* 1848-1852, p.67
46. Samuel Lewis, *A Topographical Dictionary of Scotland*, Volume 2, 1851, p.528

47. *Caithness Ordinance Survey name books*, Volume 9, 1871-1873, p.138
48. *Original Portraits and Caricature Etchings of the late John Kay*, 1877, p.100
49. *Dundee Advertiser*, 5 April 1893
50 Peter Hately Wadell, *Ossian and the Clyde: Fingal in Ireland*, 1875, p.7
51. www.orkneyjar.com/history/maeshowe/maeshowemummies.htm
52. Time Team Special 57, *Britain's Bronze Age Mummies*, Channel 4, 2014
53. https://theconversation.com/solved-the-mystery-of-britains-bronze-age-mummies-48475
54. www.bbc.co.uk/news/science-environment-14575729
55. Jo Ben, *Descripto Insuluarum Orchadiarum* (Descriptions of Orkney), c.1468
56. www.orkneyjar.com/folklore/giants/walkstones.htm
57. Jennifer Westwood, Sophia Kingshill, *The Lore of Scotland: A Guide to Scottish Legends*, 1988, p.36
58. John Spence, *Shetland Folk-Lore*, 1899, p.91
59. *Dundee Courier*, Wednesday 16 May 1951
60. Jennifer Westwood, *Albion*, Paladin Edition, 1985, p.492-493
61. www.archaeology.co.uk/articles/science-notes-survey-on-the-isle-of-arran.htm
62. Martin Martin, *A Description of the Western Islands Of Scotland, c.1695*
63. Rev. James Headrick, *View of the Mineralogy, Agriculture, and Fisheries of the Island of Arran*, 1807, p.149
64. James Beckett, *Tourist's Guide to Arran*, 1872, p.21
65. Otta Swire, *Skye: The Islands and It's Legends*, Oxford University Press, 1952, p.58
66. Otta Swire, *Inner Hebrides and their Legends*, Harper Collins, 1964, p.32

CHAPTER 10 - GODS AND GIANTS OF ANCIENT IRELAND

1. Anthony Roberts, *Atlantean Traditions in Ancient Britain*, Rider, 1977, p.4
2. Ignatius Donnelly, *Atlantis: The Antediluvian World*, 1882, p.412
3. B. Cunliffe, *Facing the ocean: the Atlantic and its peoples*, Oxford, 2001
4. Anthony Roberts, *Atlantean Traditions in Ancient Britain*, Rider, 1977, pp.31-32
5. Evans 1966, p.135
6. https://www.duchas.ie/en/cbes/4742170/4741882/4822475
7. Turtle Bunbury, *Ancient Kildare & The Kings of Leinster*, in Edmund Campion, *Historie of Ireland, Written in the Yeare 1571*, Hibernia Press, 1809. also at https://intokildare.ie/ancient-kildare-the-kings-of-leinster/
8. George Pepper, *The Irish Shield and Monthly Milesian*, 1829, p.447
9. *Book of Lecan*, verse 38, a medieval Irish manuscript c.1397-1418 in Castle Forbes, Lecan, Co. Sligo.
10. Col. Colby, *Ord Surv Ireland, Co. Londonderry Vol.1*, Dublin, 1837
11. *Transactions of the Royal Irish Academy, Volume 29*, 1892, p.557
12. James Paris du Plessis, *Short History of Human Prodigies, Dwarfs, etc*, unpublished manuscript produced between 1730 and 1733 that is preserved in the British Library in London
13. Asenath Nicholson, *Lights and Shades of Ireland*, 1850, p.192,
14. *The Morning Post*, 10 May 1832
15: Anthony Roberts, *Sowers of Thunder*, Rider, 1978, pp.31-32
16. *Schools Collection of folklore*, which has been put online at www.duchas.ie
17. *Irish News-New Zealand Tablet*, Volume 25, Issue 40, February 4, 1898 pg. 9
18. Janet and Colin Bord, *Secret Country*, Book Club Associates, 1976. p.188
19. *A Quarterly Review*, Volume 28, 1917, p.18
20. *On the Battle of Moytura*, Proceedings of the Royal Irish Academy 9, 1866, pp.545–550
21. Philip Luckombe, *A Tour Through Ireland*, 1783, p.183
22. *The Drogheda Journal or Meath and Louth Advertiser*, November 30th, 1825
23. *Kentucky Irish American*, May 24, 1902
24. https://www.jstor.org/stable/30008112?seq=1
25. 2020 BBC News article titled *DNA Study Reveals Ireland's Age of 'God-Kings'*
26. *St. Petersburg Times*, January 26, 1951, p.16
27. *Dublin Freeman Journal*, August 1812, p.25 (which also featured in the *Evening Post*, Volume LXXXIV, Issue 39, August 14, 1912, one hundred years later)
28. *Feilding Star*, Volume X1, Issue 2363, May 16 1914, P.4
29. John O'Hart, *Irish Pedigrees, or, The Origin and Stem of the Irish Nation*, 1880, p.727
30. *The Kilkenny and South-east of Ireland Archaeological Society*, 1864, p.149
31. https://sacredsites.com/europe/ireland/tower_of_cashel.html

CHAPTER 11 - SOWERS OF THUNDER

1. Anthony Roberts, *Sowers of Thunder*, Rider, 1978, p.28
2. ibid
3. John Michell, *At the Centre of the World*, Thames & Hudson, 1994, p.130

4. Anthony Roberts, *Sowers of Thunder*, Rider, 1978, p.34
5. Janet and Colin Bord, *Secret Country*, Book Club Associates, 1976. p.196
6. www.independent.ie/irish-news/sean-quinns-downfall-is-fairies-revenge-say-locals-in-cavan-26794562.html
7. Gwen Westerman and Bruce White, *Mni Sota Makoce: The Land of the Dakota*, 2012, p.18
8.www.researchgate.net/publication/316118987_Lightning_Induced_Remanent_Magnetization_at_the_Buffalo_Slough_Burial_Mound_Complex
9. Gary Lockhart, *The Weather Companion*, John Wiley & Sons, 1988, p.195
10. Janet and Colin Bord, *Secret Country*, Book Club Associates, 1976, pp.208-209
11. France Densmore, *Teton Sioux Music, Bureau of American Ethnology*. Smithsonian Institution, 1918. pp.137-138
12. http://andrewcollins.com/page/articles/thunder_people.htm
13. Andrew Collins and Greg Little, *Denosivan Origins*, Bear & Co., 2019, p.171
14. Wirt Sikes, *British Goblins*, W. Clowes & Sons, 1880, p.385
15. Both accounts from Chris Grooms, *The Giants of Wales*, Edwin Mellen press, 1993
16. trans. from Simon Jones, Straeon, *Cwm Cynllwyd*, 1989, pp.13-14 (*The Giants of Wales*, p.218)
17. Chris Grooms, *The Giants of Wales*, Edwin Mellen press, 1993, p.222
18. Ibid p.231
19. Ibid pp.91-92
20. Ibid p.238
21. Geoffrey Ashe, *A Guidebook to Arthurian Britain*, Longman, 1980, p.20
22. Rev Griffith Edwards, *History of the Parish of Llangadfan*, Montgomeryshire Collections, II, 1869, p.20
23. trans. J.O. Haws, *Y Tylwyth Teg*, Gwasg Carreg Gwalsh, 1987, pp.39-40 (*The Giants of Wales* p.100)
24. *The Mabinogion*, trans. by Gwyn and Thomas Jones, 1970, p.30
25. Rev. Walter Davies, *The English Works of Rev. Walter Davies, M.A., vol III*, W. Spurrell, 1868, p.14
26. W.A. Griffiths, Tales from Welsh History and Romance, J&j Bennett LTD, 1915, p.33
27. Chris Grooms, *The Giants of Wales*, Edwin Mellen press, 1993, pp.82-83
28. Theophilus Jones, *A History of the county of Brecknock*, Davies, 1805
29. www.reddragontales.com/single-post/2017/01/25/Did-Giants-Once-Roam-the-Welsh-Valleys
30. Anthony Roberts, *Sowers of Thunder*, Rider, 1978, p.137
31. Ibid p.35
32. William Borlase, *The Natural History of Cornwall*, 1756, p.16
33. J. wood, *A Description of Bath, vol.1*, 1765, pp.147-158
34. Gordon Strong, 'Stanton Drew', in *Megalith: Studies in Stone*, Wooden Books, 2018, p.270-271
35. Barry M. Marsden, *The Early Barrow-Diggers*, The History Press, 2011, and *Secret Country* p.208
36. *The British Critic*, Volume 42, 1813, p.544
37. Barry M. Marsden, *The Early Barrow-Diggers*, The History Press, 2011
38. Anthony Roberts, *Sowers of Thunder*, Rider, 1978, p.35
39. www.information-britain.co.uk/loredetail.php?id=80
40. John Snare, *The History and Antiquities of the Hundred of Compton, Berks*, 1844
41. Anthony Roberts, *Atlantean Traditions in Ancient Britain*, Rider, 1977, p.80
42. *The Intermountain Catholic*, October 16, 1909, p.6
43. James Beckett, *Tourist's Guide to Arran*, 1872, p.21
44. ibid
45. John Francis Campbell, *Popular Tales of the West Highlands*, 1893, pp.221-222
46. Tom Graves, *Needles of Stone*, Granada Books, 1980, pp.108-109
47. Michael Poynder, *Lost Science of the Stone Age*, Green Magic, p.49
48. https://thedailyplasma.blog/2017/10/26/lightning-and-megaliths-the-connection/comment-page-1
49. www.heraldscotland.com/news/18120526.lightning-strike-sparked-creation-stone-circle/

CHAPTER 12 - GIANT GEOMANCERS

1. Anthony Roberts, *Sowers of Thunder*, Rider, 1978, pp.XV - XVI
2. Ibid
3. John Michell's introduction in 1970 edition of, Alfred Watkins, *The Old Straight Track* p.xvi (orig. 1925)
4. Anthony Roberts, *Sowers of Thunder*, Rider, 1978, p.98
6. Danny Sullivan, *Ley Lines: A Comprehensive Guide to Alignments*, Piatkus, p.7
7. K. M. Briggs, *The Anatomy of Puck*. Routledge & Kegan Paul, 1959, p.117
8. Author Anthony Thomas discussed the reference in *Lhuyd's Parochialia*.
9. *Cardiff and Merthyr Guardian, Glamorgan, Monmouth, and Brecon Gazette*, 17 May 1851, p.4
10. Paul Newman. *The Lost Gods of Albion*. Sutton Publishing, 2000, p.102
11. Chris Grooms, *The Giants of Wales*, Edwin Mellen press, 1993, pp.187-188

12. Rosemary Ellen Guiley, *The Encyclopedia of Witches, Witchcraft and Wicca*, 1989
13. Evon Z Vogt, Ray Hyman, *Water Witching U.S.A.* 2nd ed., Chicago University Press, 1979
14. Graham Robb, *The Ancient Paths: Discovering the Lost Map of Celtic Europe*, Picador, 2014, p.285
15. *Archaeology Magazine*, Edition 284, Nov 2013, p.7
16. John Michell, *The View Over Atlantis*, Garnstone Press, 1969, p.50
17. www.cadairidriswales.com/mountain-myths-legends-of-cadair-idris
18. JWBS iv.4. Dec 1933, pp.164-5 (*The Giants of Wales*, p.205)
19. Anthony Roberts, *Sowers of Thunder*, Rider, 1978, p.77
20. trans. from *Cofiadur*, 1860, p.134, VIII, GOWp.88
21. www.themodernantiquarian.com/post/144223/fieldnotes/dolddeuli.html
22. Yuri Leitch, *The Terrestrial Alignments of Katharine Maltwood and Dion Fortune*, 2018
23. Marie Trevelyan, *Folk-lore and Folk-Stories of Wales*, p.129, 1853
24. Janet and Colin Bord, *Secret Country*, Book Club Associates, 1976. p.97
25. *Schools' Collection, Volume 969*, Wales, 1930, pp.112-113
26. *Cardiff and Merthyr Guarvdian, Glamorgan, Monmouth, and Brecon Gazette*, 17 May 1851, p.4
27. www.dunbrython.org/the-giants.html
28. Jennifer Westwood, *Albion*, Paladin Edition, 1985, p.151
29. Chris Grooms, *The Giants of Wales*, Edwin Mellen press, 1993, pp.60-61
30. Anthony Roberts, *Sowers of Thunder*, Rider, 1978, p.81
31. Anthony Mongelli, Jr., *The Craftsman's Symbology*, Lulu, p.53
32. www.craftsmansapron.com
33. www.rachelpatterson.co.uk/post/2017/10/12/coming-of-the-caileach-globe-trotting.
34. William S. Simmons, *Spirit of the New England Tribes: Indian History and Folklore*, 1984
35. Leila Azzam, *Lives of the Prophets*, Hood Hood Books, 1995
36. Louis Brenner, *Muslim Divination and the Religion of Sub-Saharan Africa. Insight and Artistry in African Divination*. ed. John Pemberton III. Smithsonian Institution Press. 2000, pp. 50–1

CHAPTER 13 - GENESIS OF THE GIANTS

1. www.nhm.ac.uk/discover/first-britons.html
2. Michael Tellinger interviewing Prof. Francis Thackary, https://youtu.be/70aRJpIY1OM
3. https://genographic.nationalgeographic.com/denisovan/
4. https://en.wikipedia.org/wiki/Cro-Magnon
5. Neil Asher Silberman, *The Oxford Companion to Archaeology*, Volume 1, 2012
6. Jack D. Forbes, *The American Discovery of Europe*, UI Press 2007, p.22
7. Vine Deloria Jr., *Red Earth, White Lies*, Fulcrum Publishing, 1997, p.63
8. Andrew Collins and Greg Little, *Denosivan Origins*, Bear & Co., 2019, p.140
9. Ibid p.136
10. Ibid p.140
11. J. Reid Moir, *The Antiquity of Man in East Anglia*, Cambridge University Press, 1927
12. https://ahotcupofjoe.net/2019/09/denisovan-origins-and-ancient-dna/
13. Andrew Collins and Greg Little, *Denosivan Origins*, Bear & Co., 2019, p.297
14. Robert Kelly & David Thomas, *Archaeology*, pg 255
15. www.ancient-origins.net/ancient-places-europe/orkney-isles-megaliths-0010828
16. ibid
17. *Multi-Proxy Characterisation of the Storegga Tsunami and Its Impact on the Early Holocene Landscapes of the Southern North Sea*, The journal Geosciences,. July 15, 2020
18. H. P. Blavatsky, *Secret Doctrine*, London & Benares, 1888, pp.760-761
19. Beckles Wilson, *Lost Lyonesse: Evidence, Records and Traditions of England's Atlantis*, 1902, p.11
20. Anthony Roberts, *Sowers of Thunder*, Rider, 1978, Footnote p.100
21. http://penelope.uchicago.edu/Thayer/E/Roman/Texts/Diodorus_Siculus/2B*.html#note36
22. www.ancient-code.com/hyperborea-atlantis-rival-land-to-the-north-and-the-homeplace-of-the-gods/
23. Hecataeus of Abdera collated all the stories about the Hyperboreans current in the fourth century BC and published a lengthy treatise on them, lost to us, but noted by Diodorus Siculus (ii.47.1–2)
24. T. P. Bridgman, Hyperboreans: Myth and history in Celtic-Hellenic contacts. Routledge, 2005, p.163-173
25. David Boyd Haycock, "Chapter 7: Much Greater, Than Commonly Imagined.", in William Stukeley: *Science, Religion and Archaeology in Eighteenth-Century England*. Boydell & Brewer, 2002
26. Aelian. *On the Nature of Animals*, Loeb Classical Library. p.357
27. See *Giants on Record*, 'Origins of the Tall Ones' chapter, Avalon Rising, 2015
28. *The Peterborough Chronicle*, 11th Century AD, Manuscript E of Anglo Saxon Chronicle
29. www.nhm.ac.uk/press-office/press-releases/ancient-dna-shows-migrants-introduced-farming-to-britain-from-eu.html

30. Comyns Beaumont, *The Riddle of Prehistoric Britain*, 1946
31. S.H. Hooke, *The Religious Institutions of Israel*, Peake's Commentary on the Bible
32. 1 Kings 18, pp.30-35
33. *The Book of Enoch*, Chapter LXI
34. Robin Heath, *Sun, Moon and Stonehenge*, Bluestone Press, 1998, and Christopher Knight, Robert Lomas, *Uriel's Machine: the Prehistoric Technology That Survived the Flood*, Century, 1999
35. *The Book of Enoch*, SE III:1 PP
36. Robin Heath, *Sun, Moon and Stonehenge*, Bluestone Press, 1998, p.147
37. *The Mabinogion*, tr. by Lady Charlotte Guest, 1877, Taliesin, p.471.
38. www.abarim-publications.com/Meaning/Enoch.html#.XlbupWT7SCU
39. www.islamichouseofwisdom.com/prophets-in-islam-/2013/1/3/idrisenoch-s-
40. Martin Sweatman, *Prehistory Decoded*, Troubador Publishing, 2018
41. Andrew Collins, *The Cygnus Key: The Denisovan Legacy, Göbekli Tepe, and the Birth of Egypt*, Bear & Company, 2018
42. Chris Grooms, *The Giants of Wales*, Edwin Mellen press, 1993, p.174
43. Christian and Barbara Joy O'Brien, *The Shining Ones*, Golden Age Project, 2001, p.416
44. Jack D. Forbes, *The American Discovery of Europe*, UI Press, 2007, p.6
45. Ibid p.33
46. Ibid p. 39
47. Michel-Gerald Boutet, *The Celtic Connection*, p. 73, quoting Bernard Assiniwi
48. Alan Wilson's books and research can be found at: www.cymroglyphics.com
49. www.historic-uk.com/HistoryUK/HistoryofWales/The-discovery-of-America-by-Welsh-Prince/
50. Jim Vieira, Hugh Newman, *Giants On Record, America's Hidden History, Secrets in the Mounds and the Smithsonian Files*, Avalon Rising, 2015, p.40
51. Seumas MacManus, *The Story of the Irish Race*, 1921
52. Christopher Knight, Robert Lomas, *Uriel's Machine: the Prehistoric Technology That Survived the Flood*, Century, 1999, p.304
53. https://bmcbiol.biomedcentral.com/articles/10.1186/1741-7007-8-15
54. Christopher Knight, Robert Lomas, *Uriel's Machine: the Prehistoric Technology That Survived the Flood*, Century, 1999, p.307
55. www.archaeology.co.uk/articles/science-notes-rare-evidence-for-neolithic-textiles-identified-on-orkney.htm
56. www.dgnhas.org.uk/tdgnhas/3045.pdf
57. www.independent.co.uk/news/world/asia/a-meeting-of-civilisations-the-mystery-of-chinas-celtic-mummies-5330366.html
58. Christopher Knight, Robert Lomas, *Uriel's Machine: the Prehistoric Technology That Survived the Flood*, Century, 1999, p.4
59. *Archaeology Magazine*, Issue 273, Dec 2012, p.9
60. *Archaeology Magazine*, issue 305, Dec 2015, p.8
61. Dolmens and a Circle of Standing Stones (article - get ref)
62. Pranidhi scene No. 5, Temple No. 9, Bäzäklik
63. Sui Shu Yiyu Zhuan, The Account of Dynasties, HanShu, www.youtube.com/watch?v=OB8eeVd7R_M
64. Sui Shu Yiyu Zhuan, *The Accounts of the Western Regions*, Shule
65. Thomas Kinsella, trans., *The Táin*, Oxford University Press, 1969, p.156
66. H.A.Basilius, *The Rhymes in Eger and Grime*, Modern Philology, vol. 35, no.2, Nov., 1937, p.129
67. 2014 article from the Ulster Medical Journal
68. Whitely Stokes, *The Destruction of Da Derga's Hostel*, 1910, p.94
69. Robert Temple, *The Sirius Mystery*, St. Martin's Press, 1976, pp.398-399
70. A. P. Fitzpatrick, *The Amesbury Archer and the Boscombe Bowmen*, Wessex Archaeology, 2013, p.24
71. www.euppublishing.com/doi/pdfplus/10.3366/saj.2018.0094, p.69
72. J.A.MacCulloch, *The Religion of the Ancient Celts*, Morrison and Gibb, 1911

GIANTS ON RECORD
By Jim Vieira and Hugh Newman
Over a 200-year period thousands of newspaper reports, town and county histories, letters, photos, diaries, and scientific journals have documented the existence of an ancient race of giants in North America. Extremely large skeletons ranging from 7 feet up to a staggering 18 feet tall have been reportedly uncovered in prehistoric mounds, burial chambers, caves, geometric earthworks, and ancient battlefields. Strange anatomic anomalies such as double rows of teeth, horned skulls, massive jaws that fit over a modern face, and elongated skulls have also been reported. Color Section.
420 pages. 6x9 Paperback. Illustrated. $19.95. Code: GOR

COVERT WARS & BREAKAWAY CIVILIZATIONS
By Joseph P. Farrell
Farrell delves into the creation of breakaway civilizations by the Nazis in South America and other parts of the world. He discusses the advanced technology that they took with them at the end of the war and the psychological war that they waged for decades on America and NATO. He investigates the secret space programs currently sponsored by the breakaway civilizations and the current militaries in control of planet Earth. Plenty of astounding accounts, documents and speculation on the incredible alternative history of hidden conflicts and secret space programs that began when World War II officially "ended."
292 Pages. 6x9 Paperback. Illustrated. $19.95. Code: BCCW

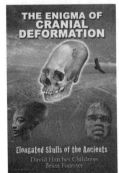

THE ENIGMA OF CRANIAL DEFORMATION
Elongated Skulls of the Ancients
By David Hatcher Childress and Brien Foerster
In a book filled with over a hundred astonishing photos and a color photo section, Childress and Foerster take us to Peru, Bolivia, Egypt, Malta, China, Mexico and other places in search of strange elongated skulls and other cranial deformation. The puzzle of why diverse ancient people—even on remote Pacific Islands—would use head-binding to create elongated heads is mystifying. Where did they even get this idea? Did some people naturally look this way—with long narrow heads? Were they some alien race? Were they an elite race that roamed the entire planet? Why do anthropologists rarely talk about cranial deformation and know so little about it? Color Section.
250 Pages. 6x9 Paperback. Illustrated. $19.95. Code: ECD

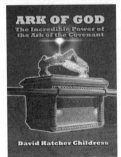

ARK OF GOD
The Incredible Power of the Ark of the Covenant
By David Hatcher Childress
Childress takes us on an incredible journey in search of the truth about (and science behind) the fantastic biblical artifact known as the Ark of the Covenant. This object made by Moses at Mount Sinai—part wooden-metal box and part golden statue—had the power to create "lightning" to kill people, and also to fly and lead people through the wilderness. The Ark of the Covenant suddenly disappears from the Bible record and what happened to it is not mentioned. Was it hidden in the underground passages of King Solomon's temple and later discovered by the Knights Templar? Was it taken through Egypt to Ethiopia as many Coptic Christians believe? Childress looks into hidden history, astonishing ancient technology, and a 3,000-year-old mystery that continues to fascinate millions of people today. Color section.
420 Pages. 6x9 Paperback. Illustrated. $22.00 Code: AOG

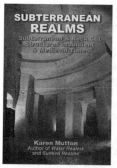

SUBTERRANEAN REALMS
By Karen Mutton

Mutton discusses such interesting sites as: Derinkuyu, an underground city in Cappadocia, Turkey that housed 20,000 people; Roman catacombs of Domitilla; Palermo Capuchin catacombs; Alexandria catacombs; Paris catacombs; Maltese hypogeum; Rock-cut structures of Petra; Treasury of Atreus, Mycenae; Elephanta Caves, India; Lalibela, Ethiopia; Tarquinia Etruscan necropolis; Hallstatt salt mine; Beijing air raid shelters; Japanese high command Okinawa tunnels; more. There are tons of illustrations in this fascinating book!

336 Pages. 6x9 Paperback. Illustrated. $19.95. Code: SUBR

WATER REALMS
Ancient Water Technologies and Management
By Karen Mutton

From the flushing toilets of ancient Crete to the qanats of Persia, aqueducts of Rome, cascading tank systems of Sri Lanka and the great baths of the Indus Valley to the eel traps of southern Australia, ancients on all continents were managing water in unique ways. Table of Contents includes: The Minoan Waterworks; Case Study—The Tunnel of Eupalinos, Samos; Sicily; Etruscan Waterworks; Aqueducts; Roman Baths; Case Study—Aqua Sulis; Case Study—The Baths of Caracalla; Flood Control Systems; Hydraulic Works in the Provinces; Case Study—The Pont Du Gard, France; Late Roman & Byzantine Technologies; The Persian Qanat System; Case Study—The Palace of Persepolis; Khmer Empire; Case Study—The Dujiangyan Irrigation System; Hohokam Water Works; Case Study—Teotihuacan; Case Study— The Puquios of Peru; Sardinia Wells; Nymphaea; Celtic Wells; Ancient Fish Traps; more. There are tons of illustrations in this fascinating book!

254 Pages. 6x9 Paperback. Illustrated. $19.95. Code: WTR

SUNKEN REALMS
A Survey of Underwater Ruins Around the World
By Karen Mutton

Mutton begins by discussing some of the causes for sunken ruins: super-floods; volcanoes; earthquakes at the end of the last great flood; plate tectonics and other theories. From there she launches into a worldwide cataloging of underwater ruins by region. She begins with the many underwater cities in the Mediterranean, and then moves into northern Europe and the North Atlantic. Places covered in this book include: Tartessos; Cadiz; Morocco; Alexandria; Libya; Phoenician and Egyptian sites; Roman era sites; Yarmuta, Lebanon; Cyprus; Malta; Thule & Hyperborea; Canary and Azore Islands; Bahamas; Cuba; Bermuda; Peru; Micronesia; Japan; Indian Ocean; Sri Lanka Land Bridge; Lake Titicaca; and inland lakes in Scotland, Russia, Iran, China, Wisconsin, Florida and more.

282 Pages. 6x9 Paperback. Illustrated. $20.00. Code: SRLM

SCATTERED SKELETONS IN OUR CLOSET
By Karen Mutton

Mutton gives us the rundown on various hominids, skeletons, anomalous skulls and other "things" from our family tree, including hobbits, pygmies, giants and horned people. Chapters include: Human Origin Theories; Dating Techniques; Evolution Fakes and Mistakes; Creationist Hoaxes and Mistakes; The Tangled Tree of Evolution; The Australopithecine Debate; Homo Habilis; Homo Erectus; Anatomically Modern Humans in Ancient Strata?; Ancient Races of the Americas; The Taklamakan Mummies—Caucasians in Prehistoric China; Strange Skulls; Dolichocephaloids (Coneheads); Pumpkin Head, M Head, Horned Skulls; The Adena Skull; The Boskop Skulls; 'Starchild'; Pygmies of Ancient America; Pedro the Mountain Mummy; Hobbits—Homo Floresiensis; Palau Pygmies; Giants; Goliath; Holocaust of American Giants?; more.

320 Pages. 6x9 Paperback. Illustrated. $18.95. Code: SSIC

THE LOST WORLD OF CHAM
The Trans-Pacific Voyages of the Champa
By David Hatcher Childress

The mysterious Cham, or Champa, peoples of Southeast Asia formed a megalith-building, seagoing empire that extended into Indonesia, Tonga, and beyond—a transoceanic power that reached Mexico and South America. The Champa maintained many ports in what is today Vietnam, Cambodia, and Indonesia and their ships plied the Indian Ocean and the Pacific, bringing Chinese, African and Indian traders to far off lands, including Olmec ports on the Pacific Coast of Central America. Topics include: Cham and Khem: Egyptian Influence on Cham; The Search for Metals; The Basalt City of Nan Madol; Elephants and Buddhists in North America; The Cham and Lake Titicaca; Easter Island and the Cham; the Magical Technology of the Cham; tons more. 24-page color section.
328 Pages. 6x9 Paperback. Illustrated. $22.00 Code: LPWC

ADVENTURES OF A HASHISH SMUGGLER
by Henri de Monfreid

Nobleman, writer, adventurer and inspiration for the swashbuckling gun runner in the *Adventures of Tintin*, Henri de Monfreid lived by his own account "a rich, restless, magnificent life" as one of the great travelers of his or any age. The son of a French artist who knew Paul Gaugin as a child, de Monfreid sought his fortune by becoming a collector and merchant of the fabled Persian Gulf pearls. He was then drawn into the shadowy world of arms trading, slavery, smuggling and drugs. Infamous as well as famous, his name is inextricably linked to the Red Sea and the raffish ports between Suez and Aden in the early years of the twentieth century. De Monfreid (1879 to 1974) had a long life of many adventures around the Horn of Africa where he dodged pirates as well as the authorities.
284 Pages. 6x9 Paperback. $16.95. Illustrated. Code AHS

NORTH CAUCASUS DOLMENS
In Search of Wonders
By Boris Loza, Ph.D.

Join Boris Loza as he travels to his ancestral homeland to uncover and explore dolmens firsthand. Throughout this journey, you will discover the often hidden, and surprisingly forbidden, perspective about the mysterious dolmens: their ancient powers of fertility, healing and spiritual connection. Chapters include: Ancient Mystic Megaliths; Who Built the Dolmens?; Why the Dolmens were Built; Asian Connection; Indian Connection; Greek Connection; Olmec and Maya Connection; Sun Worshippers; Dolmens and Archeoastronomy; Location of Dolmen Quarries; Hidden Power of Dolmens; and much more! Tons of Illustrations! A fascinating book of little-seen megaliths. Color section.
252 Pages. 5x9 Paperback. Illustrated. $24.00. Code NCD

GIANTS: MEN OF RENOWN
By Denver Michaels

Michaels runs down the many stories of giants around the world and testifies to the reality of their existence in the past. Chapters and subchapters on: Giants in the Bible; Texts; Tales from the Maya; Stories from the South Pacific; Giants of Ancient America; The Stonish Giants; Mescalero Tales; The Nahullo; Mastodons, Mammoths & Mound Builders; Pawnee Giants; The Si-Te-Cah; Tsul 'Kalu; The Titans & Olympians; The Hyperboreans; European Myths; The Giants of Britain & Ireland; Norse Giants; Myths from the Indian Subcontinent; Daityas, Rakshasas, & More; Jainism: Giants & Inconceivable Lifespans; The Conquistadors Meet the Sons of Anak; Cliff-Dwelling Giants; The Giants of the Channel Islands; Strange Tablets & Other Artifacts; more. Tons of illustrations with an 8-page color section.
320 Pages. 6x9 Paperback. Illustrated. $22.00. Code: GMOR

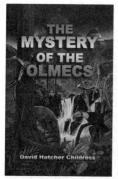

THE MYSTERY OF THE OLMECS
by David Hatcher Childress

The Olmecs were not acknowledged to have existed as a civilization until an international archeological meeting in Mexico City in 1942. Now, the Olmecs are slowly being recognized as the Mother Culture of Mesoamerica, having invented writing, the ball game and the "Mayan" Calendar. But who were the Olmecs? Where did they come from? What happened to them? How sophisticated was their culture? Why are many Olmec statues and figurines seemingly of foreign peoples such as Africans, Europeans and Chinese? Is there a link with Atlantis? In this heavily illustrated book, join Childress in search of the lost cities of the Olmecs! Chapters include: The Mystery of Quizuo; The Mystery of Transoceanic Trade; The Mystery of Cranial Deformation; more.

296 Pages. 6x9 Paperback. Illustrated. Bibliography. Color Section. $20.00. Code: MOLM

ABOMINABLE SNOWMEN:
LEGEND COME TO LIFE
The Story of Sub-Humans on Six Continents from the Early Ice Age Until Today
by Ivan T. Sanderson

Do "Abominable Snowmen" exist? Prepare yourself for a shock. In the opinion of one of the world's leading naturalists, not one, but possibly four kinds, still walk the earth! Do they really live on the fringes of the towering Himalayas and the edge of myth-haunted Tibet? From how many areas in the world have factual reports of wild, strange, hairy men emanated? Reports of strange apemen have come in from every continent, except Antarctica.

525 pages. 6x9 Paperback. Illustrated. Bibliography. Index. $16.95. Code: ABML

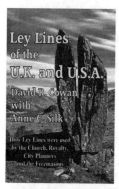

LEY LINES OF THE UK AND USA
By David R. Cowan with Anne Silk

Chapters include: Megalithic Engineering; Burial Grounds across Scotland; Following a Straight Ley Line to its Source; Saint Columba and Iona; The Royal Triangle of Great Britain; The Strange Behavior of Ley Lines; The Dance of the Dragon; Ley Lines in the USA; The Secret Knowledge of the Freemasons; Spirit Paths; The Occult Knowledge of the Nazis; How to Use Divining Rods; The Amazing Power of the Maze; more. Tons of illustrations, all in color!

184 Pages. 7x9 Paperback. All Color. Profusely Illustrated. Index. $24.00. Code: LLUK

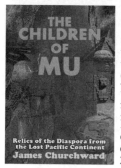

THE CHILDREN OF MU
By James Churchward

According to Churchward, the lost Pacific continent of Mu was the site of the Garden of Eden and the home of 64,000,000 inhabitants known as the Naacals. In this, his second book, first published in 1931, Churchward tells the story of the colonial expansion of Mu and the influence of the highly developed Mu culture on the rest of the world. Her first colonies were in North America and the Orient, while other colonies had been started in India, Egypt and Yucatan. Chapters include: The Origin of Man; The Eastern Lines; Ancient North America; Stone Tablets from the Valley of Mexico; South America; Atlantis; Western Europe; The Greeks; Egypt; The Western Lines; India; Southern India; The Great Uighur Empire; Babylonia; Intimate Hours with the Rishi; more. A fascinating book on the diffusion of mankind around the world—originating in a now lost continent in the Pacific! Tons of illustrations!

270 Pages. 6x9 Paperback. Illustrated. $19.95. Code: COMU

ANDROMEDA: THE SECRET FILES
The Flying Submarines of the SS
By David Hatcher Childress

Childress brings us the amazing story of the German Andromeda craft, designed and built during WWII. Along with flying discs, the Germans were making long, cylindrical airships that are commonly called motherships—large craft that house several smaller disc craft. It was not until 1989 that a German researcher named Ralf Ettl, living in London, received an anonymous packet of photographs and documents concerning the planning and development of at least three types of unusual craft—including the Andromeda. Chapters include: Gravity's Rainbow; The Motherships; The MJ-12, UFOs and the Korean War; The Strange Case of Reinhold Schmidt; Secret Cities of the Winged Serpent; The Green Fireballs; Submarines That Can Fly; The Breakaway Civilization; more. Includes a 16-page color section.
382 Pages. 6x9 Paperback. Illustrated. $22.00 Code: ASF

GODS AND SPACEMEN THROUGHOUT HISTORY
Did Ancient Aliens Visit Earth in the Past?
By W. Raymond Drake

From prehistory, flying saucers have been seen in our skies. As mankind sends probes beyond the fringes of our galaxy, we must ask ourselves: "Has all this happened before? Could extraterrestrials have landed on Earth centuries ago?" Drake spent many years digging through huge archives of material, looking for supposed anomalies that could support his scenarios of space aliens impacting human history. Chapters include: Spacemen; The Golden Age; Sons of the Gods; Lemuria; Atlantis; Ancient America; Aztecs and Incas; India; Tibet; China; Japan; Egypt; The Great Pyramid; Babylon; Israel; Greece; Italy; Ancient Rome; Scandinavia; Britain; Saxon Times; Norman Times; The Middle Ages; The Age of Reason; Today; Tomorrow; more.
280 Pages. 6x9 Paperback. Illustrated. $18.95. Code: GSTH

PYTHAGORAS OF SAMOS
First Philosopher and Magician of Numbers
By Nigel Graddon

This comprehensive account comprises both the historical and metaphysical aspects of Pythagoras' philosophy and teachings. In Part 1, the work draws on all known biographical sources as well as key extracts from the esoteric record to paint a fascinating picture of the Master's amazing life and work. Topics covered include the unique circumstances of Pythagoras' birth, his forty-year period of initiations into all the world's ancient mysteries, his remarkable meeting with a physician from the mysterious Etruscan community, Part 2 comprises, for the first time in a publicly available work, a metaphysical interpretation of Pythagoras' Science of Numbers.
294 Pages. 6x9 Paperback. Illustrated. $18.95. Code: PYOS

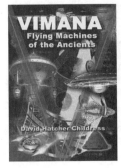

VIMANA:
Flying Machines of the Ancients
by David Hatcher Childress

According to early Sanskrit texts the ancients had several types of airships called vimanas. Like aircraft of today, vimanas were used to fly through the air from city to city; to conduct aerial surveys of uncharted lands; and as delivery vehicles for awesome weapons. David Hatcher Childress, popular *Lost Cities* author, takes us on an astounding investigation into tales of ancient flying machines. In his new book, packed with photos and diagrams, he consults ancient texts and modern stories and presents astonishing evidence that aircraft, similar to the ones we use today, were used thousands of years ago in India, Sumeria, China and other countries. Includes a 24-page color section.
408 Pages. 6x9 Paperback. Illustrated. $22.95. Code: VMA

THE GODS IN THE FIELDS
Michael, Mary and Alice-Guardians of Enchanted Britain
By Nigel Graddon

We learn of Britain's special place in the origins of ancient wisdom and of the "Sun-Men" who taught it to a humanity in its infancy. Aspects of these teachings are found all along the St. Michael ley: at Glastonbury, the location of Merlin and Arthur's Avalon; in the design and layout of the extraordinary Somerset Zodiac of which Glastonbury is a major part; in the amazing stone circles and serpentine avenues at Avebury and nearby Silbury Hill: portals to unimaginable worlds of mystery and enchantment; Chapters include: Michael, Mary and Merlin; England's West Country; The Glastonbury Zodiac; Wiltshire; The Gods in the Fields; Michael, Mary and Alice; East of the Line; Table of Michael and Mary Locations; more.

280 Pages. 6x9 Paperback. Illustrated. $19.95. Code: GIF

AXIS OF THE WORLD
The Search for the Oldest American Civilization
by Igor Witkowski

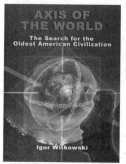

Polish author Witkowski's research reveals remnants of a high civilization that was able to exert its influence on almost the entire planet, and did so with full consciousness. Sites around South America show that this was not just one of the places influenced by this culture, but a place where they built their crowning achievements. Easter Island, in the southeastern Pacific, constitutes one of them. The Rongo-Rongo language that developed there points westward to the Indus Valley. Taken together, the facts presented by Witkowski provide a fresh, new proof that an antediluvian, great civilization flourished several millennia ago.

220 pages. 6x9 Paperback. Illustrated. $18.95. Code: AXOW

LEY LINE & EARTH ENERGIES
An Extraordinary Journey into the Earth's Natural Energy System
by David Cowan & Chris Arnold

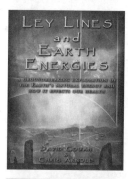

The mysterious standing stones, burial grounds and stone circles that lace Europe, the British Isles and other areas have intrigued scientists, writers, artists and travellers through the centuries. How do ley lines work? How did our ancestors use Earth energy to map their sacred sites and burial grounds? How do ghosts and poltergeists interact with Earth energy? How can Earth spirals and black spots affect our health? This exploration shows how natural forces affect our behavior, how they can be used to enhance our health and well being.

368 pages. 6x9 Paperback. Illustrated. $18.95. Code: LLEE

THE MYSTERY OF U-33
By Nigel Graddon

The incredible story of the mystery U-Boats of WWII! Graddon first chronicles the story of the mysterious U-33 that landed in Scotland in 1940 and involved the top-secret Enigma device. He then looks at U-Boat special missions during and after WWII, including U-Boat trips to Antarctica; U-Boats with the curious cargos of liquid mercury; the journey of the Spear of Destiny via U-Boat; the "Black Subs" and more. Chapters and topics include: U-33: The Official Story; The First Questions; Survivors and Deceased; August 1985—the Story Breaks; The Carradale U-boat; The Tale of the Bank Event; In the Wake of U-33; Wrecks; The Greenock Lairs; The Mystery Men; "Brass Bounders at the Admiralty"; Captain's Log; Max Schiller through the Lens; Rudolf Hess; Otto Rahn; U-Boat Special Missions; Neu-Schwabenland; more.

351 Pages. 6x9 Paperback. Illustrated. $19.95. Code: MU33

ORDER FORM

10% Discount When You Order 3 or More Items

One Adventure Place
P.O. Box 74
Kempton, Illinois 60946
United States of America
Tel.: 815-253-6390 • Fax: 815-253-6300
Email: auphq@frontiernet.net
http://www.adventuresunlimitedpress.com

ORDERING INSTRUCTIONS

✓ Remit by USD$ Check, Money Order or Credit Card

✓ Visa, Master Card, Discover & AmEx Accepted

✓ Paypal Payments Can Be Made To:
   info@wexclub.com

✓ Prices May Change Without Notice

✓ 10% Discount for 3 or More Items

SHIPPING CHARGES

United States

✓ POSTAL BOOK RATE

✓ Postal Book Rate { $4.50 First Item / 50¢ Each Additional Item

✓ Priority Mail { $7.00 First Item / $2.00 Each Additional Item

✓ UPS { $9.00 First Item (Minimum 5 Books) / $1.50 Each Additional Item

   NOTE: UPS Delivery Available to Mainland USA Only

Canada

✓ Postal Air Mail { $19.00 First Item / $3.00 Each Additional Item

✓ Personal Checks or Bank Drafts MUST BE
   US$ and Drawn on a US Bank

✓ Canadian Postal Money Orders OK

✓ Payment MUST BE US$

All Other Countries

✓ Sorry, No Surface Delivery!

✓ Postal Air Mail { $19.00 First Item / $7.00 Each Additional Item

✓ Checks and Money Orders MUST BE US$
   and Drawn on a US Bank or branch.

✓ Paypal Payments Can Be Made in US$ To:
   info@wexclub.com

SPECIAL NOTES

✓ RETAILERS: Standard Discounts Available

✓ BACKORDERS: We Backorder all Out-of-
   Stock Items Unless Otherwise Requested

✓ PRO FORMA INVOICES: Available on Request

✓ DVD Return Policy: Replace defective DVDs only

ORDER ONLINE AT: www.adventuresunlimitedpress.com

**10% Discount When You Order
3 or More Items!**

Please check: ✓

☐ This is my first order   ☐ I have ordered before

Name
Address
City
State/Province _____ Postal Code _____
Country
Phone: Day _____ Evening _____
Fax _____ Email _____

Item Code	Item Description	Qty	Total

Please check: ✓

	Subtotal ▶	
Less Discount-10% for 3 or more items ▶	
☐ Postal-Surface	Balance ▶
☐ Postal-Air Mail (Priority in USA)	Illinois Residents 6.25% Sales Tax ▶
	Previous Credit ▶
☐ UPS (Mainland USA only)	Shipping ▶
	Total (check/MO in USD$ only) ▶

☐ Visa/MasterCard/Discover/American Express

Card Number: _____

Expiration Date: _____   Security Code: _____

✓ SEND A CATALOG TO A FRIEND: